Advanced Studies on Climate Change in Urban Areas: Emerging Technologies and Strategies to Address Heat Waves and Improve Thermo-Hygrometric Comfort

Advanced Studies on Climate Change in Urban Areas: Emerging Technologies and Strategies to Address Heat Waves and Improve Thermo-Hygrometric Comfort

Guest Editors

Gianluca Pappaccogli
Ferdinando Salata

MDPI

Basel • Beijing • Wuhan • Barcelona • Belgrade • Novi Sad • Cluj • Manchester

Guest Editors

Gianluca Pappaccogli
Dipartimento di Scienze e Tecnologie Biologiche ed Ambientali
University of Salento
Lecce
Italy

Ferdinando Salata
Department of Astronautics, Electronic and Energy Engineering (DIAEE)
Sapienza University of Rome
Rome
Italy

Editorial Office
MDPI AG
Grosspeteranlage 5
4052 Basel, Switzerland

This is a reprint of the Special Issue, published open access by the journal *Atmosphere* (ISSN 2073-4433), freely accessible at: https://www.mdpi.com/journal/atmosphere/special_issues/225JB2200P.

For citation purposes, cite each article independently as indicated on the article page online and as indicated below:

Lastname, A.A.; Lastname, B.B. Article Title. *Journal Name* **Year**, *Volume Number*, Page Range.

ISBN 978-3-7258-7719-5 (Hbk)
ISBN 978-3-7258-7720-1 (PDF)
https://doi.org/10.3390/books978-3-7258-7720-1

Contents

atmosphere

MDPI

Article

LUCIDiT: A Lean Urban Comfort Intelligent Digital Twin for Quick Mean Radiant Temperature Assessment

Michele Baia, Giacomo Pierucci and Carla Balocco *

Department of Architecture, University of Florence, Via della Mattonaia 8, 50121 Florence, Italy; michele.baia@unifi.it (M.B.); giacomo.pierucci@unifi.it (G.P.)
* Correspondence: carla.balocco@unifi.it

Abstract

The intensification of Global Warming and Urban Heat Island phenomena necessitates advanced, computationally effective tools for evaluating outdoor thermal comfort and microclimatic dynamics by means of Mean Radiant Temperature assessment. However, existing high-resolution physical models often suffer from prohibitive computational costs. This research proposes LUCIDiT (Lean Urban Comfort Intelligent Digital Twin), a physically based modeling framework implemented for a quick mean radiant temperature assessment inside complex urban morphologies. The method integrates a simplified balance of mutual radiative heat exchanges with recursive time-series filtering to account for the thermal inertia of different urban materials, alongside greenery heat exchange due to evapotranspiration. This architecture creates an operational urban comfort digital twin that reduces computational times by orders of magnitude for large-scale mappings, without sacrificing physical accuracy. Validation against drone-acquired thermographic data and the established Urban Multi-scale Environmental Predictor model demonstrates high reliability and coherence with the real physical phenomena and context. The application to an urban pilot site in Florence reveals that strategic interventions, such as substituting impervious surfaces with irrigated greenery and arboreal canopies, can mitigate radiant loads by up to 20 °C. Findings show that the proposed urban comfort digital twin can be a robust, scalable instrument for designing evidence-based climate adaptation strategies and quick testing mitigation scenarios to enhance urban resilience.

Keywords: outdoor thermal comfort; urban comfort digital twin; mean radiant temperature; built-up/greenery and radiative integrated modeling; Urban Heat Island

1. Introduction

The Global Warming (GW) phenomenon and the increasing frequency of extreme weather events intensify the Urban Heat Island (UHI) effect, leading to severe physiological and psychological discomfort in densely populated areas [1–14]. The complex interplay between urban morphology, impervious surfaces with low albedo, anthropogenic heat, and reduced green infrastructure synergistically traps solar radiation, creating UHI and thermal "cul-de-sacs" characterized by elevated daytime peaks and reduced night-time cooling [15–21]. To preserve the usability of urban spaces and mitigate thermal stress, it is crucial to devise robust, computationally efficient modeling tools capable of mapping outdoor thermal comfort and evaluating mitigation strategies [22–25].

In this context, Urban Digital Twins (UDTs) have been suggested as a transformative paradigm for urban planning and environmental management. A UDT is a highly detailed, virtual replica of the physical city, integrating static morphological information—such as 3D building geometries, street canyons, and material properties—with dynamic data streams from real-time fixed meteorological stations or sensor networks.

Within the domain of urban microclimate, UDTs provide a powerful, interactive framework capable of simulating complex thermodynamic interactions and visualizing spatial thermal distributions. Moreover, they can be used as a virtual testbed for predictive analysis, allowing the evaluation of mitigation climate change scenarios (i.e., "what and/or if") such as the integration of green infrastructures, planting trees with higher canopy expansion, or the application of high-albedo pavements compared to any building/urban regeneration solution. In particular, the UDTs effectively bridge the gap between complex environmental physics and regenerative resilient urban design integrating multi-source data into a cohesive digital environment.

To effectively enhance the use of these digital environments, the existing literature provides comprehensive approaches for modeling urban-scale comfort using indices like the Universal Thermal Climate Index (UTCI) and Physiological Equivalent Temperature (PET) [26–31]. For example, a recent notable research [32] has demonstrated the feasibility and efficacy of an open-source UDT capable of simulating the Mean Radiant Temperature (MRT) in an interactive 3D environment by integrating the well-established Urban Multi-scale Environmental Predictor (UMEP) model [33].

While highly effective and widely recognized as a robust benchmark for microclimatic simulations, such state-of-the-art models like UMEP typically rely on complex explicit 3D geometries, extensive boundary condition datasets, and significant computational overhead to solve transient heat exchanges. This computational burden heavily limits their scalability for large urban areas.

To overcome these computational limitations and capture the multi-scale complexity of the urban environment, recent advancements have increasingly integrated Machine Learning (ML) techniques. Several studies have successfully employed algorithms such as Random Forest, CatBoost, and Deep Neural Networks (DNN) to predict MRT and Land Surface Temperature (LST), quantifying the impact of different road morphologies, shading configurations, and material properties [34–43].

However, this shift introduces a significant methodological dichotomy. On one hand, purely data-driven ML models offer rapid predictions, but their "black-box" nature often limits generalizability outside their specific training domains. On the other hand, pure ML models can struggle to maintain thermodynamic consistency especially for evaluating novel urban resilience regeneration scenarios, such as introducing new highly reflective paving and building materials or new vegetation layouts, effective integration of Nature-Based Solutions (NBS), that is, de facto, the core of any UDT. Furthermore, explicit physically based models ensure rigorous thermodynamic consistency, but with higher computational cost.

Therefore, a critical gap exists between the computational speed of data-driven algorithms and the physical reliability of explicit models. To bridge this specific gap, our research proposes LUCIDiT (Lean Urban Comfort Intelligent Digital Twin).

Faced with the challenges of poor data availability and prohibitive calculation times, LUCIDiT is designed as an effective, immediate, and highly scalable tool. By using basic physical and thermodynamic equations rather than relying solely on data-driven inferences, LUCIDiT guarantees the reliability and robustness required for suitable scenario testing. At the same time, its primary novelty lies in an innovative and lean computational architecture that achieves execution speeds comparable to ML models without reducing

 https://doi.org/10.3390/atmos17030305

physical coherence. Specifically, LUCIDiT integrates a simplified multi-physics thermal balance with particular attention to mutual radiative heat exchanges and while bypassing computationally expensive explicit differential equations connected to thermo-physical behavior over time of different materials. Thermal capacity and inertia of each material are successfully emulated using an Exponential Weighted Moving Average (EWMA) [44]. This approach allows for a rigorous yet simplified evaluation of the phase shift and damping of the thermal wave as a function of material properties (mass, specific heat, thermal diffusivity, density). Furthermore, the model treats urban greenery as an active thermodynamic component, solving equations for the evapotranspiration cooling effect via a parameterized Penman–Monteith approach [45,46]. By overcoming the traditional bottlenecks associated with the reliance on hard-to-source specific data, LUCIDiT enables the generation of ultra-high-resolution diurnal thermal maps with very high computational efficiency, preserving physical accuracy while facilitating rapid large-scale scenario testing.

The structure of the manuscript is organized as follows. Section 2 presents the modeling framework behind LUCIDiT and its implementation using readily accessible real-world data. Section 3 details the validation of the proposed model against drone-derived thermographic measurements and provides a comparative performance analysis with a well-established open-source model (UMEP). Section 4 demonstrates the application of LUCIDiT for the rapid analysis of future urban regeneration scenarios. Finally, Section 5 outlines the conclusions and future developments.

2. Materials and Methods

2.1. The Radiative Physical Model

The estimation of outdoor thermal comfort under clear-sky conditions, heavily relies on the MRT, which integrates the complex radiative exchanges between the human body and the urban environment. To facilitate the rapid estimation of MRT within complex urban canyons, a physically based simplified radiative model is developed. This model computes the total radiative flux absorbed by a reference pedestrian, approximated as a vertical cylinder, by accounting for the interplay between solar geometry, urban morphology, and material properties. The MRT is subsequently derived from the Stefan–Boltzmann law, incorporating both shortwave solar radiation and longwave thermal emission from the surrounding environment:

$$\epsilon_{body}\sigma \mathrm{MRT}^4 = \sigma \sum_i F_i \epsilon_i T_i^4 + \alpha_k I_{sw}. \quad (1)$$

The left-hand side of Equation (1) represents the total long-wave radiation emitted by the human body to achieve equilibrium with an ideal, isothermal enclosure. $\sigma = 5.67 \cdot 10^{-8}$ Wm^{-2}K^{-4} is the Stefan–Boltzmann constant, $\epsilon_{body} = 0.97$ is the average emissivity of human body. The two components on the right-hand side of Equation (1) represent the short and long-wave radiation, e.g., the total amount of radiation received by the human body. The short-wave component, $\alpha_k I_{sw}$, quantifies the solar radiation adsorbed by the subject, where α_k is the absorption coefficient for a standard pedestrian and I_{sw} is the total incident solar radiation on the subject. The term $\sum_i F_i \epsilon_i T_i^4$ accounts for the long-wave radiation, received from surrounding urban surfaces, including building facades, ground materials, vegetation, and the sky vault. Each surface i is characterized by its emissivity ϵ_i and a surface temperature T_i which is dynamically computed based on the local configuration and solar exposure, due to the inertia of different materials. The relative influence of these surfaces on the pedestrian, is weighted by the view factors F_i. Figure 1 shows the schematic view of the described balance (Equation (1)).

https://doi.org/10.3390/atmos17030305

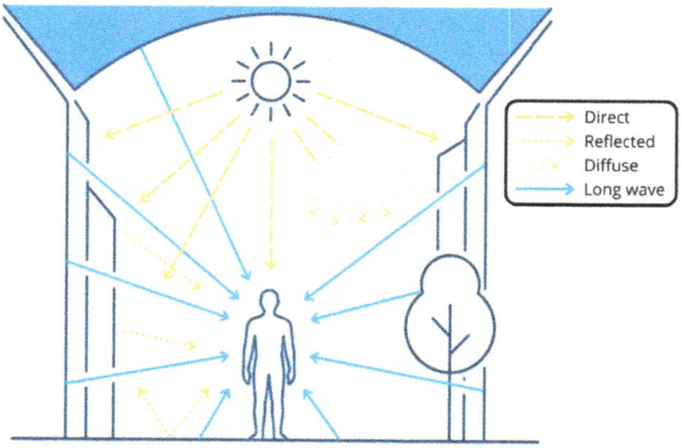

Figure 1. Schematic view of contributions used in the urban energy balance. In yellow color shortwave contribute, dashed line for direct radiation, dotted line for reflected radiation and dashed curve for diffuse radiation. In blue color is the longwave contribution.

The geometric coefficients F_i represent the fraction of the pedestrian's field of view occupied by each environmental component, effectively linking the urban morphology—such as canyon narrowness or open spaces—to the thermal sensation.

Two parameters are defined to characterize the complexity of the urban morphology: the Sky View Factor (SVF) and the density of trees (gf_{trees}), representing, respectively, the portion of sky viewed and the amount of arboreal vegetation surrounding the pedestrian. By exploiting these two parameters, the normalized view factors F_i can be defined as follows:

$$F_{ground} = 0.5 \tag{2}$$

$$F_{sky} = 0.5 \cdot \mathrm{SVF} \cdot (1 - \tau_{trees} \cdot gf_{trees}) \tag{3}$$

$$F_{wall} = 0.5 \cdot (1 - \mathrm{SVF}) \cdot (1 - \tau_{trees} \cdot gf_{trees}) \tag{4}$$

$$F_{veg} = 0.5 - F_{wall} - F_{sky} \tag{5}$$

With this parametrization, it is assumed that the leaf cover of the trees is a screen in front of everything, which, therefore, dampens the portion of sky and buildings with a transmittance factor τ_{trees}. The assignment of a fixed value of 0.5 to the ground view factor (F_{ground}) is firmly rooted in standard biometeorological practices for assessing pedestrian thermal comfort (e.g., ISO 7726 and VDI 3787 Part 2 [47,48]), which are structurally implemented in state-of-the-art benchmark models such as RayMan pro3.1, ENVI-met 5.8, and SOLWEIG/UMEP 4.4. In this framework, the pedestrian is approximated as a vertically oriented cylinder or sphere at a reference height of 1.1 m. Consequently, on a flat surface, the lower hemisphere of the subject's radiant field of view is entirely occupied by the ground (F_{ground} = 0.5).

Notably, this geometric relation remains robust even in densely built urban canyons, as surrounding obstructions primarily affect the upper hemisphere—redistributing the view factors between the sky and vertical walls—without significantly altering the pedestrian's view of the street surface.

However, it is important to note that this assumption is theoretical. Inside highly complex urban topographies featuring slopes, staircases, or immediate low-level obstructions, the actual F_{ground} may deviate from this theoretical constant.

 https://doi.org/10.3390/atmos17030305

Recent empirical analyses by Zhengrong Li et al. [49] have quantified the critical sensitivity of MRT to the ground view factor, demonstrating that a deviation of 0.1 in F_{ground} can induce an MRT calculation error of up to 5 °C, and deviations as small as 0.01 can still shift the MRT by over 1 °C. While the constant 0.5 value provides a reliable baseline for the flat topographies simulated in this study, future iterations of the LUCIDiT framework will aim to implement a dynamically variable F_{ground}. This could be achieved by incorporating a parameterized dependence on the SVF or high-resolution 3D morphological features [19], further enhancing the model's accuracy in capturing the thermal nuances of complex urban terrains.

The longwave radiative exchange is modeled by summing the thermal emissions from four primary surfaces: sky, ground, building walls, and vegetation.

Surface temperatures are estimated using a simplified heat balance rather than full computational fluid dynamics. Following the approach suggested in [19], ground and wall temperatures are modeled as deviations from the air temperature:

$$T_{wall(ground)} = T_{air} + \mathrm{d}T_{wall(ground)}. \tag{6}$$

However, to address the limitation of static coupling, we introduce a time-dependent radiative term that accounts for the thermal inertia of different urban materials. The surface temperature deviation is calculated as proportional to an inertial solar irradiance ($I_{inertial}$):

$$\mathrm{d}T_i = K_i \cdot \frac{\mathrm{SAF} \cdot I_{inertial}}{I_{max}} \cdot (1 - \alpha_i), \tag{7}$$

where α_i is the albedo of the surface material i, K_i is a scaling coefficient [°C], I_{max} = 1000 Wm^{-2} a normalizing factor. This value is selected as it represents a widely accepted standard reference for peak Global Horizontal Irradiance (GHI) at sea level under clear sky conditions at medium low latitudes (exactly like the one that concerns Florence), aligning with the reference irradiance used in Standard Test Conditions (STC) for solar applications [50,51]. It effectively scales the radiative forcing for typical urban summer conditions in temperate climates.

In geographical regions characterized by high elevation, tropical latitudes, or extremely high atmospheric transparency, the actual solar irradiance can systematically exceed 1000 Wm^{-2} and the I_{max} parameter within the LUCIDiT framework can be adjusted to reflect the local climatological upper bounds to prevent underestimation of the thermal forcing.

SAF is the Solar Access Factor, a geometric proxy introduced to easy integrate the shading effect, defined as:

$$\mathrm{SAF} = \max(1 - a \cdot S_d, \mathrm{SAF}_{\mathrm{min}}), \tag{8}$$

with a = 0.5 and $\mathrm{SAF}_{\mathrm{min}}$ = 0.5. SAF depends directly form the shadow density (S_d).

This parameter, included in [0, 1], represents the amount of direct solar radiation shielded by buildings or trees. This is dynamically updated from the solar angle, projecting the shape of the obstructions onto the surface at height z = 1.5 m. Shadows generated by buildings give values of S_d = 1, while the shadow density from trees depends on specific vegetative parameter, the Leaf Area Index (LAI), and the Sun elevation β (the angle between the Sun's rays and a horizontal plane), following a Beer–Lambert exponential decay [52]:

$$S_d = 1 - \exp\left\{\frac{k_e \cdot \mathrm{LAI}}{\sin(\beta)}\right\}, \tag{9}$$

with the attenuation factor k_e = 0.6.

https://doi.org/10.3390/atmos17030305

The term $I_{inertial}$ is not the instantaneous solar radiation, but a historical convolution of the radiation that affects the surface. To simulate heat storage and release without solving differential equations, EWMA is employed [44]. This acts as a recursive low-pass filter on the radiation time series:

$$I_{inertial}(t) = \gamma \cdot I_{global}(t) + (1 - \gamma) \cdot I_{inertial}(t-1). \quad (10)$$

The decay factor γ ($0 < \gamma \leq 1$), directly related to the thermal time constant τ, determines the thermal memory of the system.

Radiative contributions from the sun are obtained starting from the decomposition of GHI into direct and diffuse components using the semi-empirical model proposed by [53]. Based on solar elevation β and the clearness index, the model determines the diffusive fraction k_d. Consequently, the Diffuse Horizontal Irradiance (DHI) and the Direct Normal Irradiance (DNI) are calculated as follows:

$$\mathrm{DHI} = \mathrm{GHI} \cdot k_d, \quad (11)$$

$$\mathrm{DNI} = \frac{\mathrm{GHI} - \mathrm{DHI}}{\sin(\beta)}. \quad (12)$$

For the walls, I_{global} = DNI e $\gamma = 0.25$, corresponding to a value $\tau = 2 - 3$ h. For the soil, a primary composition of two classes of different materials, asphalt and greenery is assumed. Greenery concentration is characterized using a density fraction gf_{ground}. The resulting temperature of the soil will be an average of the surface temperature deviation of the two material (obtained from Equation (7)) weighted by the gf_{ground}:

$$T_{ground} = T_{air} + \left(1 - gf_{ground}\right) \cdot \mathrm{d}T_{asphalt} + gf_{ground} \cdot \mathrm{d}T_{greenery}. \quad (13)$$

For this surface, the main assumption is: an incident solar radiation I_{global} = GHI, $\gamma_{asphalt} = 0.25$, corresponding to a value $\tau = 5 - 6$ h and $\gamma_{ground} = 0.8$, corresponding to a value $\tau \lesssim 1$ h. The scaling factor K_i is chosen to represent the potential maximum temperature rise for the material under global solar exposure. Following the literature experimental/simulation evidence [54,55], a value of K_{wall} = 25 °C, and $K_{asphalt}$ = 45 °C and K_{ground} = 7 °C (otherwise until 18 °C for extreme limit conditions) is imposed. These values physically correspond to a maximum increase in surface temperatures with respect to the air temperatures of 15 °C, 30 °C, and 5 °C (otherwise until 15 °C), respectively.

The bifurcated choice for the K_{ground} coefficient is introduced to distinguish between healthy, irrigated grass—where evapotranspiration has a significant impact on the adiabatic saturation effect of the air and therefore on local cooling, reducing surfaces' warming—and the condition in which the herbaceous green cover is dry and/or subjected to water stress, and does not trigger cooling processes, thus reaching higher surface temperatures [56]. Unless explicitly specified, in this study, the greenery, all types of green cover, is considered healthy, both biophysically and structurally, and well-irrigated. Essentially, the European directives [57] on plant health and biosecurity are also taken into account, which require all plants (entire plants, fruit trees, lawns, and green cover, etc.) imported into the EU to be accompanied by a phytosanitary certificate (exemptions: bananas, coconuts, dates, pineapples, and durians, which are not easily direct introduced into urban areas, like Florence due to their climate, morphology, and hydrogeology).

Deep sky is modeled as a heat sink, with an apparent temperature derived from ambient air temperature adjusted by a cloud-cover proxy calculated from the diffuse fraction ratio:

$$T_{sky} = T_{air} - 12 \cdot \left(1 - \frac{\mathrm{DHI}}{\mathrm{GHI}}\right). \quad (14)$$

https://doi.org/10.3390/atmos17030305

In this, a maximum cooling effect up to 12 °C with respect to ambient temperatures can be assumed.

Vegetation is modeled as life system; indeed, trees temperature is evaluated taking in account the evapotranspiration effect of the vegetative system, modeled as a Penman–Monteith equation [45,46]. This model calculates the energy balance of the canopy by solving for the latent heat flux (λE) and the sensible heat flux. The standard aerodynamic resistance (r_a), which governs the heat and mass transfer is considered, estimating the water vapor mass as a function of wind speed and surface roughness, this last one parameterized as a constant (r_a = 20 sm^{-1}) [58]. This parameterization decouples the model from explicit wind speed field inputs, a climatic parameter that usually is mapped through urban canyon modeling with computationally expensive Computational Fluid Dynamics (CFD) simulations or dense sensor networks.

The canopy temperature is derived from the residual sensible heat flux after accounting for evapotranspiration:

$$T_{veg} = T_{air} + \frac{(R_n - \lambda \cdot E) \cdot r_a}{\rho \cdot C_p}, \tag{15}$$

where $\rho \cdot C_P$ is the volumetric air heat capacity and R_n is the net solar radiation, estimated as a balance between shortwave absorption (accounting for albedo $\alpha = 0.2$) and a net longwave loss (assumed as average constant value −60 Wm^{-2}).

The physiological regulation of transpiration is modeled via the canopy stomatal resistance (r_s), which follows a multiplicative Jarvis-type approach [59]. This formulation accounts for the limiting factors of available solar energy and atmospheric demand:

$$r_s = \frac{r_{s,min}}{f_{light} \cdot f_{VDP}}, \tag{16}$$

where $r_{s,min}$ represents the minimum stomatal resistance (set to 100 sm^{-1}). The scaling functions f_{light} and f_{VPD} modulate stomatal aperture, based on GHI and Vapor Pressure Deficit (VDP), respectively, simulating the physiological closure of stomata during periods of low light or high evaporative demand.

For the short-wave component, the total solar flux I_{sw} needs to be estimated. This is highly sensitive to the spatial distribution of shadows, which is modeled through a shadow density parameter (S_d).

The total shortwave flux (I_{sw}) incident on the pedestrian aggregates four distinct contributions: direct beam radiation, diffuse sky radiation, and reflected radiation from both vertical walls and the horizontal ground:

$$I_{sw} = I_{dir} + I_{diff} + I_{ref}^{wall} + I_{ref}^{ground}. \tag{17}$$

The direct component is modulated by the projected area factor of the human body and the probability of shading, which is dynamically determined by the presence of buildings and vegetation:

$$I_{dir} = \text{DNI} \cdot (1 - S_d) \cdot f_p, \tag{18}$$

where the term $(1 - S_d)$ represents the pedestrian's exposition to direct sunlight. This parameter is introduced to modulate the impact of direct beam radiation and differentiate the conditions of complete solar exposure from shaded areas. The parameter f_p is the projected area factor, which accounts for the cylindrical geometry of the standing subject as a function of the Sun elevation β [60], neglecting the shadows produced by the 3D cylinder/person:

https://doi.org/10.3390/atmos17030305

$$f_p = 0.308 \cdot cos\left(\beta \cdot \left(0.998 - \frac{\beta^2}{50000}\right)\right). \quad (19)$$

The scattered radiation is weighted by the portion of sky observed by the pedestrian:

$$I_{diff} = \mathrm{DHI} \cdot F_{sky}. \quad (20)$$

The reflected radiation contributions from the ground and walls are subsequently calculated by modulating the incident flux by the surface albedo and the specific view factors for the surface *i* = *ground*, *wall*:

$$I_{ref}^{(i)} = I_{incident}^{(i)} \cdot \mathrm{SAF} \cdot \alpha_i \cdot F_i, \quad (21)$$

where SAF is the surrounding sun factor defined as Equation (8), with a = 0.6 and SAF_{min} = 0.4. In this way it is assumed that when the pedestrian is in the shade, the surfaces from which it receives the reflected radiation are also placed in the shade, but with a variable portion of these remaining exposed to the sun.

The ground albedo is the average between the albedo of two main materials composing it, i.e., asphalt and greenery, weighted by the green fraction of the greenery gf_{ground}.

The incident horizontal solar radiation on the ground is GHI, i.e., $I_{incident}^{(ground)}$ = GHI, the total radiation incident on vertical walls $I_{incident}^{(wall)}$ must be explicitly derived.

The latter one is defined as the sum of the vertical component of the direct beam, diffuse sky radiation and the radiation reflected from the ground onto the walls:

$$I_{incident}^{(wall)} = I_{verical\_dir} + I_{verical\_diff} + I_{vertical\_ref\_ground}. \quad (22)$$

These vertical components are approximated by the following group of three similar equations:

$$I_{verical\_dir} = 0.5 \cdot \mathrm{DNI} \cdot cos(\beta), \quad (23)$$

$$I_{verical\_diff} = 0.5 \cdot \mathrm{DHI}, \quad (24)$$

$$I_{verical\_ref\_ground} = \mathrm{GHI} \cdot \alpha_{ground} \cdot F_{ground}. \quad (25)$$

The factor of 0.5 applied to the direct and diffuse vertical component, relies on the statistical assumption that within a random urban orientation, approximately half of the vertical surfaces are exposed to the sun while the other half are in shade.

2.2. Mapping and Implementation on Real Urban Scenario

To demonstrate the applicability of LUCIDiT model on real data, a geospatial pipeline is developed to translate real urban geometries into the input feature of the model. This process integrates static georeferenced datasets with dynamic climatological series to generate all the inputs required by the model.

The mapping process utilizes OpenStreetMap (OSM) [61] and, where provided, municipal cadastral data to reconstruct the building footprints and their respective heights, which are essential for determining the local urban morphology. This geometric information is discretized into a high-resolution grid (typically 1 × 1 m), where each cell is treated as a potential pedestrian location. To optimize computational efficiency, static morphological features are precomputed for each grid cell.

The SVF, as obstruction made from the buildings, is derived using a simplify sector-based ray-tracing algorithm that scans the surrounding environment to identify obstruction angles caused by urban elements, integrating these into a view factor representing the

 https://doi.org/10.3390/atmos17030305

visible sky dome. Simultaneously, vegetation density is modeled as a continuous field rather than a binary classification; an inverse distance weighting approach smooths the discrete presence of trees and grass to assign effective green fraction values (gf_{trees}, gf_{ground}) to each cell, based on the proximity factor.

Meteorological boundary conditions are established using data from meteorological station, or, if not provided, Typical Weather Year (TWY) datasets retrieved via the PVGIS database [62]. For every simulation time step, the pipeline extracts air temperature T_{air}, relative humidity RH and GHI while solving the solar geometry to determine Sun elevation β. These dynamic variables drive the shadow-casting algorithm, which projects the footprints of buildings and trees onto the ground plane based on the instantaneous solar position.

The resulting shadow polygons are intersected with the grid centroids to compute the Shadow density S_d. Building shadows are modeled as binary obstructions blocking direct radiation, whereas vegetation shadows are treated probabilistically; within tree shadows, intensity follows an exponential decay function of the local green fraction gf_{trees} and LAI to simulate canopy transmittance (Equation (9)).

Finally, surface albedos are assigned, with ground albedo dynamically interpolated between vegetative and asphalt reference values. This procedure yields a complete, time-resolved dataset for the entire domain, serving as the input for LUCIDiT model to generate high-resolution MRT maps.

The use of LUCIDiT model allows for generating a spatially continuous map of the MRT in a fraction of the time required by traditional CFD or ray-tracing simulations. This efficiency enables the rapid transition from a static urban representation, to a dynamic UDT capable of simulating various microclimatic conditions and using all easily accessible and available data/information in an integrated approach.

3. The Case Study

To ensure the reliability of the surrogate model, a validation phase is conducted by comparing the model's predictions against experimental data collected during a field measurements campaign in an urban area of Florence (Firenze Nova district) and, consequently with UMEP, an open source tool for QGIS (v.3.44.7 Solothurm), a validated model from [33].

The experimental investigation has been carried out for the EU LIFE-ESCAPOS project implementation, on a representative and significant urban area, from the UHI formation point of view, used as a primary pilot site.

The proposed method lays the foundation for one possible approach to the Dynamic Control Volume (DCV) modeling, a thermodynamic approach specifically designed for the multi-scale analysis of urban energy and entropy flows [63]. The studied area (43°48′ 23.472″ N, 11°13′ 40″ E; elevation 50 m a.s.l.) covers approximately 0.15 km^2 and is characterized by a heterogeneous urban fabric, which includes varying building heights and vegetation densities, providing a challenging benchmark for the UDT.

The core of this urban site is Lippi's Garden, an expansive green space featuring a different arboreal canopy. Existing greenery includes mature specimens (exceeding 20 m in height) and younger trees. This green infrastructure is integrated into a residential context, where building geometries generate urban canyon effects (buildings' height does not exceed 25 m due to the presence of the airport site few kilometers away).

The district is further characterized by the presence of a nursery school in the north-western sector and a large industrial facility to the south-west (Figure 2).

Due to its geographical location, Florence occupies a transitional zone between a humid subtropical (Cfa) and a hot-summer Mediterranean (Csa) climate [64,65]. The region is characterized by temperate, humid winters and hot, often sultry summers frequently

https://doi.org/10.3390/atmos17030305

marked by convective phenomena. Long-term meteorological records (1991–2020) provided by LAMMA Consortium [66] indicate an annual mean temperature of 15.5 °C, which rises to a seasonal average of 24.5 °C during the summer months. Historical temperature extremes for the city range from a minimum of −10.2 °C to a maximum of 41.3 °C. Relative humidity typically fluctuates around 70%, peaking at 80% in winter and reaching its minimum average (approx. 65%) in July. Prevailing winter winds are predominantly south-westerly (SW), with average velocities ranging between 10–12 km/h.

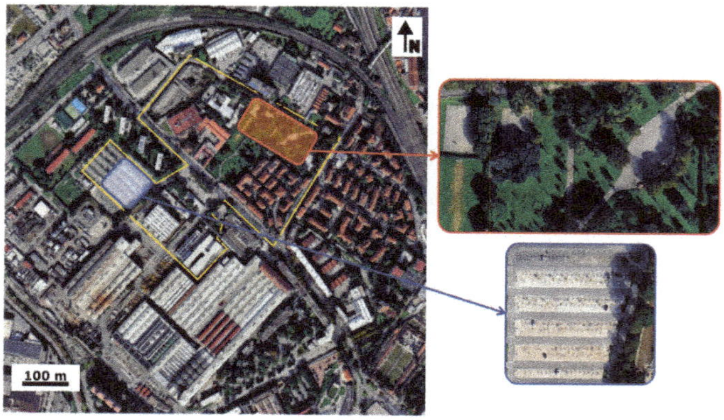

Figure 2. Top view of the study area in the North–East part of Florence—Google Maps web version source (enclosed by yellow line). In particular: red rectangle identifies drone flight zone above the Lippi's Garden, blue rectangle the parking lot where the LUCIDiT model can be applied for suitable regenerative future scenarios.

In the first subsection we report the results obtained from the validation of LUCIDiT model with the data obtained from the drone measurement, in the second the comparison between LUCIDiT and UMEP models.

3.1. *LICIDiT vs. Drone*

The validation of the surrogate model in a real-world operational context, is obtained by means of model predictions compared against the high-resolution thermographic data, acquired for the Lippi's Garden. The experimental campaign is carried out exploiting drone surveys techniques. In particular, the Unmanned Aerial System (UAS) is equipped with a gimbal-stabilized thermal sensor 13.5 mm focal length, spectral sensitivity in the 8–14 µm range). Data acquisition is carried out through automated flight missions, enabling high-resolution coverage thanks to the match of hundreds of georeferenced thermal and RGB frames, aiming at the generation of an overall ground surface temperature map across the study area (referred to a mission conducted on 12 August 2025). The data post-processing procedure and aggregation used, is that suggested in [67]. Figure 3 shows the area considered for the analysis, as RGB image obtained by processing the photographs provided by drone flights (Figure 3a), and the digital reconstruction derived with the previously presented pipeline (Figure 3b).

To map the surface radiosity detected by the drone into a Mean Radiant Temperature (MRT_{drone}), directly comparable with the model's output, a simplified longwave radiation balance is applied. Specifically, the MRT_{drone} is derived by aggregating the radiative

 https://doi.org/10.3390/atmos17030305

contributions from the surrounding environment—including sky, buildings, trees, greenery and ground—according to the fundamental relationship, as follows:

$$\epsilon_{body}\mathrm{MRT}^4_{\mathrm{drone}} = \sum \epsilon_i F_i T_i. \tag{26}$$

In this formulation, the ground surface temperature (T_{ground}) is directly derived from the drone measurements. The temperatures of trees and buildings are approximated as equal to the ambient air temperature. Although this assumption may seem broad and lax in relation to building facades, its impact is numerically and physically negligible in the investigated Lippi's Garden scenario, where the geometric view factor of buildings ($F_{buildings}$) approaches zero. The sky temperature (T_{sky}) is parameterized according to the previously defined physical model (Equation (14)).

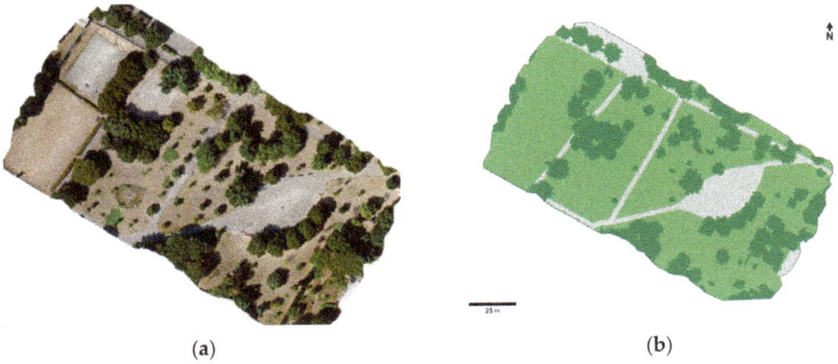

Figure 3. The studied area by drone in RGB (**a**) and in virtual representation of UDT (**b**).

Furthermore, constant emissivity values are assigned to the different radiating surfaces: 0.94 for the ground (representing an average between vegetation and asphalt), 0.80 for the effective sky vault, 0.97 for trees, 0.90 for walls, and 0.97 for the reference pedestrian body (ϵ_{body}). The view factors (F_i) are maintained consistent with those derived from the morphological analysis algorithm.

The surrogate model, driven by boundary meteorological variables acquired from a local weather station, is specifically tuned to emulate the actual surveyed site conditions by setting the ground thermal scaling factor to K_{ground} = 18 °C, in agreement with the hypothesis made for the scale factor for dry grass. Complementing this thermal tuning, the surface albedos within the model, are explicitly calibrated to reflect the specific thermal conditions of the local materials during summer: the asphalt albedo is set to 0.15, representative of an aged, dry, and light-colored pavement, while the grass albedo is fixed to 0.26, typical of a dry, unirrigated lawn.

Figure 4 shows the spatial comparison between the model predictions (Figure 4a), the drone-derived MRT obtained using Equation (26) (Figure 4b), and the difference between the LUCIDiT prediction and the drone-derived estimation of MRT (Figure 4c). It can be noted that a good agreement is achieved between the modeled data and the measurement-derived results, with a mean error of −0.3 °C and a standard deviation of 5.6 °C, and a detailed observation of the spatial error distribution reveals two interesting points.

The first concerns a close inspection of the spatial map, which indicates that the drone-derived MRT values in shaded areas tend to be systematically lower than those predicted by the model. The second concerns the visual and statistical analysis, which highlights a broad simplification for the classification of different surfaces. Indeed, while the surrogate model successfully captures thermal gradients at the macroscale, its rigid

https://doi.org/10.3390/atmos17030305

categorization of horizontal surfaces into just two basic material classes—asphalt and dry grass—oversimplifies physical morphologies.

Consequently, the model generates thermal maps that do not take into account of highly intricate micro-variability of surface materials, localized variability of moisture levels, and subtle shading variations that are intrinsically present in the physical urban fabric, which instead are accurately captured by the drone's high-resolution thermal sensors.

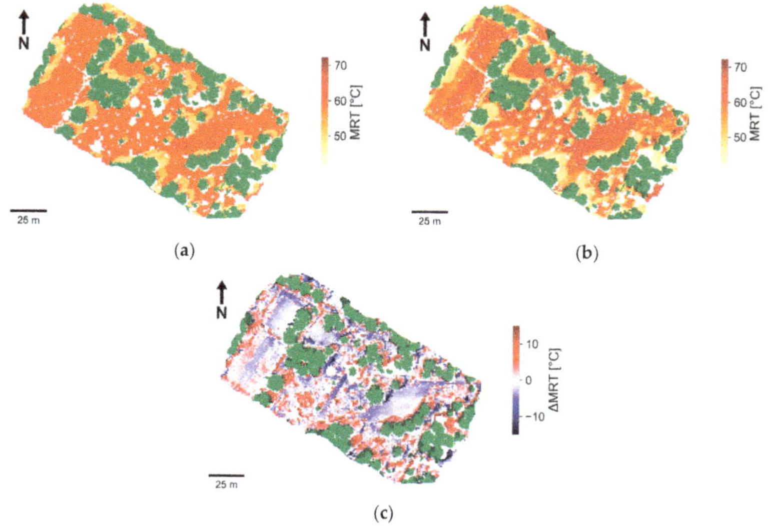

Figure 4. MRT maps for the real data test area predicted by LUCIDiT (**a**), Drone measurement (**b**) and the difference between LUCIDiT and Drone estimation (**c**). Areas in green color indicate the trees obstruction.

3.2. LUCIDiT vs. UMEP

The efficiency, performance, and specific behavioral traits of the LUCIDiT framework are evaluated by comparing its spatial and temporal predictions with those obtained with the established UMEP model. The area considered for comparison, the extent of which is approximately 38 ha, is shown in Figure 2. Meteorological inputs used for simulations are extracted from the TWY dataset.

Figure 5 provides the spatial comparison between the two model predictions at 3 p.m. on August 12, the same day of the drone flight, with a maximum air temperature of 32.5 °C. In Figure 5a the MRT map predicted drone by LUCIDiT, in Figure 5b the UMEP prediction, and in Figure 5c the difference between our model and UMEP prediction.

The raw comparison yields a Mean Absolute Error (MAE) of 3.6 °C and a Mean Bias Error (MBE) of −0.9 °C.

However, this global metric is heavily skewed by highly localized peak deviations. To assess the baseline agreement of the radiative solvers across the bulk of the urban domain, a filtered metric is computed by excluding the absolute differences greater than 15 °C. When isolating the general domain from these specific boundary and microclimatic divergences—the physical reasons for which are detailed below—the baseline MAE drops to 1.5 °C with a positive bias of 0.7 °C.

LUCIDiT exhibits a wider variability in its spatial predictions, with extreme localized differences bounded within a range of [−20, 20] °C. These peak deviations do not indicate a

 https://doi.org/10.3390/atmos17030305

systemic model failure but rather highlight fundamental differences in how the two models handle spatial discretization and applied thermodynamics.

Specifically, positive discrepancies (i.e., LUCIDiT predicting MRT values up to 20 °C higher than UMEP) are predominantly clustered along building perimeters and within narrow urban canyons. This is driven by two factors: firstly, the analytical parameterization of shortwave and longwave radiation trapping in LUCIDiT, which is strictly scaled by (1-SVF), resulting in pronounced thermal loads in deep geometries. Secondly, the boundary effects between LUCIDiT's continuous vector raytracing and UMEP's discrete raster grid generate sharp pixel-wise discrepancies along shadow terminators. A slight sub-pixel offset can result in one model computing full direct solar irradiance while the other computes deep shade, inherently generating an instantaneous ΔMRT of 15–20 °C.

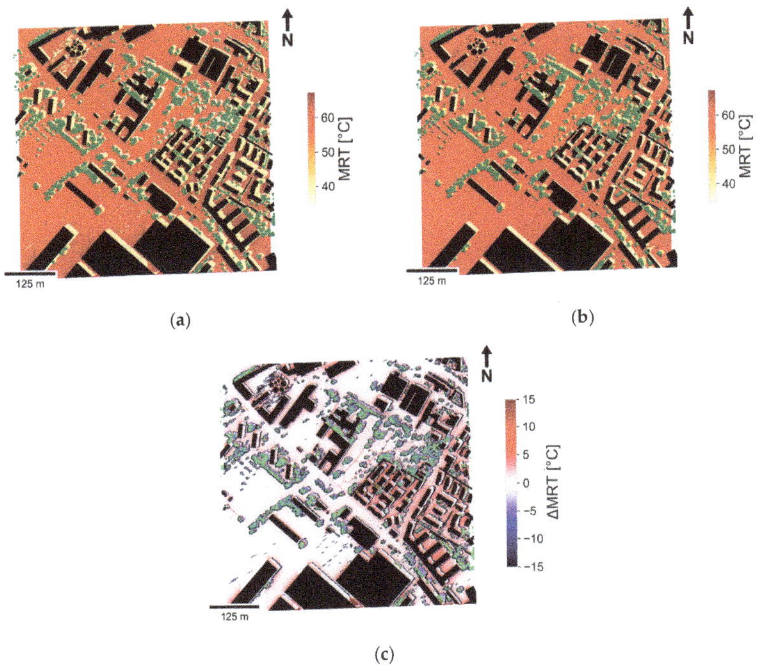

Figure 5. MRT maps for the models test area predicted by LUCIDiT (**a**), UMEP (**b**), and the difference between LUCIDiT and UMEP predictions (**c**). Black areas indicate buildings; green areas indicate the trees obstruction. The spatial domain is titled due to the latitude–longitude orientation.

Conversely, the extreme negative deviations (LUCIDiT predicting MRT up to 20 °C cooler) are strictly localized beneath and immediately surrounding vegetation canopies.

Because LUCIDiT treats trees as active thermodynamic entities—incorporating evapotranspiration cooling mechanisms—rather than mere passive optical filters for shortwave radiation, the longwave emission from the foliage is substantially lower.

This formulation is particularly important because yields a deep, localized microclimatic mitigation, highlighting the model's high sensitivity to nature-based solutions.

The temporal dynamics and the aforementioned divergences are effectively captured in the boxplots of Figure 6, which shows the hourly evolution across the entire day of August 12. The diurnal trends are accurately respected by LUCIDiT, capturing a broader daytime statistical dispersion of MRT extremes compared to UMEP, before converging tightly with UMEP's baseline during the nighttime cooling phase.

 https://doi.org/10.3390/atmos17030305

Notably, the implementation of a delayed radiative forcing approach for surface temperature estimation, allows LUCIDiT to successfully replicate the nighttime cooling trends demonstrated by UMEP, which instead relies on explicitly solving transient heat exchange models for urban surfaces.

Consequently, LUCIDiT achieves comparable prognostic capabilities at a fraction of the computational cost: executed on equivalent hardware, the UMEP simulation required approximately 2.5 h, whereas the LUCIDiT pipeline converged in merely 90 s.

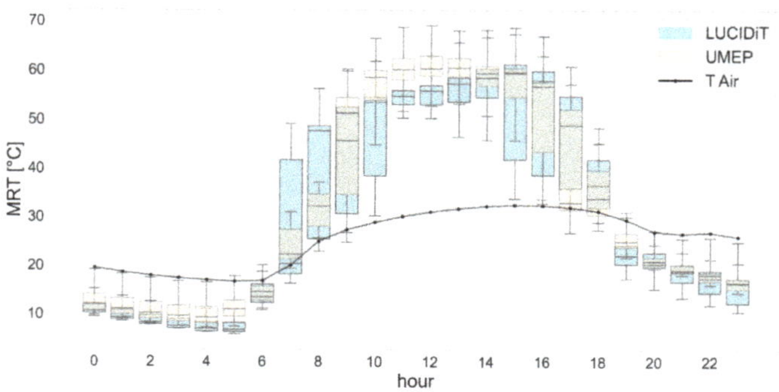

Figure 6. Comparison of MRT calculation by LUCIDiT and UMEP for the extreme day of the TWY.

4. Results and Discussion

The previous sections (Sections 3.1 and 3.2) demonstrate a fulfilling coherence between the LUCIDiT model and the local experimental results obtained by drone flights, such as between the same proposed method and the UMEP tool for QGIS.

With the objective of testing the power and the capability of the model, LUCIDiT is applied in a specific zone of the pilot area for the impact assessment of some suitable urban interventions. The evaluation of the MRT is achieved in the area of the parking lot (Figure 7a in RGB and Figure 7b in the UDT) simulating: (i) the current layout, (ii) substitution of the asphalt with grass, and (iii) grass and planting of four new rows of trees (8 m height with a diameter of 5 m for each tree, the distance between the trees of each row is 8 m and the spacing among the rows is 16 m). These different solutions are analyzed in the extreme day of the TWY, which corresponds to 24 August 2023 (at 3 p.m.). The weather inputs for the simulation are T_{air} 37.7 °C, RH 21.7% and GHI 613 Wm^{-2}.

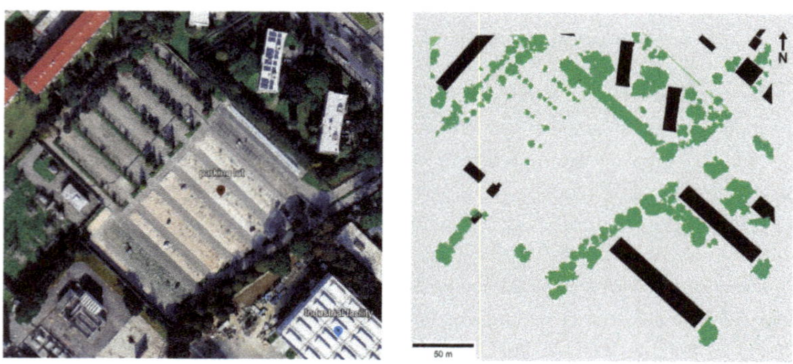

(**a**) (**b**)

Figure 7. The area studied for different suitable solutions in RGB (**a**) and in the UDT (**b**). In the UDT images, buildings are in black, and trees in green color.

 https://doi.org/10.3390/atmos17030305

The current configuration exhibits a critical thermal scenario as shown in Figure 8: the parking lot is mainly paved with asphalt or dry ground and the radiative contributions, due to shadows' absence, drive MRT to values between 60 °C and 65 °C (with an average value of 63 °C).

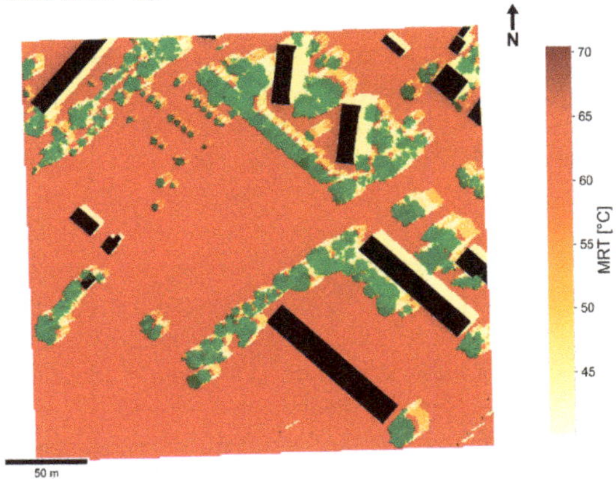

Figure 8. MRT map for the parking lot in the existing scenario. Black areas identify buildings; green areas the trees obstruction. The spatial domain is titled due to the latitude–longitude orientation.

Building upon this framework, a vegetated lawn is modeled into the central portion of the parking area (of about 5000 m^2). The changes of the MRT values are mapped as a difference ΔMRT, relative to the initial baseline configuration (Figure 9a).

Thanks to the increasing in the albedo and the scaling coefficient K_{ground} (set at 7 °C for a well-irrigated healthy turfgrass), a drop in MRT is achieved for about 6 °C, even without integrating shade effects. Furthermore, the inclusion of trees results in a more significant decrease in MRT values. This enhancement is a direct consequence of the combined effect of evapotranspiration and shading, with the latter playing a crucial role in a local reduction in the radiant load on the ground. A maximum temperature reduction of 20 °C is observed in this case (Figure 9b). Again, LUCIDiT demonstrates high versatility and computational efficiency. Once the boundary conditions—including georeferencing, soil morphology, and material properties—are defined, each scenario requires approximately 10 s to generate high-resolution MRT maps.

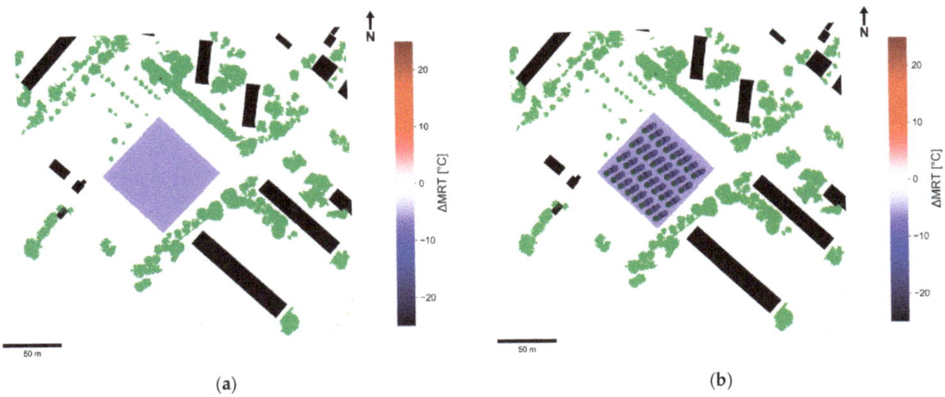

Figure 9. Difference of MRT in respect to the initial scenario: (**a**) adding turfgrass and (**b**) adding turfgrass and trees. Grey areas indicate buildings, and green areas color the trees obstruction. The spatial domain is titled due to the latitude–longitude orientation.

https://doi.org/10.3390/atmos17030305

A limitation observed during the spatial validation is the primary strict categorization of horizontal surfaces into broad macro-classes (e.g., standard asphalt versus dry grass), which provides a strong simplification of the real complex micro-variability of different materials. To address this, future development of LUCIDiT framework will be designed to incorporate an oriented much more granular, pixel-specific material classification—such as spatially variable different albedo, emissivity, and thermal admittance matrices derived from high-resolution satellite or multi-spectral drone imagery.

The authors believe that the integration of this level of granularity will not compromise the model "lean" computational advantage. Because LUCIDiT processes morphological and thermophysical properties as a priori input matrices, shifting from constant class-based values to heterogeneous spatial arrays will only slightly increase the initial memory allocation, while the execution time of the fundamental physical equations remain unaltered. Another simplification aspect of the proposed model is that it neglects fluid dynamic inside the urban context that which would require CFD simulations with greater computational costs. A CFD approach would undoubtedly enhance the method's accuracy, by integrating local air-flow and velocity fields into the heat exchange analysis, but its application would involve greater complexity, both due to the multiplicity of necessary data required and to slow solutions due to the multi-physics coupling. Indeed, e.g., standard fixed meteorological stations are not suitable to provide the high-resolution data required for CFD-multi-physics modeling, needing to complex, hyper-local experimental data for its validation.

Furthermore, CFD-enhanced tools require highly specialized personnel, significant computational costs and resources, leading to substantially longer simulation times which would not allow the rapid-assessment by means of LUCIDiT.

In particular, convective heat exchange phenomena at the urban scale becomes significant when natural convection occurs in turbulent flow regimes, such as long, straight avenues aligned with prevailing winds (wind channels) or narrow high-rise streets. In these specific geometries, LUCIDiT will systematically underestimate convective cooling, leading to a conservative overestimation of surface temperatures and MRT. Conversely, in highly complex, deep, and with air stagnant and/or corto-circuit phenomena canyon geometries (e.g., narrow historical alleys), the lack of air mixing might lead to an underestimation of the perceived thermal stress.

In the investigated case, used for LUCIDiT application, hyperlocal air motion is mainly induced by very low mean air velocity values, ensuring that the radiative heat exchange is strong and prevalent. Indeed, the real measured thermal conditions show high external air temperatures, high material surface temperatures, and an average air velocity never exceeding 2 ms^{-1}, that is a value usually associated to convection in turbulent flow regimes and modeled with CFD simulations [68,69].

5. Conclusions

This research proposes a novel methodology, based on LUCIDiT, for the dynamic assessment of MRT at the urban scale. Grounded in applied physics and thermodynamics, the algorithm effectively quantifies radiative heat exchange and the microclimatic influence of different materials of built-up and green areas. By bypassing computationally heavy explicit solvers, LUCIDiT drastically reduces execution times while maintaining high reliability—as demonstrated through validation against experimental drone data and established benchmark models like UMEP. The minimal requirements for input data—which in many cases are already readily accessible through modern open-source repositories—makes the tool operational for non-expert users, such as urban planners, municipal authorities, and various stakeholders.

 https://doi.org/10.3390/atmos17030305

The potential and effectiveness of LUCIDiT are further supported and corroborated by the results of the sensitivity analysis (Appendix A), which confirm the expected thermodynamic behaviors: an exponential cooling effect associated with increasing the LAI (consistent with the Lambert-Beer law) and the linear, yet distinct, impacts of ground and wall albedos. While these parameters act as independent design variables that can be actively modified for climate mitigation, intrinsic thermal properties—such as the thermal scaling coefficient (K_i) and time decay constant (γ)—were intentionally maintained as empirical "thermal footprints" to preserve physical realism. Arbitrarily modifying them would be mathematically equivalent to simulating the properties of fictitious, non-existent materials. However, a theoretical exploration of these parameters presents a highly valuable future prospect for advanced materials engineering, particularly for assessing the microclimatic impact of advanced next-generation urban surfaces with customized thermal inertia.

Future perspectives for LUCIDiT involve the development of a platform that, starting from local urban data to global databases on urban morphology (such as the Global Building Atlas [70]), facilitates extensive studies on urban comfort and design of effective, integrated climate adaptation interventions and strategies to mitigate UHIs across multiple scales, from local neighborhoods to global urban assessments. Moreover, additional intensive drone flight missions will be designed with a view to developing the method for the real radiative phenomena measurements, in several areas of the city, more densely built-up and with different environmental morphology, aiming at capillary validation of the model in contexts where the built-up is more predominant and/or strictly combined greenery and NBs.

The authors are currently investigating the application of their proposed tool to generate extensive datasets for training ML algorithms. This studied approach, aims to identify large-scale statistical correlations—while maintaining a rigorous link to the physics and thermodynamics of the underlying phenomena—to further accelerate outdoor thermal comfort assessment in urban environments.

Ultimately, LUCIDiT effectively bridges the gap between explicit 3D microclimatic modeling and computational efficiency. By providing a versatile, upgradeable, and fast framework for MRT evaluation, it empowers the implementation of intelligent UDTs, enabling rapid "what-if" scenario testing and the immediate comparison of climate adaptation strategies across different urban morphologies and climatic conditions, making it an effective and attractive tool for planners and decision-makers.

Author Contributions: M.B.; Writing, Methodology, Investigation, Validation, Software, Data curation; G.P.; Writing, Methodology, Investigation, Validation, Software, Data curation; C.B.; Writing, Review and Editing, Supervision, Funding Acquisition, Conceptualization. All authors have read and agreed to the published version of the manuscript.

Funding: This work was supported by the EU Project, LIFE Energy + LIFE Climate—Project 101157553-LIFE23-CCA-IT-LIFE ESCAPOS "Environmental energy for Strategic CApillary urban POlicieS", funded by the European Union.

Institutional Review Board Statement: Not applicable.

Informed Consent Statement: Not applicable.

Data Availability Statement: The original contributions presented in this study are included in the article. Further inquiries can be directed to the corresponding author.

Conflicts of Interest: The authors declare no conflicts of interest and that they have no competing financial interests or personal relationships that could have appeared to influence the work reported in this paper.

https://doi.org/10.3390/atmos17030305

Abbreviations

The following abbreviations are used in this manuscript:

CFD	Computational fluid dynamics
DCV	Dynamical control volume
DHI	Diffuse horizontal irradiance [Wm^{-2}]
DNI	Direct normal irradiance [Wm^{-2}]
DNN	Deep neural network
EWMA	Exponential weighted moving average
EU	European union
GHI	Global horizontal irradiance [Wm^{-2}]
GW	Global warming
LAI	Leaf area index [-]
LST	Land surface temperature
LUCIDiT	Lean urban comfort intelligent digital twin
MAE	Mean absolute error
MBE	Mean bias error
ML	Machine learning
MRT	Mean radiant temperature [°C]
NBS	Nature-based solutions
OSM	OpenStreetMap
PET	Physiological environmental temperature [°C]
PIML	Physics-informed machine learning
PVGIS	Photovoltaic geographical information system
RH	Relative humidity [%]
SAF	Solar access factor [-]
SHAP	Shapley additive explaining
SVF	Sky view factor [-]
STC	Standard test conditions for solar applications
TWY	Typical weather year
UAS	Unmanned aerial system
UDI	Urban digital twin
UHI	Urban Heat Islands
UMEP	Urban multi-scale environmental predictor
UTCI	Universal thermal climate index [°C]
VDP	Vapor pressure deficit [Pa]
α	Albedo [-]
β	Sun elevation [°]
γ	Decay factor of EWMA [-]
ϵ	Emissivity [-]
λ	Latent heat of vaporization [Jkg^{-1}]
ρ	Density [$Kg\ m^{-3}$]
σ	Stefan Boltzmann constant [$Wm^{-2}K^{-4}$]
τ	Thermal time constant [s^{-1}]
τ_{trees}	Tree transmittance [-]
a	Dispersion coefficient of SAF [-]
C_p	Specific heat [Jkg^{-1}]
E	Evapotranspiration rate [$kgm^{-2}s^{-1}$]
F_i	View factor [-]
f_{light}	Light scaling function of stomatal aperture [-]
f_p	Projected area factor [-]
f_{VDF}	Evaporation scaling function of stomatal aperture [-]
I_{dir}	Direct irradiance on pedestrian [Wm^{-2}]

https://doi.org/10.3390/atmos17030305

I_{diff}	Diffuse irradiance on pedestrian [Wm^{-2}]
$I_{incident}$	Incident solar irradiance on surface [Wm^{-2}]
$I_{inertial}$	Inertial solar irradiance [Wm^{-2}]
I_{max}	Maximum reference irradiance [Wm^{-2}]
I_{ref}	Reflected irradiance from surface on pedestrian [Wm^{-2}]
I_{sw}	Short wave irradiance on pedestrian [Wm^{-2}]
$I_{vertical\_dir}$	Direct irradiance on vertical surface [Wm^{-2}]
$I_{vertical\_diff}$	Diffuse irradiance on vertical surface [Wm^{-2}]
$I_{vertical\_ref\_ground}$	Reflected irradiance from ground on vertical surface [Wm^{-2}]
k_d	Diffusive fraction [-]
k_e	Attenuation factor [-]
K_i	Scaling coefficient of surface [°C]
gf_{trees}	Trees density [-]
gf_{ground}	Greenery density [-]
r_a	Aerodynamic resistance [sm^{-1}]
R_n	Net solar radiation on tree [Wm^{-2}]
r_s	Stomatal resistance [sm^{-1}]
S_d	Shadow density [-]
T_{air}	Air temperature [°C]
T_i	Surface temperature [°C]

Appendix A. Sensitivity Analysis

To evaluate the impact of the key parameters included in the modeling framework, a sensitivity analysis is performed on the Leaf Area Index (LAI) and the albedo of both the ground and the walls. The main aim is to assess how these parameters' variation, affect the Mean Radiant Temperature (MRT) predicted by the model.

For a straightforward evaluation, all baseline model parameters are kept constant while the specific parameters selected for the sensitivity analysis are individually varied, calculating the resulting MRT values at strategic locations of interest.

Specifically, to evaluate the impact of the LAI, a location directly beneath a tree canopy is selected to quantify how this parameter dictates the shadow density S_d (Equation (9)).

To assess the influence of the ground albedo, which alters both the ground surface temperature and the reflected shortwave radiation, a location in an open, fully sunlit space is chosen. This fact effectively minimizes the interference from vegetation and the canyoning effects of surrounding buildings. Finally, to evaluate the impact of the wall albedo, a location adjacent to a single wall, but situated in its shadow, is considered. This isolates the weight of the wall's albedo on the MRT—primarily driven by longwave and reflected radiation—by removing the direct radiative effects of the sunlit ground and surrounding vegetation.

The obtained results are provided in Figure A1. Specifically, Figure A1a shows the sensitivity analysis for the LAI, Figure A1b for the ground albedo, and Figure A1c for the wall albedo.

It can be noted that increasing the LAI from values corresponding to bare trees (LAI = 0.5) to very dense canopies (LAI = 8), causes the MRT to decrease exponentially. This behavior consistently follows the increase in shadow density, also shown in Figure A1, which is modeled with a decreasing exponential trend according to the Lambert–Beer law (Equation (9)).

Conversely, the albedo variations result in a linear increase in the MRT. This trend emerges from the thermodynamic balance between the reflected shortwave radiation and the emitted longwave radiation; both of them depend linearly on the albedo parameter,

https://doi.org/10.3390/atmos17030305

but with opposite slopes. However, it is noteworthy that while an increase in the ground albedo from $\alpha = 0.05$ to $\alpha = 0.5$ leads to a substantial MRT increase on the order of ten degrees, the impact of the wall albedo is significantly less pronounced, resulting in an MRT increase on the order of only one degree.

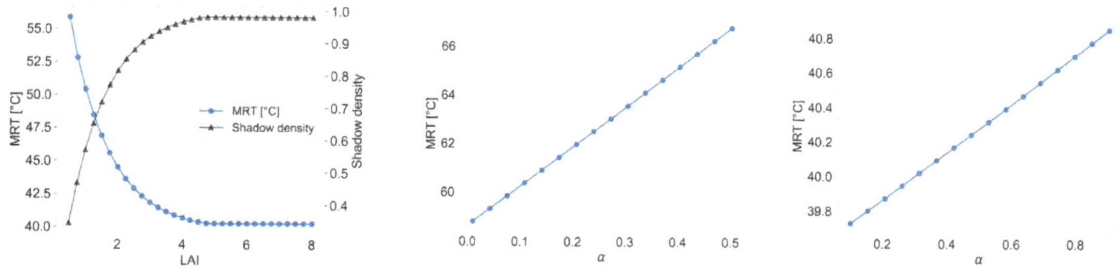

Figure A1. Sensitivity analysis to quantify the impact of LAI (**a**), soil albedo (**b**), and wall albedo (**c**) on MRT predictions.

References

1. Huang, J.; Yu, H.; Guan, X.; Wang, G.; Guo, R. Accelerated dryland expansion under climate change. *Nat. Clim. Change* **2016**, *6*, 166–171. [CrossRef]
2. Yuan, W.; Zheng, Y.; Piao, S.; Ciais, P.; Lombardozzi, D.; Wang, Y.; Ryu, Y.; Chen, G.; Dong, W.; Hu, Z.; et al. Increased Atmospheric Vapor Pressure Deficit Reduces Global Vegetation Growth. *Sci. Adv.* **2019**, *5*, eaax1396. [CrossRef] [PubMed]
3. Hugonnet, R.; McNabb, R.; Berthier, E.; Menounos, B.; Nuth, C.; Girod, L.; Farinotti, D.; Huss, M.; Dussaillant, I.; Brun, F.; et al. Accelerated Global Glacier Mass Loss in the Early Twenty-First Century. *Nature* **2021**, *592*, 726–731. [CrossRef]
4. Johnson, C.N.; Balmford, A.; Brook, B.W.; Buettel, J.C.; Galetti, M.; Guangchun, L.; Wilmshurst, J.M. Biodiversity Losses and Conservation Responses in the Anthropocene. *Science* **2017**, *356*, 270–275. [CrossRef]
5. Díaz, S.; Settele, J.; Brondízio, E.S.; Ngo, H.T.; Agard, J.; Arneth, A.; Balvanera, P.; Brauman, K.A.; Butchart, S.H.M.; Chan, K.M.A.; et al. Pervasive Human-Driven Decline of Life on Earth Points to the Need for Transformative Change. *Science* **2019**, *366*, eaax3100. [CrossRef]
6. Intergovernmental Panel On Climate Change (Ipcc). *Climate Change 2021—The Physical Science Basis: Working Group I Contribution to the Sixth Assessment Report of the Intergovernmental Panel on Climate Change*, 1st ed.; Cambridge University Press: Cambridge, UK, 2023.
7. Diffenbaugh, N.S.; Singh, D.; Mankin, J.S.; Horton, D.E.; Swain, D.L.; Touma, D.; Charland, A.; Liu, Y.; Haugen, M.; Tsiang, M.; et al. Quantifying the Influence of Global Warming on Unprecedented Extreme Climate Events. *Proc. Natl. Acad. Sci. USA* **2017**, *114*, 4881–4886. [CrossRef]
8. Mora, C.; McKenzie, T.; Gaw, I.M.; Dean, J.M.; Von Hammerstein, H.; Knudson, T.A.; Setter, R.O.; Smith, C.Z.; Webster, K.M.; Patz, J.A.; et al. Over Half of Known Human Pathogenic Diseases Can Be Aggravated by Climate Change. *Nat. Clim. Change* **2022**, *12*, 869–875. [CrossRef]
9. Vicedo-Cabrera, A.M.; Scovronick, N.; Sera, F.; Royé, D.; Schneider, R.; Tobias, A.; Astrom, C.; Guo, Y.; Honda, Y.; Hondula, D.M.; et al. The Burden of Heat-Related Mortality Attributable to Recent Human-Induced Climate Change. *Nat. Clim. Change* **2021**, *11*, 492–500. [CrossRef]
10. Watts, N.; Adger, W.N.; Agnolucci, P.; Blackstock, J.; Byass, P.; Cai, W.; Chaytor, S.; Colbourn, T.; Collins, M.; Cooper, A.; et al. Health and Climate Change: Policy Responses to Protect Public Health. *Lancet* **2015**, *386*, 1861–1914. [CrossRef] [PubMed]
11. Obradovich, N.; Migliorini, R.; Paulus, M.P.; Rahwan, I. Empirical Evidence of Mental Health Risks Posed by Climate Change. *Proc. Natl. Acad. Sci. USA* **2018**, *115*, 10953–10958. [CrossRef] [PubMed]
12. Dunne, J.P.; Stouffer, R.J.; John, J.G. Reductions in Labour Capacity from Heat Stress under Climate Warming. *Nat. Clim. Change* **2013**, *3*, 563–566. [CrossRef]
13. Chen, L.; Ng, E. Outdoor Thermal Comfort and Outdoor Activities: A Review of Research in the Past Decade. *Cities* **2012**, *29*, 118–125. [CrossRef]
14. Sherwood, S.C.; Huber, M. An Adaptability Limit to Climate Change Due to Heat Stress. *Proc. Natl. Acad. Sci. USA* **2010**, *107*, 9552–9555. [CrossRef]

https://doi.org/10.3390/atmos17030305

15. Stewart, I.D.; Oke, T.R. Local Climate Zones for Urban Temperature Studies. *Bull. Am. Meteorol. Soc.* **2012**, *93*, 1879–1900. [CrossRef]
16. Ghoniem, M.; Fahmy, A.; Kamel, T.M. Evaluating Urban Surface Materials: Albedo Optimization and Shading for Pedestrian Thermal Comfort in Hot-Arid Environments. *Architecture* **2025**, *13*, 4331–4352. [CrossRef]
17. Yang, J.; Yang, Y.; Sun, D.; Jin, C.; Xiao, X. Influence of Urban Morphological Characteristics on Thermal Environment. *Sustain. Cities Soc.* **2021**, *72*, 103045. [CrossRef]
18. Iamarino, M.; Beevers, S.; Grimmond, C.S.B. High-resolution (Space, Time) Anthropogenic Heat Emissions: London 1970–2025. *Int. J. Climatol.* **2012**, *32*, 1754–1767. [CrossRef]
19. Santamouris, M. Recent Progress on Urban Overheating and Heat Island Research. Integrated Assessment of the Energy, Environmental, Vulnerability and Health Impact. Synergies with the Global Climate Change. *Energy Build.* **2020**, *207*, 109482. [CrossRef]
20. Cui, S.; Wang, X.; Yang, X.; Hu, L.; Jiang, Z.; Feng, Z. Mapping Local Climate Zones in the Urban Environment: The Optimal Combination of Data Source and Classifier. *Sensors* **2022**, *22*, 6407. [CrossRef]
21. Chen, J.; Dong, Z.; Shi, R.; Sun, G.; Guo, Y.; Peng, Z.; Deng, M.; Chen, K. Urban Multi-Scenario Land Use Optimization Simulation Considering Local Climate Zones. *Remote Sens.* **2024**, *16*, 4342. [CrossRef]
22. Chen, C.Y.; Chuang, Y.H.; Chen, H.W.; Teng, C.-S. Optimizing Ecological Corridors for Urban Sustainability by Using Remote Sensing and Decision Modeling. *Int. J. Environ. Sci. Technol.* **2025**, *22*, 10013–10028. [CrossRef]
23. Hu, M.; Ghorbany, S.; Yao, S.; Wang, C. Micro-Urban Heatmapping: A Multi-Modal and Multi-Temporal Data Collection Framework. *Buildings* **2024**, *14*, 2751. [CrossRef]
24. Guan, S.; Chen, Y.; Wang, T.; Hu, H. Mitigating Urban Heat Island through Urban-Rural Transition Zone Landscape Configuration: Evaluation Based on an Interpretable Ensemble Machine Learning Framework. *Sustain. Cities Soc.* **2025**, *123*, 106272. [CrossRef]
25. Henn, K.A.; Peduzzi, A. Surface Heat Monitoring with High-Resolution UAV Thermal Imaging: Assessing Accuracy and Applications in Urban Environments. *Remote Sens.* **2024**, *16*, 930. [CrossRef]
26. Zhang, S.; Zhang, X.; Niu, D.; Fang, Z.; Chang, H.; Lin, Z. Physiological Equivalent Temperature-Based and Universal Thermal Climate Index-Based Adaptive-Rational Outdoor Thermal Comfort Models. *Build. Environ.* **2023**, *228*, 109900. [CrossRef]
27. Sharmin, T.; Steemers, K.; Humphreys, M. Outdoor Thermal Comfort and Summer PET Range: A Field Study in Tropical City Dhaka. *Energy Build.* **2019**, *198*, 149–159. [CrossRef]
28. Kicovic, D.; Vuckovic, D.; Markovic, D.; Jovic, S. Assessment of Visitors' Thermal Comfort Based on Physiologically Equivalent Temperature in Open Urban Areas. *Urban Clim.* **2019**, *28*, 100466. [CrossRef]
29. Höppe, P. The Physiological Equivalent Temperature–a Universal Index for the Biometeorological Assessment of the Thermal Environment. *Int. J. Biometeorol.* **1999**, *43*, 71–75. [CrossRef] [PubMed]
30. Vanos, J.; Guzman-Echavarria, G.; Baldwin, J.W.; Bongers, C.; Ebi, K.L.; Jay, O. A Physiological Approach for Assessing Human Survivability and Liveability to Heat in a Changing Climate. *Nat. Commun.* **2023**, *14*, 7653. [CrossRef] [PubMed]
31. Li, N.; Shu, H.; Sun, D. Human Thermal Comfort Indicator in High-Temperature Environments in Deep Mining. *J. Saf. Sustain.* **2025**, *2*, 134–141. [CrossRef]
32. Lopez-Cabeza, V.P.; Videras-Rodriguez, M.; Gomez-Melgar, S. An Open-Source Urban Digital Twin for Enhancing Outdoor Thermal Comfort in the City of Huelva (Spain). *Smart Cities* **2025**, *8*, 160. [CrossRef]
33. Lindberg, F.; Grimmond, C.S.B.; Gabey, A.; Huang, B.; Kent, C.W.; Sun, T.; Theeuwes, N.E.; Järvi, L.; Ward, H.C.; Capel-Timms, I.; et al. Urban Multi-Scale Environmental Predictor (UMEP): An Integrated Tool for City-Based Climate Services. *Environ. Model. Softw.* **2018**, *99*, 70–87. [CrossRef]
34. Zhong, X.; Voogt, J.; Bailey, B.; Zhang, X.; Krayenhoff, E.S. Evaluating High-Resolution Mean Radiant Temperature within an Urban Street Canopy: Resolving Spatiotemporal Variations with LiDAR/Thermal Infrared Scanning and Data-Driven Simulation. *Build. Environ.* **2025**, *287*, 113819. [CrossRef]
35. Zhu, S.; Yan, Y.; Zhao, B.; Wang, H. Assessing the Impact of Adjacent Urban Morphology on Street Temperature: A Multisource Analysis Using Random Forest and SHAP. *Build. Environ.* **2025**, *267*, 112326. [CrossRef]
36. Conigliaro, E.; Monti, P.; Leuzzi, G.; Cantelli, A. A Three-Dimensional Urban Canopy Model for Mesoscale Atmospheric Simulations and Its Comparison with a Two-Dimensional Urban Canopy Model in an Idealized Case. *Urban Clim.* **2021**, *37*, 100831. [CrossRef]
37. Tian, G.; Ma, Y.; Chen, Y.; Wan, M.; Chen, S. Impact of Urban Canopy Characteristics on Turbulence Dynamics: Insights from Large Eddy Simulation. *Build. Environ.* **2024**, *250*, 111183. [CrossRef]
38. Nazarian, N.; Krayenhoff, E.S.; Martilli, A. A One-Dimensional Model of Turbulent Flow through "Urban" Canopies (MLUCM v2.0): Updates Based on Large-Eddy Simulation. *Geosci. Model Dev.* **2020**, *13*, 937–953. [CrossRef]

https://doi.org/10.3390/atmos17030305

39. Mateen, S.A.; Giometto, M.G.; Biswal, B.; Tapper, N.J.; Parlange, M.B. Large Eddy Simulation Based Evaluation of an Urban Canopy Model. *Bound. Layer Meteorol.* **2025**, *191*, 34. [CrossRef]
40. Snaiki, R.; Merabtine, A. Recent Advances on Machine Learning Techniques for Urban Heat Island Applications: A Review and New Horizons. *Sustain. Cities Soc.* **2025**, *134*, 106943. [CrossRef]
41. Briegel, F.; Wehrle, J.; Schindler, D.; Christen, A. High-Resolution Multi-Scaling of Outdoor Human Thermal Comfort and Its Intra-Urban Variability Based on Machine Learning. *Geosci. Model Dev.* **2024**, *17*, 1667–1688. [CrossRef]
42. Jia, S.; Wang, Y.; Hien Wong, N.; Liang Tan, C.; Chen, S.; Weng, Q.; Ming Mak, C. Estimation of Mean Radiant Temperature across Diverse Outdoor Spaces: A Comparative Study of Different Modeling Approaches. *Energy Build.* **2024**, *310*, 114068. [CrossRef]
43. Rashtian, Z.; Manesh, M.T.; Tahsildoost, M.; Zomorodian, Z.S. Data-Driven Real-Time Visualization of Urban Heat Islands Using Mean Radiant Temperature for Urban Design. *Energy Build.* **2025**, *333*, 115470. [CrossRef]
44. Nicol, F.; Humphreys, M. Derivation of the Adaptive Equations for Thermal Comfort in Free-Running Buildings in European Standard EN15251. *Build. Environ.* **2010**, *45*, 11–17. [CrossRef]
45. Monteith, J.L. Evaporation and the Environment. In *Proceedings of the 19th Symposium of the Society for Experimental Biology*; Cambridge University Press: New York, NY, USA, 1965.
46. Grimmond, C.S.B.; Oke, T.R. An Evapotranspiration-interception Model for Urban Areas. *Water Resour. Res.* **1991**, *27*, 1739–1755. [CrossRef]
47. *ISO 7726:2025*; Ergonomics of the Thermal Environment—Instruments for Measuring and Monitoring Physical Quantities. International Organization for Standardization (ISO): Geneva, Switzerland, 2025.
48. VDI. *VDI 3787-2*; VDI 3787 Part 2: Human Biometeorological Assessment of Climate for Urban and Regional Planning. VDI: Verein Deutscher Ingenieure: Düsseldorf, Germany, 2008.
49. Li, Z.; Feng, X.; Chen, W.; Fang, Z. Quantifying the Effect of Ground View Factor and Ground Temperature on Outdoor Mean Radiant Temperature. *Sustain. Cities Soc.* **2022**, *84*, 104030. [CrossRef]
50. *IEC 61215-1:2021*; Terrestrial Photovoltaic (PV) Modules—Design Qualification and Type Approval—Part 1: Test Requirements. International Electrotechnical Commission (IEC): Geneva, Switzerland, 2021.
51. *IEC 60904-3:2019*; Photovoltaic Devices—Part 3: Measurement Principles for Terrestrial Photovoltaic (PV) Solar Devices with Reference Spectral Irradiance Data. International Electrotechnical Commission (IEC): Geneva, Switzerland, 2019.
52. Campbell, G.S.; Norman, J.M. *An Introduction to Environmental Biophysics*; Springer: New York, NY, USA, 1998.
53. Erbs, D.G.; Klein, S.A.; Duffie, J.A. Estimation of the Diffuse Radiation Fraction for Hourly, Daily and Monthly-Average Global Radiation. *Sol. Energy* **1982**, *28*, 293–302. [CrossRef]
54. Pomerantz, M.; Akbari, H.; Chen, A.; Taha, H.; Rosenfeld, A.H. *Paving Materials for Heat Island Mitigation*; Report LBL-38074; Lawrence Berkeley National Laboratory: Berkeley, CA, USA, 1997.
55. Santamouris, M. Cooling the Cities—A Review of Reflective and Green Roof Mitigation Technologies to Fight Heat Island and Improve Comfort in Urban Environments. *Sol. Energy* **2014**, *103*, 682–703. [CrossRef]
56. Santamouris, M. Using Cool Pavements as a Mitigation Strategy to Fight Urban Heat Island—A Review of the Actual Developments. *Renew. Sustain. Energy Rev.* **2013**, *26*, 224–240. [CrossRef]
57. Regulation (EU) 2016/2031 of the European Parliament of the Council of 26 October 2016 on Protective Measures against Pests of Plants, Amending Regulations (EU) No 228/2013, (EU) No 652/2014 and (EU) No 1143/2014 of the European Parliament and of the Council and Repealing Council Directives 69/464/EEC, 74/647/EEC, 93/85/EEC, 98/57/EC, 2000/29/EC, 2006/91/EC and 2007/33/EC. Available online: https://eur-lex.europa.eu/eli/reg/2016/2031/oj (accessed on 14 March 2026).
58. Jones, H.G. *Plants and Microclimate: A Quantitative Approach to Environmental Plant Physiology*; Cambridge University Press: Cambridge, UK, 2014.
59. Jarvis, P. The Interpretation of the Variations in Leaf Water Potential and Stomatal Conductance Found in Canopies in the Field. *Philos. Trans. R. Soc. Lond. B Biol. Sci.* **1976**, *273*, 593–610. [CrossRef]
60. Fanger, P.O. *Thermal Comfort. Analysis and Applications in Environmental Engineering*; CABI Digital Library: Leusden, The Netherlands, 1970.
61. Open Street Map. Available online: https://Www.Openstreetmap.Org (accessed on 16 January 2026).
62. Photovoltaic Geographical Information System (PVGIS). Available online: https://Joint-Research-Centre.Ec.Europa.Eu/Photovoltaic-Geographical-Information-System-Pvgis_en (accessed on 16 January 2026).
63. Balocco, C.; Grazzini, G. Thermodynamic Parameters for Energy Sustainability of Urban Areas. *Sol. Energy* **2000**, *69*, 351–356. [CrossRef]
64. Beck, H.E.; Zimmermann, N.E.; McVicar, T.R.; Vergopolan, N.; Berg, A.; Wood, E.F. Present and Future Köppen-Geiger Climate Classification Maps at 1-Km Resolution. *Sci. Data* **2018**, *5*, 180214. [CrossRef]

65. Kottek, M.; Grieser, J.; Beck, C.; Rudolf, B.; Rubel, F. World Map of the Köppen-Geiger Climate Classification Updated. *Meteorol. Z.* **2006**, *15*, 259–263. [CrossRef] [PubMed]
66. Consorzio LAMMA. Available online: https://Www.Lamma.Toscana.It/Clima-e-Energia/Climatologia/Clima-Firenze (accessed on 16 January 2026).
67. Pierucci, G.; Baia, M.; Balocco, C. Urban Area Sustainability Analysis by Means of Integrated Microclimatic Measurement Techniques Combined with Thermal Comfort Modelling—A Pilot Project Application. *Energies* **2025**, *19*, 217. [CrossRef]
68. Palyvos, J.A. A Survey of Wind Convection Coefficient Correlations for Building Envelope Energy Systems' Modeling. *Appl. Therm. Eng.* **2008**, *28*, 801–808. [CrossRef]
69. Defraeye, T.; Blocken, B.; Carmeliet, J. Convective Heat Transfer Coefficients for Exterior Building Surfaces: Existing Correlations and CFD Modelling. *Energy Convers. Manag.* **2011**, *52*, 512–522. [CrossRef]
70. Zhu, X.X.; Chen, S.; Zhang, F.; Shi, Y.; Wang, Y. GlobalBuildingAtlas: An Open Global and Complete Dataset of Building Polygons, Heights and LoD1 3D Models. *Earth Syst. Sci. Data* **2025**, *17*, 6647–6668. [CrossRef]

https://doi.org/10.3390/atmos17030305

atmosphere

MDPI

Article

Remote Sensing Monitoring of Summer Heat Waves–Urban Vegetation Interaction in Bucharest Metropolis

Maria Zoran *, Dan Savastru and Marina Tautan

ITC Department, National Institute of R&D for Optoelectronics, 409 Atomistilor Street, MG5, 077125 Magurele, Romania
* Correspondence: mzoran@inoe.ro; Tel.: +40-723024834

Abstract

Through a comprehensive analysis of urban vegetation summer seasonal and interannual patterns in the Bucharest metropolis in Romania, this study explored the response of urban vegetation to heat waves' (HWs) impact in relation to multi-climatic parameters variability from a spatiotemporal perspective during 2000–2024, with a focus on summer HWs periods (June–August), and particularly on the hottest summer 2024. Statistical correlation, regression, and linear trend analysis were applied to multiple long-term MODIS Terra/Aqua and MERRA-2 Reanalysis satellite and in situ climate data time series. To support the decline in urban vegetation during summer hot periods due to heat stress, this study found strong negative correlations between vegetation biophysical observables and urban thermal environment parameters at both the city center and metropolitan scales. In contrast, during the autumn–winter–spring seasons (September–May), positive correlations have been identified between vegetation biophysical observables and a few climate parameters, indicating their beneficial role for vegetation growth from 2000 to 2024. The recorded decreasing trend in evapotranspiration from 2000 to 2024 during summer HW periods in Bucharest's metropolis was associated with a reduction in the evaporative cooling capacity of urban vegetation at high air temperatures, diminishing vegetation's key function in mitigating urban heat stress. The slight decline in land surface albedo in the Bucharest metropolis due to increased urbanization may explain the enhanced air temperatures and the severity of HWs, as evidenced by 41 heat wave events (HWEs) and 222 heat wave days (HWDs) recorded during the summer (June–August) period from 2000 to 2024. During the severe 2024 summer heat wave episodes in the south-eastern part of Romania, a rise of 5.89 °C in the mean annual land surface temperature and a rise of 6.76 °C in the mean annual air temperature in the Bucharest metropolitan region were observed. The findings of this study provide a refined understanding of heat stress's impact on urban vegetation, essential for developing effective mitigation strategies and prioritizing interventions in vulnerable areas.

Keywords: heat waves; urban vegetation; climate variables; land surface albedo; aerosol optical depth; Bucharest metropolitan area

1. Introduction

Global climate change, associated with increasingly frequent and intense heat wave events that kill thousands of people each year, is considered a global priority for sustainable development in terms of adaptation and mitigation [1,2]. Heat waves, defined as consecutive days of abnormally high temperatures exceeding local or regional norms, are among the most extreme climate events and negatively impact people's health, urban ecosystems, and infrastructure. HWs can amplify urban heat island (UHI) effects, exacerbating air and land-surface temperatures and triggering major impacts on urban vegetation ecosystems [3,4].

Urban vegetation (green areas, including parks, gardens, and urban forests) is gaining increasing significance in the context of sustainable urban development. As future projections predict, by 2030, the increase in the frequency, duration, seasonality, and intensity of HWs in the Euro–Mediterranean regions will lead to a decrease in summer precipitation rates [5]. As a consequence, in the context of forecasted climate change, the state of urban vegetation and its heat- and water-stress vulnerability will increasingly affect the urban microclimate, and adaptation strategies [6,7] will be adopted. During summer seasons, urban vegetation faced significant complex environmental heat and water stressors over the past 25 years in the south-eastern part of Romania. Thermal stress, exacerbated by the HWs and UHI effects, elevated day–night time temperatures, and changes in evapotranspiration rates, may alter the cooling efficiency of urban vegetation [8]. However, the interaction between air pollution and high air temperature adds a significant complexity to global warming, particularly the alarming rise during compound climate events, such as heat–ozone extremes driven by the increased temperatures in synergy with non-carbon pollutants [9,10].

Several studies investigating the physical mechanisms driving HWs in the Northern Hemisphere found that large-scale atmospheric dynamics and persistent synoptic anticyclones are responsible for prolonged HWs through subsidence, light winds, and warm-air advection [11–17]. Also, prolonged droughts can decrease soil moisture and increase surface sensible heat fluxes, amplifying HW [18–20]. Other studies have found that strong adiabatic warming within anticyclones and increased solar radiation due to clear skies, which trigger increased surface sensible heating, play a crucial role in producing high near-surface temperatures during HWs [21,22]. According to these studies, the surface sensible-heat flux plays a dominant role in increasing the daytime urban planetary boundary layer (PBL) height during HWs, which become deeper and hotter, providing crucial information for extreme-temperature weather climatology [23,24]. PBL climatology, its characteristics, and its relationship with meteorological parameters play a significant role during HWs, being affected by solar radiation and by deeper PBL during summer seasons [25,26]. Another important parameter in describing the radiative properties of the Earth's surface, urban land surface albedo (LSA), an essential primary climate variable (ECV), is a crucial tool of climate change at the local urban scale [27]. Moreover, climate parameters at both local and regional scales, and their seasonal variability, affect the interaction between heat waves and vegetation [28–31]. Heat stress induced in vegetation by high air and land surface temperatures in synergy with water stress can impact the possible physiological or biochemical mechanisms of urban green, being a critical limiting factor of vegetation growth and development [32–35], from the cellular to whole-vegetation scale, reducing photosynthesis, stomatal conductance, growth, reproduction, and leading to premature leaves deaths depending on the heat stress episode's duration, frequency, and severity [36–39].

Also, during heatwaves, air pollution negatively impacts urban vegetation health through the retention of particulate matter (PM2.5 and PM10) by dry and wet atmospheric deposition onto leaf surfaces, increasing light reduction, blocking stomata, and reducing photosynthesis [40–42]. Absorption of surface-level gaseous air pollutants (nitrogen dioxide—NO_2, ozone—O_3, and sulfur dioxide—SO_2, carbon monoxide—CO, formaldehyde—HCHO,

 https://doi.org/10.3390/atmos17010109

volatile organic compounds—VOCs) can cause chlorophyll degradation and induce oxidative stress in plants [43–45]. CO_2 promotes urban vegetation health through the fertilization effect, while O_3 causes oxidative damage to vegetation [46,47]. Urban aerosols exert dual effects on urban vegetation by altering surface radiation for photosynthesis.

Satellite-based remote sensing monitoring of HW events allows the regional-to-global analysis of urban vegetation ecosystem responses to climate change, and may prove helpful in identifying trends in urban vegetation health, for achieving UN SDGs goals 3, 11, and 13: "Ensure healthy lives and promote well-being for all at all ages", "Make cities and human settlements inclusive, safe, resilient and sustainable", and "Take urgent action to combat climate change and its impacts [48].

While several studies investigated the role of urban vegetation in air pollution removal capacity [49–53], or mitigating heatwaves and urban heat island influenced by high levels of air pollutants in large cities by conducting extensive air quality measurements and using analytical methods [54–56], limited research focused on the response of urban vegetation to summer heat stress and climate variability from a spatiotemporal perspective [57–60]. The year 2024 was the warmest on record at both the global and European scales, with the global mean surface temperature exceeding the pre-industrial level by more than 1.5 °C, and Europe was the fastest-warming continent at twice the global average [61].

In the Bucharest metropolitan area in Romania, urban vegetation has faced increasingly complex environmental stressors over the past 25 years, including the 2024 summer HWs, which were the warmest on record. This study explored the impact of heatwaves on urban/periurban vegetation in Bucharest from a spatiotemporal perspective during the 2000–2024 period, focusing on summer (June–August) months and heatwave periods in relation to climate seasonality. Time series of in situ monitoring data have been used in synergy with biophysical variables generated by the NASA MODIS Terra/Aqua satellites, the NCEP/NCAR and Giovanni MERRA-2 reanalysis datasets, and analytical processing methods, along with statistical cross-correlation and regression analysis. The present study addresses the need to assess the impacts of multiple climatic parameters on biophysical observables of urban vegetation during the 2000–2024 period, for the interannual and summer seasons (June–August), in the Bucharest city center and the metropolitan area. The main objective of this study was to comparatively quantify the effects on vegetation of the main environmental parameters (mean air temperature at 2 m height (TA) and its maximum value (TAmax), land surface temperature (LST), air relative humidity (RH), wind speed intensity (w), surface solar irradiance (SI), total aerosol optical depth at 550 nm (AOD), land surface albedo (LSA)) on vegetation parameters (normalized difference vegetation index (NDVI), leaf area index (LAI), the fraction of absorbed photosynthetically active radiation (FPAR), evapotranspiration, and net primary productivity (NPP)). The number of summer HW events (HWEs) and the number of HW days (HWDs) per summer season have been considered.

2. Materials and Methods

2.1. Study Area

The study site, Bucharest metropolitan region, is situated in the Romanian Plain, on the banks of the Dâmbovița River, a small tributary of the Danube, and the south-eastern part of Romania, and south-eastern Europe. It is bounded by latitudes 44.33° N and 44.66° N and 25.90° E and 26.20° E longitudes. Its center is placed at 44.4355381° N Latitude and 26.100049° E Longitude (Figure 1). The Bucharest metropolitan area, with approximately 2.2 million inhabitants and a 1811 km^2 area, includes Bucharest city, the capital of Romania,

 https://doi.org/10.3390/atmos17010109

with 1.8 million people and a 240 km^2 area, of which urban green infrastructure (forests, parks, small green gardens) covers almost 10% of its area [62].

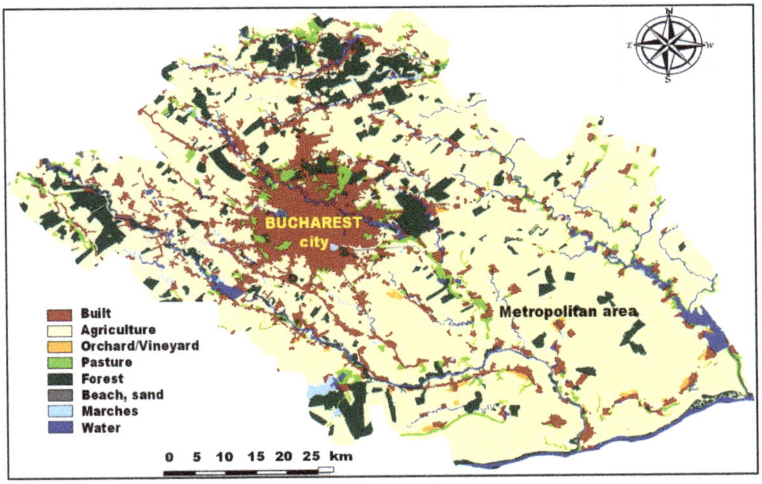

Figure 1. Bucharest city in the metropolitan test area.

The periurban areas, distributed across five counties, cover approximately 543 km^2 and have a population of approximately 0.81 million. The most important relief units are Vlăsiei Plain and Mostiştea Plain, with low altitudes ranging between 5 and 150 m. In the Bucharest metropolitan area, significant human-related impacts associated with urbanization and climate change have altered natural vegetation land cover, reducing its extent and altering its floristic structure and composition, thereby favoring aridization processes. Over the last 30 years, the metropolitan vegetation land cover in Bucharest has decreased, from 4839 ha in 1993 to 4506 ha [63], encompassing multiple urban-to-rural transition areas and comprising different vegetation types (shrubs, grass, forest, crops, etc.). The rapid urbanization of Bucharest after 1990, along with urban expansion, reduced vegetation cover and increased impervious surfaces, contributed to higher air and land surface temperatures associated with HWs and UHIs recorded during the summer (June–August) period. According to the Copernicus Urban Atlas in 2018, land use land cover (LULC) distribution (km^2) of different surface land cover classes in the Bucharest metropolis was: 53% agricultural area, 33.6% artificial area, 10.7% natural areas, 2.4% water, and 0.2% wetland [64]. Urban green infrastructure in Bucharest's metropolis, including periurban forests, provides recognized cultural ecosystem services that are important for enhancing social well-being and resilience [65]. With a diverse landscape, including the spatial distribution of different land cover types and flat plains, Bucharest is one of the most polluted cities in Romania and Europe, ranking 9th out of 96 most polluted cities in Europe in 2022 [66].

With a temperate continental climate and Western European Climate Circulation, Mediterranean Cyclones, and the East-European Anticyclone, Bucharest's climate is characterized by very hot summers, particularly during frequent heat waves recorded since 2000, and cold, humid winters. The annual mean air temperature belongs to the (10.6 to 11.5) °C range, for periurban to urban areas, while the mean annual rainfall belongs to the (613.2 to 578.9) mm range for periurban to urban areas. Frequent HWs occur during the

https://doi.org/10.3390/atmos17010109

June–August period, and sometimes extend from May to September. Co-occurrence with UHI events results in this area experiencing significantly higher summer temperatures.

2.2. Data Sources

To analyze interannual spatiotemporal urban vegetation greenness and summer heatwaves—urban vegetation interaction in Bucharest metropolitan area during (June–August) periods of 2000–2024, this study used available observational, satellite remote sensing, and reanalysis data provided by various free sources. The climate and other environmental time series datasets were supplied by different monitoring networks and satellite platforms:

1. *Meteorological driving factors.* Air temperature (TA) at 2 m height and its maximum values TAmax, air relative humidity (RH), air pressure (p), wind speed intensity (w) and direction, planetary boundary layer height (PBL), and solar surface irradiance (SI) have been provided by MERRA-2 Version 2 (Modern-Era Retrospective Analysis for Research and Applications) [67], C3S (Copernicus Climate Change Service) [68], and the online database NASA's Center (GES DISC) Geospatial Interactive Online Visualization and Analysis Infrastructure (GIOVANNI) V4.28 via its portal [69], as well as from NOAA Physical Sciences Laboratory [70].
2. *Air Pollution load.* Time series of total aerosol optical depth (AOD) at 550 nm were supplied by MERRA-2 Version 2 (Modern-Era Retrospective Analysis for Research and Applications) on the GIOVANNI platform.
3. *MODIS LST data.* From the Terrestrial Ecology Subsetting & Visualization Services (TESViS) Global Subsets Tool at the ORNL DAAC, MODIS Terra/Aqua data [71]. Among the available LST products developed through the different retrieval algorithms based on TIR sensors from different satellite missions (MODIS, LandsatTM/ETM+/OLI, AVHRR, SENTINEL, AMSR-E, AATSR, VIRR), this study considered MODIS Terra/Aqua to be a suitable data source for LST monitoring due to its high observation frequency, moderate spatial resolution, and free availability [72,73]. DAAC, MOD11A2 LST_Day_1 km and MOD11A2 LST_Night_1 km collected within 8 days [74], provide time series LST data for the Bucharest metropolitan area. The root mean square error (RMSE) of the MODIS LST data is within 2.0 °C and exhibits high accuracy in the major global cities [75].
4. *Vegetation MODIS NDVI data.* MOD13Q1 MODIS/Terra Vegetation Indices NDVI/EVI 16-Day L3 Global 250 m SIN Grid V06116-day, *MODIS NDVI data*, from TESViS.
5. *Vegetation MODIS LAI/FPAR data.* MOD15A2H MODIS/Terra Leaf Area Index/FPAR 8-Day with 500 m spatial resolution, mainly for their capacity to detect anthropogenic and climate impacts on urban vegetation land cover changes, from TESViS.
6. *Vegetation Evapotranspiration MODIS data.* MOD16A2ET_500 m, at 8-day data for evapotranspiration monitoring, from TESViS.
7. *Vegetation Annual Net Primary Production data.* MOD17A3HGF v061, MODIS/Terra Net Primary Production Gap-Filled Yearly L4 Global 500 m SIN Grid, from TESViS.
8. *Land Surface Albedo MODIS LSA data.* MCD43A1 MODIS/Terra + Aqua BRDF/Albedo Model Parameters Daily L3 Global—500 m V061, from TESViS.

2.3. Methodology

This study aims to provide insight into the spatiotemporal evolution of the urban vegetation ecosystem in one of the most dynamic urban–rural systems in south-eastern Europe, namely the Bucharest metropolitan area. Our analysis focused on annual and summer (June–August) datasets spanning the entire period of 2000–2024. For this research, a heat

https://doi.org/10.3390/atmos17010109

wave analysis has been performed using time-series data from the Copernicus GIOVANNI platform. A robust methodology was used to investigate, in a comprehensive manner, the complex interplay between summer intense HWs and heat stress on urban green land cover, in relation to environmental variables, incorporating satellite and in situ time-series datasets and an approach less commonly employed in urban vegetation contexts.

In climate science, HWs are defined as extended periods of excessive heat [76,77], characterized by different measures for each of these features (intensity, maximum amplitude, duration, frequency, and spatial extent) [78–80]. This study uses the definition of HWs as persistent periods of at least 3 consecutive days of high air temperature anomalies, during which the daily maximum TA exceeds the 90th percentile in a given region [81]. Also, as a main characteristic of HWs, the heat wave duration (HWD) is defined as n, the total number of heat wave days in the event [82].

Cross-correlation statistical analysis was adopted for the similarity between two time series data of the outdoor daily, monthly or annual averages of meteorological multiparameters (TA, RH, SI, PBL, LST, LSA), and vegetation parameters (NDVI, LAI, FPAR, ET, NPP) in Bucharest capital of Romania (both center defined by a 6.5 km $\times$ 6.5 km area, and metropolitan area defined by a 4.5 km $\times$ 40.5 km area). NDVI values range from -1.0 to 1.0, representing vegetation for positive values; clouds, water, and snow for negative values; and bare soils and rocks for values close to zero [83]. When vegetation dries out or becomes sick, its color usually turns from green to brown. This change is particularly significant in the near-infrared region of the spectrum, where absorption increases when vegetation is damaged. Therefore, NDVI can accurately detect chlorophyll and is related to plant health [84].

Cross-correlation analysis, nonparametric test coefficients, trend analysis, and linear regression have been used to assess the dependence between dependent and independent variables in the time series of mean environmental parameters. The strength of relationships between variables was described by linear regression values, ranging from 0 to 1, with values closer to 1 signifying stronger connections.

Because climate observables exhibit non-normal distributions during extreme events such as HWs, the Kolmogorov–Smirnov test of normality has been applied to assess the normality of daily or monthly mean time-series datasets [85]. This method is efficient for analyzing the diverse characteristics of vegetation phenology across different spatio-temporal scales of summer seasonal variation, and especially in complex urban environments. The Spearman rank correlation coefficient, r, ranging from -1 to 1, with higher absolute values indicating stronger correlations, is used to measure the strength and direction of both linear and nonlinear relationships between variables. Significant correlations correspond to a p-value between 0.01 and 0.05. These statistical methods were applied to investigate associations between independent variables (explanatory variables), such as NDVI, LAI, FPAR, ET, NPP, and AOD, and the dependent variable (response variable), which represents TA, LST, LSA, HWE, and HWD. The main novelty of this research is the introduction of correlations between urban vegetation and environmental parameters under HWs, which can help understand trends in climate change and temperature variation in the present and future periods. To determine the statistical significance of the correlation, we used the p-value ($p < 0.01$). ORIGIN 10.0 software version 2021 for Microsoft Windows, and ENVI 5.7 were used for data processing. All time series datasets were preprocessed using the TIMESAT 3.2 software. Several preprocessing levels were tested, including smoothing and phenometrics computation.

https://doi.org/10.3390/atmos17010109

3. Results

Climate change intensifies heat stress on urban vegetation ecosystems, creating an urgent need to understand how climate variables influence vegetation health. The present study explored NDVI, LAI, FPAR, and NPP urban vegetation health biophysical variables' temporal relationship with different climate drivers, such as TA, TAmax, LSA, PBL height, RH, w, SI, and AOD, for a period spanning twenty-five years between 2000 and 2024 in the Bucharest metropolitan area. In a comparative analysis, the heat-stress impact of summer HWs on urban vegetation was examined from June to August.

3.1. Spatiotemporal Patterns of the Urban Vegetation During 2000–2024 in Bucharest

Urban vegetation NDVI daily means exhibit annual patterns in Figure 2 (2000–2024 at 16-day intervals), with clear seasonality: maximum values during summer and minimum during winter.

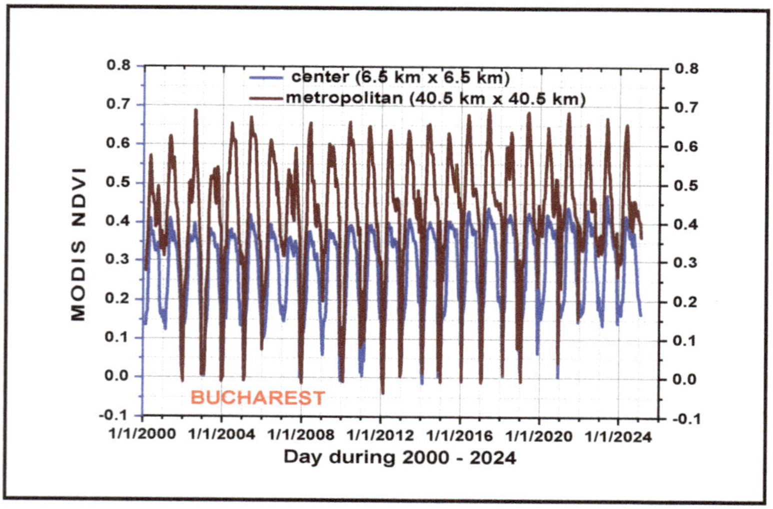

Figure 2. Spatio-temporal patterns of daily means NDVI for Bucharest city center area and metropolitan area during the 2000–2024 period.

For the analyzed period, the daily mean NDVI registered values of (0.2736 ± 0.11848) in the range (−0.00273 to 0.47214) for Bucharest center, defined by a (6.5 km × 6.5 km) area, corresponding to classified as slightly dense and stressed vegetation. At the Bucharest metropolitan scale, MODIS-derived NDVI registered a daily mean value of (0.42705 ± 0.15913) in the range (−0.0386 to 0.6922) for the defined area of (40.5 km × 40.5 km), corresponding to classified as moderately dense vegetation. During all seasons, at the metropolitan scale, LAI recorded a daily mean value of (0.91962 ± 0.68978) in the range of (0.1–2.8473), while FPAR recorded a daily mean value (0.3688 ±0.1761) in the range of (0.0734–0.7392).

The results of this analysis also suggest stress vegetation in both central and metropolitan areas and reveal a range from sparse to rich periurban vegetation influenced by diverse urban microclimates. The time of maximum greenness is similarly distributed from early June to early August, findings which are consistent with previous studies [86,87].

Annual computed mean time-series of the NDVI vegetation index show an overall increasing trend from 2000 to 2024 (Figure 3), explained by climate change at mid-latitudes, and therefore, vegetation appears more active. Similar results were reported in the scientific literature [88,89]. Urban vegetation satellite monitoring in the central part of the city and

 https://doi.org/10.3390/atmos17010109

metropolitan area is essential for efficient planning and management, especially in the south-eastern part of Romania, where HW events are more frequent.

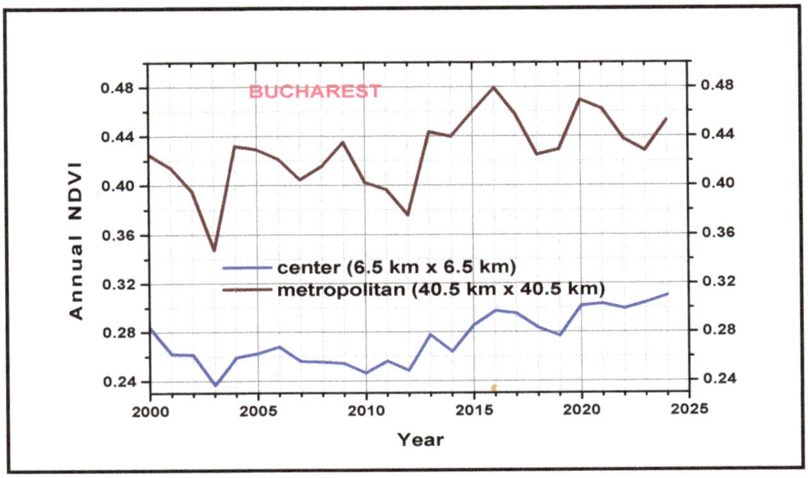

Figure 3. Annual patterns of daily means NDVI in Bucharest city center and metropolitan areas during the 2000–2024 period.

3.2. Variability in Heatwaves and Urban Vegetation Fluctuations over Summer (June-August) Seasons of 2000–2024

Summer 2024 recorded the most extreme and longest heatwaves in the South-eastern part of Europe, especially in July 2024, with persistent and stable drought conditions mostly in the Balkans and Romania [90,91]. From the NOAA NCEP/NCAR reanalysis Europe map in Figure 4, it is clear that an extreme high-air-temperature composite anomaly of more than 3.5 °C at 1000 mb was observed over South-eastern Europe and Bucharest, marked by a star during the summer months of 2024 [90].

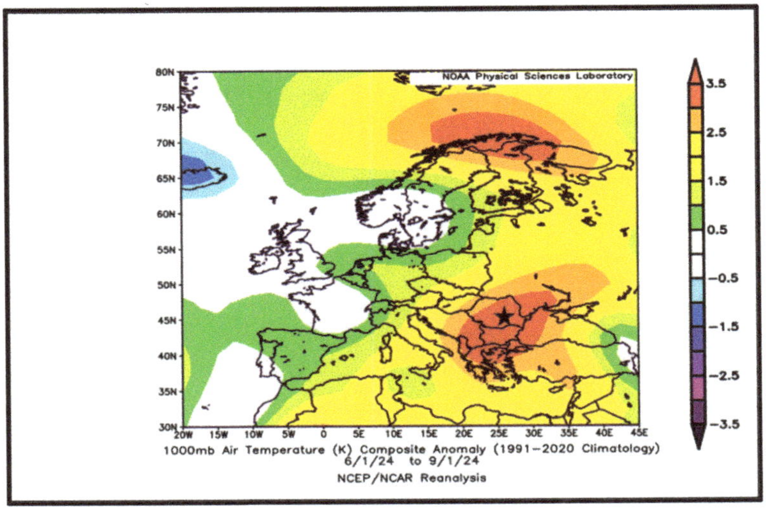

Figure 4. Air temperature composite anomaly map for Europe during the summer season (June–August) 2024.

https://doi.org/10.3390/atmos17010109

Between 2000 and 2024 summer (June–August) seasons, in Bucharest, a total of 41 independent summer heat wave events (HWE) were identified, corresponding to 222 heat wave days (HWD) with considerable variability in both duration and intensity, of which 2024 exhibited the strongest hot period with 5 HWE and 43 HWD. Neither duration nor intensity exhibited significant temporal trends. The most severe HWs associated with higher mean values of air TA max, lower mean air relative humidity RH, and lower values of mean vegetation indices NDVI, LAI, FPAR, and ET have been recorded in Bucharest during mostly summer seasons of 2000, 2012, and 2024 (Table 1).

Table 1. The most significant heatwaves recorded during the summer (June–August) periods in the Bucharest metropolis, with their main characteristics and corresponding mean values of TAmax, RH, and vegetation indices (NDVI, LAI, and FPAR).

Year (June–August)	HWE	HWD	Mean TA Max (°C)	Mean RH (%)	Mean NDVI	Mean LAI	Mean FPAR
2000	5	22	32.77	38.17	0.43805	1.11454	0.41469
2012	5	29	33.37	44.21	0.51408	1.57357	0.5270
2024	5	43	34.89	44.06	0.4313	0.98797	0.40939

Lower values of vegetation parameters recorded in 2000, 2012, and 2024 may indicate that stress induced by higher maximum air temperatures, in synergy with water stress during HWs, can affect the physiological or biochemical mechanisms underlying urban green infrastructure, a critical limiting factor for vegetation growth and development.

The increasing trends of the summer annual mean HWD and TAmax over the 2000–2024 period are clearly illustrated in Figure 5.

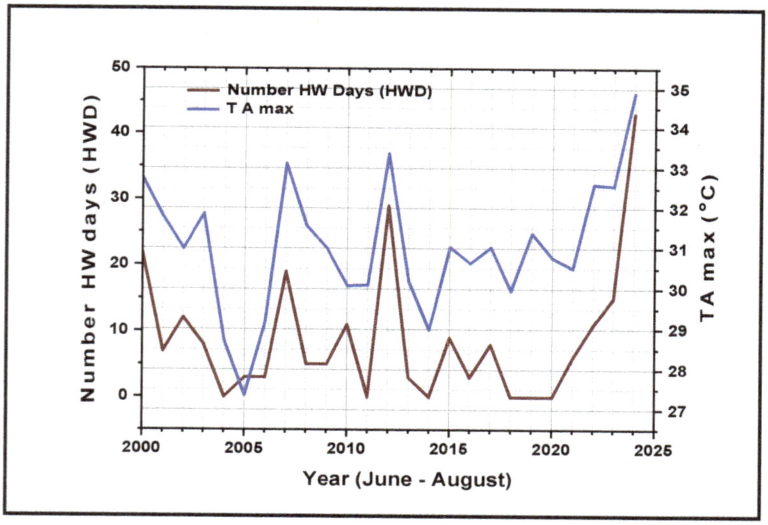

Figure 5. Temporal patterns of summer annual heat wave number (HWD) and annual mean maximum air temperature at 2 m height (TAmax) in Bucharest between 2000 and 2024.

3.3. Planetary Boundary Layer Height During Summer (June–August) Seasons of 2000–2024 and Its Impact on Urban Vegetation

Although large-scale persistent synoptic circulation anomalies are considered key factors in generating HWs, urban vegetation–atmosphere interactions and feedbacks seem

https://doi.org/10.3390/atmos17010109

to play a crucial role in HW intensity. During the summer (June–August) 2000–2024 periods, PBL height in Bucharest displayed a mean of (1170.81 ± 146.20 m in the range of (900.26–1547.63) m, and was positively correlated with daily mean TA, TAmax, and LST. A good correlation between PBL height and air temperature at 2 m height indicates (r = 0.55; $p < 0.01$), while better positive correlations are found for TAmax (r = 0.75; $p < 0.01$) and daily mean land surface temperature LST (r = 0.76; $p < 0.01$). However, the high temperatures during the HWs extend to a higher range, indicating stronger thermal convection and a higher PBL during these periods, results consistent with other studies [25,92]. Also, compared with the same months during 2000–2024, HWs significantly increased PBL height during the June–August period, results similar to those of other studies [93,94].

A strong anticorrelation between PBL height and key biogeophysical variables characterizing vegetation structure and functioning (NDVI, LAI, FPAR, ET, and annual mean NPP) in the Bucharest metropolitan area is shown in Table 2 and Figure 6. During the summer (June–August) 2000–2024 periods, PBL height in Bucharest displayed a mean of (1170.81 ± 146.20 m in the range of (900.26–1547.63) m, and was inversely correlated with vegetation observables NDVI (r = −0.71; $p < 0.01$), LAI (r = −0.74; $p < 0.01$), FPAR (r = −0.79; $p < 0.01$), and ET (r = −0.95; $p < 0.01$). It was also reflected in the annual mean NPP.

Table 2. Spearman rank correlation coefficients and p-values between the monthly means of vegetation variables and the main climate variables in the Bucharest metropolis during the June–August analysis period (2000–2024).

Vegetation/ Environmental Factors	PBL (m)	TA (°C)	TAmax (°C)	LST (°C)	LSA	RH (%)	SI (W/m^2)	w (m/s)	AOD	HWD
NDVI	−0.71 *	−0.76 *	−0.85 *	−0.83 *	−0.41 **	0.71 *	−0.68 *	−0.29 **	−0.72 *	−0.67 *
LAI	−0.74 *	−0.58 *	−0.76 *	−0.82 *	−0.27 ***	0.72 *	−0.63 *	−0.31 **	−0.71 *	−0.68 *
FPAR	−0.79 *	−0.66 *	−0.86 *	−0.86 *	−0.30 ***	0.78 *	−0.68 *	−0.30 **	−0.74 *	−0.68 *
ET (kg m^{-2} s^{-1})	−0.95 *	−0.55 *	−0.80 *	−0.28 ***	−0.37 **	0.87 *	−0.61 *	−0.42 *	−0.85 *	−0.60 *
NPP (g C m^{-2} year^{-1})	−0.77 *	−0.61 *	−0.80 *	−0.79 *	−0.10 ***	0.82 *	−0.57 *	−0.30 ***	−0.69 *	−0.64 *

Note: NDVI (normalized difference vegetation index), LAI (leaf area index), FPAR (photosynthetically active radiation), NPP (net primary production), PBL (planetary boundary layer height), TA (air temperature at 2 m height), TAmax (maximum air temperature at 2 m height), LST (land surface temperature), LSA (land surface albedo), RH (air relative humidity), SI (surface solar irradiance), w (wind speed intensity), AOD (total aerosol optical depth at 550 nm), HWD (total number of HW days per summer study period); * and ** indicate $p < 0.01$ and $p < 0.051$, respectively, and *** indicates $p > 0.05$.

3.4. Land Surface Albedo Variability and Impact on Vegetation in Bucharest Metropolitan Region

This study analyzed the spatiotemporal variations in urban land surface albedo in the Bucharest metropolitan area in Romania from 2000 to 2024, and assessed the relationships between the spatial and temporal patterns of the albedo and associated influencing factors (vegetation, air pollution, and climate parameters) on a seasonal and interannual basis using satellite products and reanalysis data.

As Figure 7 shows, the monthly mean LSA fluctuated over the 2000–20,024 period and exhibited a clear downward trend after 2020, which may explain the observed climate warming in the Bucharest metropolitan region. The derived MODIS multiannual mean LSA in the Bucharest metropolitan area in the 2000–2024 analyzed period was (0.2547 ± 0.04962), in the range of (0.13467–0.50673) for all seasons.

https://doi.org/10.3390/atmos17010109

Also, Spearman correlation coefficients between LSA and NDVI in Bucharest for all seasons during the entire analyzed period 2020–2024 show negative correlations for both the center city/metropolitan area, as follows: r = −0.56 ($p < 0.01$)/r = −0.36 ($p < 0.01$).

The 25-year albedo showed a weak decline trend, mainly due to a significant summer (June–August) reduction, reaching 30% at (0.17833 ± 0.0013) in the range (0.17493–0.18135). This research found an inverse correlation between LSA and NDVI, which explains the essential role of urban green infrastructure in reducing land surface temperatures, particularly during hot summers, heat waves, and urban heat islands.

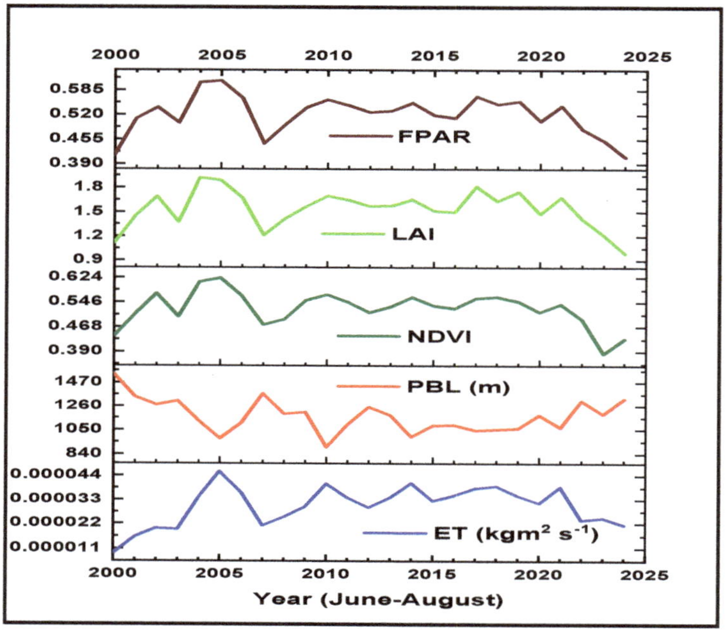

Figure 6. Temporal patterns of the planetary boundary layer (PBL) height and urban vegetation parameters in the Bucharest metropolitan area during the summers (June–August) of the 2000–2024 period.

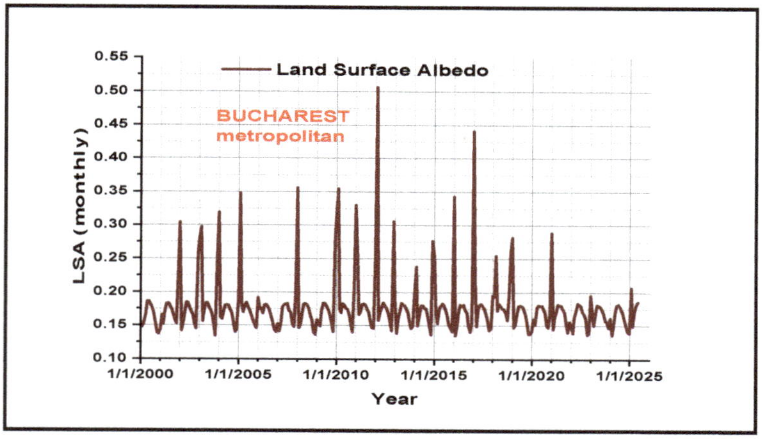

Figure 7. Temporal pattern of the monthly mean land surface albedo (LSA) in the Bucharest metropolitan area during the 2000–2024 period.

https://doi.org/10.3390/atmos17010109

As a consequence, the reduction in LSA enhanced TA, a result that may explain the high summer temperatures of 41 HWE and 222 HWD recorded during the summer (June–August) from 2000 to 2024. A clear anticorrelation was found between LSA and vegetation ET (r = −0.37; $p < 0.05$) during June–August seasons (Table 2). Also, significant decreases in LSA and ET were found throughout the summer (June–August) over a 25-year study duration (Figure 7).

Lower land surface albedo values recorded after 2020 in the Bucharest metropolitan region are associated with greater absorbed solar radiation and higher air and land surface temperatures, while reduced urban vegetation land cover and increased impermeable surfaces decrease ET. It leads to a higher proportion of energy being converted into sensible heat, driving higher temperatures and enhancing urban heat stress during HWs and the urban heat island effect.

Table 2 shows that the decreasing trends in ET and the strong anticorrelation between air temperature TA and its maximum TAmax, as well as between TA and evapotranspiration ET during summer HWs of the 2000–2024 period, are associated with reduced evaporative cooling effects, likely due to stomatal closure at high air temperatures. This effect will diminish vegetation's key function in mitigating urban heat stress.

This study also found an anticorrelation between LSA and urban NDVI (r = −0.41; $p < 0.05$), in good agreement with other studies exploring LSA variability and its impact on vegetation, which found that over vegetation-covered areas, interannual variation in albedo was more sensitive to NDVI [92,93].

Our results reveal a stronger negative correlation between LSA and LST (r = −0.80; $p < 0.01$) during the summer time in both Bucharest city areas with dominant impervious surfaces, hurting the urban thermal environment. Broadband albedo, which measures urban surface properties, also depends on the atmospheric conditions.

3.5. Air and Land Surface Temperature Variability and Their Impacts on Urban Vegetation During Summer Seasons

Daily and monthly mean surface air temperature (TA) and its maximum value, TAmax, measured at around 2 m above the surface, have significant impacts on both human and natural ecosystems, affecting urban vegetation and biodiversity, agriculture, and energy demand. Due to the synergy of HWs and UHIs during hot summers, this study found in Bucharest's center, with increased artificial built areas, higher temperatures at 2 m height of 3.6 °C compared to the adjacent, less urbanized periurban zones. The findings of this study support the need to increase urban green spaces, such as street trees, parks, and green roofs, to provide shading and evaporative cooling and mitigate urban heat island effects, especially during heat waves in the Bucharest metropolis.

Also, as one of the most important Earth System variables and a key parameter in urban climate research, land surface temperature (LST) reflects the interaction between solar radiation and the land surface, influencing thermal radiation from the vegetated areas or the ground. In urban areas, LST provides a crucial indicator of HWs and UHIs' effects, vegetation ecosystem health, energy consumption, and responses to climate change [94,95]. Also, this study analyzed the interannual variations in LST data and their relationships with air temperature, AT. The rank correlation analysis shows that, at the pixel-scale, during summer seasons (June–August) 2000–2024, air temperature at 2 m height, TA, and LST present a strong positive correlation (r = 0.96; $p < 0.01$). As the physical climate system is highly sensitive to changes in land surface albedo, urban/periurban vegetation systems could significantly feed back into projected climate change scenarios. As Table 2 shows, during the June–August summer analysis period from 2000 to 2024, TA, TAmax, LST, and

https://doi.org/10.3390/atmos17010109

HWD display strong inverse correlations with the vegetation variables NDVI, LAI, FPAR, and NPP, which means that urban vegetation's growth is limited during persistent HWs.

Table 2 and Figure 8 clearly show existing anticorrelations between TA, TAmax, LST, and ET, and LSA. Computed for the Bucharest metropolitan area the monthly mean TAmax recorded during the summer (June–August) 2000–2024 periods was (31.06 ± 1.64) °C in the range of (27.36–34.89) °C, while evapotranspiration ET registered very low mean value of ($3.02654 \times 10^{-5} \pm 9.02518 \times 10^{-6}$) kg m^{-2} s^{-1} in the range of (8.99333×10^{-6}–4.58667×10^{-5}) kg m^{-2} s^{-1}.

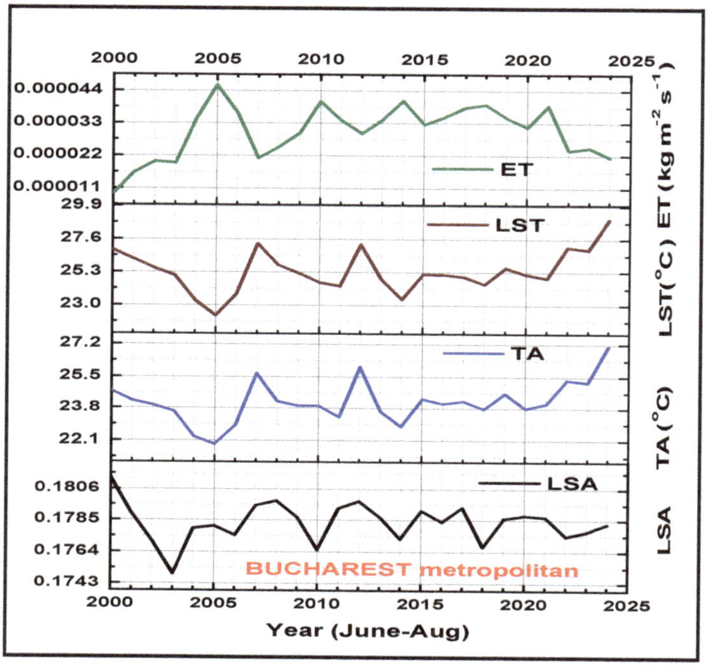

Figure 8. Temporal pattern of the monthly mean land surface albedo (LSA), air temperature at 2 m height (AT), land surface temperature (LST), and evapotranspiration (ET) in the Bucharest metropolitan area during the summers 2000–2024 period.

This finding confirms a decrease in urban vegetation's cooling effect during summer heat stress associated with HWs and higher temperatures. However, urban vegetation's evapotranspiration (ET) has significant cooling effects, helping mitigate the summer urban thermal environment and reduce heat exposure risks in large metropolitan areas [96,97]. Despite increased air temperatures during HWs, urban vegetation continued to cool through evapotranspiration, but further studies are needed to identify the best-suited vegetation species to optimize cooling.

3.6. Air Relative Humidity, Surface Solar Irradiance, and Aerosol Optical Depth Impact on Urban Vegetation During Summer Seasons

This research found that higher values of air relative humidity RH enhance vegetation's growth and evapotranspiration, and surface heat exchange, thereby strengthening vegetation's cooling effect. This finding is consistent with other studies [98–103]. Knowing the relationship between evapotranspiration and heat stress during HWs is essential for effective water resource management, irrigation planning, and climate change assessments

https://doi.org/10.3390/atmos17010109

in urban vegetation systems [104–106]. As key thermal and water-related factors influencing urban microclimates, air relative humidity and solar radiation have high impacts on urban vegetation growth during summer hot periods.

The results of this study revealed that among several climate factors during frequent HWs recorded in the summers 2000–2024 in the Bucharest metropolitan region, solar surface irradiance SI was an important influential factor limiting the urban vegetation growth. From Figure 9 and Table 2, it is evident that increased solar radiation during heatwaves affects urban vegetation growth, as indicated by lower values of NDVI, LAI, and FPAR.

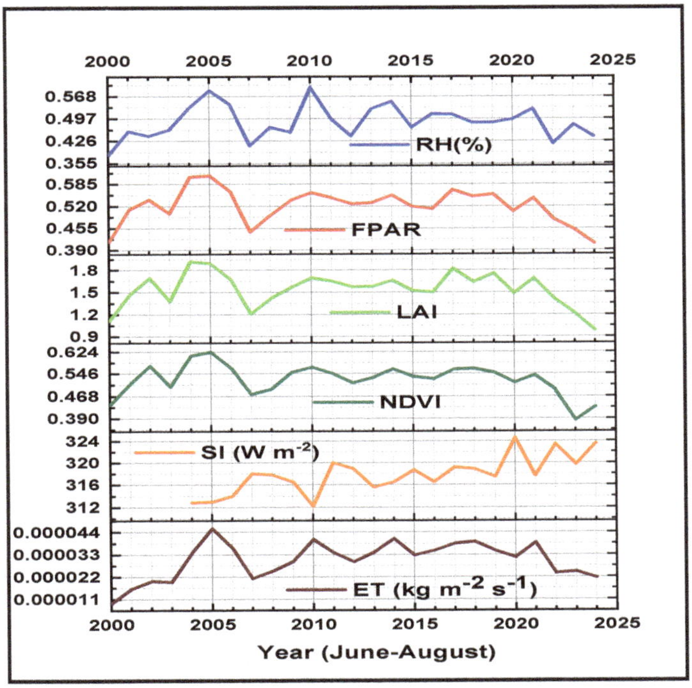

Figure 9. Temporal patterns of the monthly mean air relative humidity (RH), surface solar radiation (SI), and vegetation characteristics NDVI, LAI, FPAR, and ET, in the Bucharest metropolitan area during the summer (June–August) 2000–2024 periods.

Figure 9 and Table 2 clearly show decreasing trends in the monthly mean summer (June–August) time series of urban vegetation parameters NDVI, LAI, FPAR, and ET, associated with decreasing air relative humidity and increasing solar radiation during heatwaves, which affect urban vegetation growth.

During the summers 2000–2024 period, the total aerosol load over Bucharest metropolitan area AOD at 550 nm had an increased monthly mean value (0.26413 ± 0.05526) in the range (0.19767–0.39567), being negative correlated with all monthly mean vegetation indices NDVI (r = −0.61; $p < 0.01$), LAI (r = −0.71; $p < 0.01$), FPAR (r = −0.74; $p < 0.01$), ET (r = −0.83; $p < 0.01$). Positive correlations between AOD and air temperature TA (r = 0.59; $p < 0.01$), TAmax (r = 0.76; $p < 0.01$), and HWD (r = 0.73; $p < 0.01$) confirm that increased air pollution in synergy with air temperatures during summer HWs may have negative impacts on urban vegetation health. Environmental parameters primarily govern the diurnal and seasonal variations in vegetation characteristics, including NDVI, LAI, FPAR, NPP,

https://doi.org/10.3390/atmos17010109

and ET, in Bucharest's metropolis. According to Table 2, higher levels of air temperature at 2 m height (AT), ATmax, land surface temperature (LST), planetary boundary layer height (PBL), wind speed intensity (w), surface solar irradiance (SI), and total aerosol optical depth at 550 nm (AOD) decrease urban vegetation's cooling effect during heatwaves and vegetation's growth. Long-term changes in vegetation are essential indicators for understanding the response of ecosystems to climate change. NDVI and other vegetation parameters could be used for urban ecosystem monitoring and as an additional criterion for assessing the impact of extreme climatic events.

In a comparison with Table 2, Table 3 shows that while during summer seasons, vegetation greenness indicators NDVI were inversely correlated with TA, TA max, LST, PBL heights, and SI, during September–May months, corresponding to autumn–winter–spring seasons of 2000–2024, these turned to positive correlations in both the Bucharest center area as well as at the metropolitan scale. Also, opposite correlations were found for air relative humidity RH, which was positively correlated with NDVI during summer seasons and negatively correlated during autumn–winter–spring seasons. In all seasons, land surface albedo LSA, total aerosol optical depth (AOD), and wind speed intensity (w) were negatively correlated with all vegetation parameters, highlighting their damaging effects on urban vegetation.

Table 3. Spearman cross correlation coefficients between monthly mean NDVI variables in Bucharest center city/metropolitan area, and main climate variables during autumn–winter–spring seasons, September–May analysis period 2000–2024.

NDVI Vegetation//Environmental Factors	PBL (m)	TA (°C)	TAmax (°C)	LST (°C)	LSA	RH (%)	SI (W/m^2)	w (m/s)	AOD
NDVI Bucharest Center area (6.5 km × 6.5 km)	0.78 *	0.86 *	0.74 *	0.89 *	−0.58 **	−0.65 *	0.85 *	−0.24 **	−0.63 *
NDVI Bucharest metropolitan area (40.5 km × 40.5 km)	0.62 *	0.52 *	0.51 *	0.67 *	−0.35 **	−0.47 *	0.67	−0.29 **	−0.51 *

Note: NDVI (normalized difference vegetation index), LAI (leaf area index), FPAR (photosynthetically active radiation), NPP (net primary production), PBL (planetary boundary layer height), T (air temperature at 2 m height), TAmax (maximum air temperature at 2 m height), LST (land surface temperature), LSA (land surface albedo), RH (air relative humidity), SI (surface solar irradiance), w (wind speed intensity), AOD (total aerosol optical depth at 550 nm); * and ** indicate $p < 0.01$ and $p < 0.05$, respectively.

3.7. Hot Summer 2024 Analysis

Due to rapid urbanization and extensive infrastructure development, Bucharest is highly vulnerable to extreme heat events [107]. By tracking NOAA positive temperature anomalies relative to the 1991–2020 climatology, Figure 3 shows the extreme hot summer of 2024, when temperatures exceeded 3.5 °C. Cumulative positive anomalies highlight the intensity and duration of extreme heat events over time, with high values suggesting the development of severe heat wave conditions. The temporal variations in the monthly mean air temperature at 2 m height (TA), its maximum mean value (TAmax) and its minimum mean value (TAmin) in the Bucharest metropolitan area during the summer (June–August) 2024 period are illustrated in Figure 10 that display very high temperatures (daily mean TAmax = 34.89 °C, daily mean TA = 27.11 °C, and daily mean TAmin = 20.07 °C especially during July observed when some typical heat wave processes took place. This period was important due to 5 significant heat wave events, totaling 43 HW days. Urban vegetation characteristics, NDVI, LAI, and FPAR in the Bucharest metropolitan area registered lower values compared with other summer periods when there have not been HWs (Figure 5).

https://doi.org/10.3390/atmos17010109

Also, during the severe 2024 summer heat wave episodes in the south-eastern part of Romania, this study identified increases of 5.89 °C in mean annual LST and 6.76 °C in mean annual TA in the Bucharest metropolitan region.

Understanding the relationship between HWs' interaction with urban vegetation in the Bucharest area, in a comparative analysis for both the city center and the metropolitan area, and urban heat during HWs is crucial for mitigating the thermal environment and improving environmental quality.

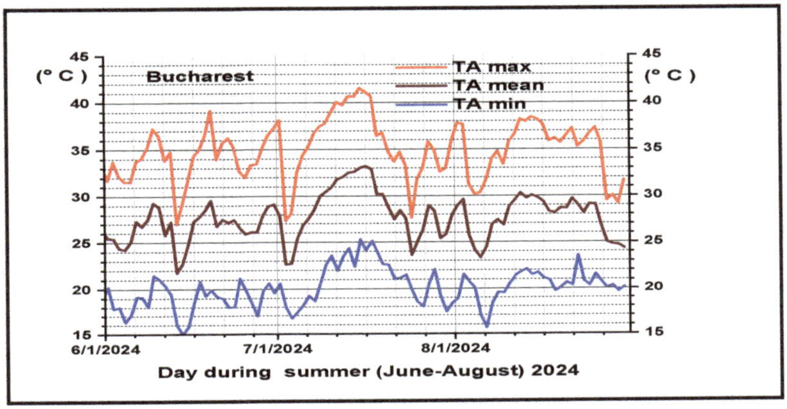

Figure 10. Temporal pattern of the daily mean air temperature at 2 m height TA, its maximum mean value TAmax, and its minimum mean value TAmin in the Bucharest metropolitan area during the summer (June–August) 2024 period.

4. Discussion

Successive record-breaking summer temperatures, both globally and in south-eastern Europe, including Romania, raise the urgent question of how to better protect vulnerable people's health and urban vegetation ecosystems. Also, compound drought–air pollution–heat wave events are associated with several risks, including vegetation fire risk during summer extreme hot temperatures [85]. Several recent studies analyzed the impacts of HWs on vegetation in large metropolitan areas.

As essential metrics for assessing urban vegetation greenness and health from satellite sensors, the normalized difference vegetation index (NDVI) indicating photosynthetic activity, is a strong indicator for essential climate variables (ECVs) like as leaf area index (LAI), and fraction of absorbed photosynthetic radiation (FPAR) provided by satellite images, which show their condition based on how much light of different wavelengths plants absorb and reflect without in situ measurement data. In urban ecosystems with high spatial heterogeneity, vegetation responses to climate and air pollution variability at the city scale exhibit diverse spatiotemporal patterns. As a global index used to analyze the characteristics of vegetation land cover and its dynamic spatiotemporal patterns, NDVI is a key parameter for assessing vegetation stability and its complex interaction with climate and anthropogenic stressors. LAI and FPAR indicators were used to characterize the effects of climate parameters and urbanization on vegetation health and growth, and to assess their changes. Urban vegetation phenology in LAI is the primary driver of seasonal variation in FPAR, which, in turn, drives photosynthesis. While FPAR absorbed by urban vegetation is a key biophysical variable for monitoring vegetation growth, productivity, and the energy balance of urban ecosystems, evapotranspiration (ET) is a key process in the water cycle. Moreover, monitoring variations in these indices over time allows scientists

 https://doi.org/10.3390/atmos17010109

to track seasonal development, detect early signs of stress, and plan interventions before damage becomes critical. During summer hot periods, a sudden decline might indicate drought stress, disease or pest outbreaks, nutrient deficiencies, or harvesting.

Urban vegetation in the Bucharest metropolitan area is sensitive to climate change, being significantly affected by variations in climate parameters and exhibiting considerable differences in distribution patterns during summer heat waves and during the autumn–winter–spring seasons. However, during the 2000–2024 period, as reported in other studies in Europe [102–106], vegetation greenness in Bucharest's metropolitan area, as measured by NDVI, showed a slight increasing trend after 2000, with 2000 identified as a turning point. This was explained by climate change in European mid-latitudes, and therefore, vegetation is observed to be more active.

There is an important interplay between deeper and hotter planetary boundary layer PBL, especially during the daytime HWs, and its synergies with the enhanced urban heat island (UHI) effect, and urban vegetation, which can alter how heat is mixed and how it interacts with large-scale synoptic weather patterns, affecting turbulence and air quality [107–109]. It was demonstrated that heatwaves are driven by persistent large-scale circulation anomalies, and that land–atmosphere interactions intensify and propagate them. In urban areas, various vegetation types play a vital role in land–atmosphere interactions by modulating energy and water exchange through different pathways. Similar to other studies, the present research found a strong anticorrelation of urban greenness during the summer seasons with PBL heights between 2000 and 2024 in Bucharest [110–112].

While HWs are significantly associated with high air and land surface temperatures, it is also crucial to examine changes in the other meteorological factors, especially air relative humidity, wind speed intensity, and solar surface irradiance, which are critical controls of urban thermal comfort and heat stress on vegetation [113–116]. A serious impact on local and regional climate due to the alteration of the effective surface albedo, a change in land surface albedo can have a positive (cooling) or negative (warming) effect on climate change [117,118].

Land surface albedo (LSA), air temperature at 2 m (TA), and land surface temperature (LST) are thermophysical parameters that define the urban thermal environment, including urban heat islands (UHIs) and heat waves (HWs) [119,120]. These parameters are correlated such that materials with low albedo (low reflectivity of solar radiation) exhibit higher heat absorption and, consequently, higher TA and LST values, intensifying the effects of UHI and HWs. While Bucharest center predominantly consists of impervious surfaces such as roofs, parking lots, streets, and sidewalks, which have low albedo, absorbing significant amounts of heat, in contrast, periurban metropolitan counties with increased vegetation land cover mitigate heat through natural shading and evapotranspiration. However, the ongoing conversion of green spaces with concrete and asphalt exacerbates the effects of urban heat stress on the vegetation ecosystem [121,122]. However, long-term vegetation dynamics that significantly impact the climate are reflected by the dramatic variations in urban land surface albedo. The slight decline in land surface albedo in the Bucharest metropolis due to increased urbanization may explain the enhanced air temperatures and the severity of HWs, as evidenced by 41 heat wave events and 222 heat wave days recorded during the summer (June–August) period from 2000 to 2024. Bucharest, densely packed with heat-retaining concrete and limited green space, is highly vulnerable to summer heatwaves and the urban heat island (UHI) effect. Land surface albedo and heat waves in cities are closely linked because low-albedo urban surfaces (like dark, impermeable materials) absorb more solar radiation, significantly increasing urban heat island (UHI) effects and intensifying heat waves. Conversely, increasing the albedo of urban surfaces—making

https://doi.org/10.3390/atmos17010109

them more reflective—is a key strategy to reduce daytime heat absorption, lower surface temperatures, and mitigate the negative impacts of heat waves on urban environments and human thermal comfort [123,124].

Quantifying the impact of TA, TAmax, and LST during HWs on urban vegetation in the highly heterogeneous metropolis like Bucharest is challenging. However, this research found that the normalized difference vegetation index (NDVI) is the most significant indicator of urban vegetation affecting LST with a significant cooling effect, particularly in summer. The total effects of all vegetation indicators are anticorrelated with LST, with the greatest impact in spring and the least in winter. Also, seasonal metropolitan morphology has a high impact on the urban thermal environment during HWs [125,126].

The interplay between urban heat environments and air pollution is deeply interconnected; their synergistic effects, associated with stagnant synoptic conditions and lower PBL heights, reduced wind speeds, and reduced turbulent flux, reduce the dispersion of air pollutants [127].

Like other scientific publications [128], this study found that heat waves and high air temperatures can severely stress urban vegetation in Bucharest, thereby reducing the cooling effect it provides to the metropolis. This effect is explained by vegetation stomatal closure, which conserves water and reduces evapotranspiration, thereby diminishing its cooling effect. However, this stress cycle can damage vegetation health, limit its growth, and make cities more vulnerable to extreme heat. It was demonstrated that local and regional background climate conditions, vegetation types, and spatial configuration influence the cooling effectiveness of urban vegetation [129]. In densely built-up urban areas with limited vegetation cover, there is a significant lack of summer thermal regulation during the day, especially in the early morning and late evening, thus exacerbating thermal stress. Other studies considered that thermal impacts of heat waves vary by urban green space type, with broad-leaved forests showing the strongest cooling [130]. Also, urban impervious-surface expansion reduces ET, offsetting the increase from anthropogenic heat. The replacement of natural landscapes like forests and grasslands by built impervious surfaces is leading to changes in meteorological factors, such as air temperature, humidity, precipitation, and wind speed near the ground in urban areas, and further affecting the energy and water balance. An additional heat load during heat waves must account for the heat generated by densely populated areas and frequent human activities. However, the mechanisms linking urban vegetation to climate drivers are very complex at temperate latitudes; therefore, a targeted investigation is required to detect possible trends in evapotranspiration and in urban surface morphology and roughness, to support understanding of the dynamics of urban vegetation during heat waves. Several studies found that heat stress negatively affects vegetative photosynthesis, thereby affecting tree growth and, eventually, survivability [122–124].

The mechanisms underlying the interactions among urban vegetation health and growth, heat stress, and HWs are of immense importance for assessing vegetation cooling effects and developing mitigation measures. Another aspect to consider is the existing differences among plant and tree species in their tolerance and adaptation to climate change [82,84,125]. Understanding the interplay between summer heatwaves, urban vegetation, air pollution, and climate variables is essential for developing evidence-based public health strategies and mitigating environmental health risks.

Like other studies in the field [126–128], this study has revealed the mechanism of a bidirectional influence between environmental and climate factors and urban vegetation greenness, emphasizing the green space protection needs during heat wave periods in

https://doi.org/10.3390/atmos17010109

reducing air pollutant emissions, promoting environmental technology innovation, and safeguarding ecological measures

Study Limitations and Future Directions

Despite providing valuable insights into the impact of summer heatwave variability on Bucharest metropolitan vegetation, this study has several limitations that underscore the need for a more complex approach. Our results might not fully align with those spatial and contextual variations. Also, MODIS Terra/Aqua datasets have coarse spatial resolution; higher-resolution imagery, such as Worldview-3 (WV-3) satellite imagery [129] or ESA Copernicus Sentinel imagery, will be preferred in urban greenness studies.

Recent studies highlight that urban vegetation (forest trees, parks, yards, streets, etc.) structure and vertical morphology, particularly mean tree height, canopy aggregation, and spatial continuity, can have a greater impact on LST than total green land cover, with effects modulated by urban form in seasonal LST patterns [130]. Conventional metrics used in this study for urban greenness —NDVI, LAI, and FPAR—derived from MODIS satellite observations do not capture vertical vegetation structure or spatial configuration, offering limited insight into intraurban spatiotemporal HWs—vegetation interactions and their cooling effects during the day.

However, the use of remotely sensed LST and LSA datasets to quantify the heat waves–urban vegetation interaction may introduce uncertainties, because the derived MODIS satellite LST and LSA reflect emitted radiation from the urban surface of an object seen from the point of view of the thermal sensor, which is not a good proxy for the air temperature TA perceived by people, compared to in situ TA monitoring, which is important to be considered for the realistic mitigation strategies of alleviating urban heat stress on vegetation and people during synergies of HWs with UHIs through increasing vegetation land cover [131].

In general, urban vegetation plays a significant role in the interaction between heatwaves, urban heat islands, and vegetation, lowering daytime temperatures, but the magnitude of their cooling effect is notably amplified when using only satellite datasets [132,133]. In the context of increasing frequency of globally deadly heatwaves, declining urban vegetation land cover, and rising imperviousness, which intensify decision-makers' concerns, there is an urgent need, especially in European Southern cities, to improve the adaptive capacity of biosystems.

One significant hypothesis that can be made is that the conditions for vegetation growth and health during hot summers and heat waves have deteriorated (i.e., high air and land surface temperatures, low precipitation, and heat stress), caused by climate change at mid-latitudes, and, therefore, vegetation can be observed from satellites to be more affected. Similar patterns have been observed in the other parts of the World [134–136], and further studies are needed to confirm and characterize this change across the different biogeographical regions and urban landscapes.

These insights are essential for developing human and alth urban vegetation health adaptation strategies during summer heat stress and informing sustainable metropolitan environmental management practices in the face of ongoing climatic changes [137–140].

5. Conclusions

This study aims to explore the response of urban vegetation to the impacts of heat waves in relation to multi-climatic parameter variability, from a spatiotemporal perspective during 2000–2024, with a focus on summer HWs (June–August), particularly the hottest summer of 2024 in the Bucharest metropolitan area, Romania. In the frame of worsening

https://doi.org/10.3390/atmos17010109

heat wave severity under future warming, this research revealed the interannual and inter-seasonal relationships between HWs and urban vegetation parameters NDVI, LAI, FPAR, ET, and their interplay with environmental variables and background climate conditions, with implications for the management of urban resources.

The results of this study highlight the importance of considering land surface albedo, land surface temperature, aerosol optical depth, surface solar irradiance, and other climate variables when assessing mitigation strategies to reduce summer extreme heat in urban areas. In the frame of sustainable urban development, optimizing the structure of green space and the built environment in cities helps mitigate the adverse impacts of extreme summer heat on public health and the environment.

Urban heat stress–vegetation-induced biophysical feedbacks have been considered for moderating increases in land-surface air temperature and reducing warming during summer heat waves and under urban heat island effects. Recognized for the cooling effect associated with changes in surface convection and reduced transmissivity of shortwave radiation, evapotranspiration from urban vegetation has shown high spatiotemporal variability in response to the urban thermal environment in Bucharest. The impacts of heat waves on urban vegetation evapotranspiration during hot summers limit the cooling benefits of urban vegetation. In such conditions, the preservation of urban ecosystems becomes a priority for urban decision-makers, who need scientific information to develop mitigation and adaptation strategies for sustainable planning and design of green-blue infrastructure, thereby guaranteeing the conservation of biodiversity in cities.

Author Contributions: M.Z.: Conceptualization; methodology, supervision, review, editing, and validation. D.S.: Methodology, validation, software. M.T.: methodology, writing. All authors have read and agreed to the published version of the manuscript.

Funding: This research received no external funding.

Institutional Review Board Statement: Not applicable.

Informed Consent Statement: Not applicable.

Data Availability Statement: The original contributions presented in this study are included in the article. Further inquiries can be directed to the corresponding author.

Acknowledgments: This study was supported by the Romanian Ministry of Education and Research, Research, Development, and Innovation Plan 2022–2027, CONTRACT PN 23 05 NUCLEU. We are very thankful for the NASA MODIS products, the MERRA-2-derived meteorological and solar parameters product provided by the Copernicus Atmosphere Monitoring Service (CAMS), and the GIOVANNI products.

Conflicts of Interest: The authors declare no conflicts of interest.

References

1. Li, Y.; Zhong, C.; Deng, Z.; Jiang, Z. Weather-Regime-Based Heatwave Risk Typing and Urban Climate Resilience Assessment in New Delhi (1997–2016). *Atmosphere* **2025**, *16*, 1179. [CrossRef]
2. Adeyeri, O.E.; Ajadi, S.A.; Ishola, K.A.; Mustapha, S.O. The societal impact of heatwave intensification and heat stress on African Urban populations. *Soc. Impacts* **2025**, *6*, 100148. [CrossRef]
3. Jiang, S.; Zhao, Y.; Zhao, L.; Strebel, D.; Derome, D.; Ürge-Vorsatz, D.; Carmeliet, J.; Peng, J. Nonlinear impacts of urban size and vegetation cover on global surface urban heat: Insights from 6022 cities. *Remote Sens. Environ.* **2026**, *333*, 115154. [CrossRef]
4. European Environment Agency. Climate-ADAPT. 2022. Available online: https://climate-adapt.eea.europa.eu/en/ (accessed on 30 August 2025).
5. European Environment Agency. Climate-ADAPT. Precipitation Sum, 2011–2099. 2022. Available online: https://climate-adapt.eea.europa.eu/en/metadata/indicators/precipitation-sum-2011-2099#details (accessed on 30 August 2025).

https://doi.org/10.3390/atmos17010109

6. Garcia de Cezar, M.; Cheviron, B.; Tomas, S.; Lutete, M.M.; Rodler, A.; Aprin, L. Modelling the impact of irrigated vegetation on urban microclimate: A review. *Sci. Total Environ.* **2025**, *997*, 180166. [CrossRef] [PubMed]
7. Vaňo, S.; Duchková, H.; Bašta, P.; Jančovič, M.; Geletič, J.; Lorencová, E.K.; Suchá, L. From scenarios to strategies: Integrated methodology for addressing urban heat vulnerability in an uncertain future. *Sustain. Cities Soc.* **2025**, *121*, 106202. [CrossRef]
8. Wang, Y.; Lin, C.; Lau, A.K.H.; Fung, J.K.H.; Chong, K.C.; Ren, C. Assessing the impact of compound heatwave-ozone event with varied definitions on mortality risk in Hong Kong. *Environ. Int.* **2025**, *201*, 109595. [CrossRef]
9. Liang, C.; Yuan, J.; Tang, X.; Kan, H.; Cai, W.; Chen, J. The influence of humid heat on morbidity of megacity Shanghai in China. *Environ. Int.* **2024**, *183*, 108424. [CrossRef]
10. Stafoggia, M.; Michelozzi, P.; Schneider, A.; Armstrong, B.; Scortichini, M.; Rai, M.; Achilleos, S.; Alahmad, B.; Analitis, A.; Åström, C.; et al. Joint effect of heat and air pollution on mortality in 620 cities of 36 countries. *Environ. Int.* **2023**, *181*, 108258. [CrossRef]
11. Zou, J.; Zhong, G.; Wang, L.L.; Katal, A.; Gaur, A.; Yan, S.; Albettar, M.; Marey, A. Urban heatwaves and mortality: A socioeconomic and environmental study using a novel building heat vulnerability index. *Cities* **2026**, *169*, 106558. [CrossRef]
12. Pei, X.; Wu, J.; Li, Y. Spatiotemporal assessment of heatwaves adaptation in Chinese cities and urban agglomerations: An integrated ND-GAIN framework and multi-model approach. *J. Clean. Prod.* **2025**, *520*, 146150. [CrossRef]
13. Xu, R.; Sun, H.; Zhong, Z.; Zheng, Y.; Liu, T.; Li, Y.; Liu, L.; Luo, L.; Wang, S.; Lv, Z.; et al. Ozone, heatwave, and cardiovascular disease mortality: A population-based case-crossover study. *Environ. Sci. Technol.* **2024**, *58*, 171–181. [CrossRef]
14. Du, H.; Yan, M.; Liu, X.; Zhong, Y.; Ban, J.; Lu, K.; Li, T. Exposure to concurrent heatwaves and ozone pollution and associations with mortality risk: A nationwide study in China. *Environ. Health Perspect.* **2024**, *132*, 047012. [CrossRef] [PubMed]
15. Huang, Q.; Reeder, M.J.; Jakob, C.; King, M.J. Disentangling the near-surface physical processes from the synoptic-dynamics in heatwaves using a single column model and trajectory analysis. *Q. J. R. Meteorol. Soc.* **2025**, *151*, e70006. [CrossRef]
16. Cassou, C.; Terray, L.; Phillips, A.S. Tropical Atlantic influence on European Heatwaves. *J. Clim.* **2005**, *18*, 2805–2811. [CrossRef]
17. Wang, W.; Zhou, W.; Li, X.; Wang, X.; Wang, D. Synoptic-scale characteristics and atmospheric controls of summer heatwaves in China. *Clim. Dyn.* **2016**, *46*, 2923–2941. [CrossRef]
18. Zschenderlein, P.; Fink, A.H.; Pfahl, S.; Wernli, H. Processes determining heatwaves across different European climates. *Q. J. R. Meteorol. Soc.* **2019**, *145*, 2973–2989. [CrossRef]
19. Ventura, S.; Miró, J.R.; Peña, J.C.; Villalba, G. Analysis of synoptic weather patterns of heatwave events. *Clim. Dyn.* **2023**, *61*, 4679–4702. [CrossRef]
20. Hirschi, M.; Seneviratne, S.I.; Alexandrov, V.; Boberg, F.; Boroneant, C.; Christensen, O.B.; Formayer, H.; Orlowsky, B.; Stepanek, P. Observational evidence for soil-moisture impact on hot extremes in Southeastern Europe. *Nat. Geosci.* **2011**, *4*, 17–21. [CrossRef]
21. Miralles, D.G.; Teuling, A.J.; Van Heerwaarden, C.C.; Vila Guerau de Arellano, J. Mega-Heatwave temperatures due to combined soil desiccation and atmospheric heat accumulation. *Nat. Geosci.* **2014**, *7*, 345–349. [CrossRef]
22. Zhang, Y.; Wang, L.; Santanello, J.A.; Pan, Z.; Gao, Z.; Li, D. Aircraft observed diurnal variations of the planetary boundary layer under heatwaves. *Atmos. Res.* **2020**, *235*, 104801. [CrossRef]
23. Matsueda, M. Predictability of Euro-Russian blocking in summer of 2010. *Geophys. Res. Lett.* **2011**, *38*, L06801. [CrossRef]
24. Huang, Q.; Reeder, M.J.; Jakob, C.; King, M.J.; Su, C.-H. The life cycle of the heatwave boundary layer identified from commercial aircraft observations at Melbourne Airport (Australia). *Q. J. R. Meteorol. Soc.* **2023**, *149*, 3440–3454. [CrossRef]
25. Wang, Y.; Xu, T.; Shi, G.; Yang, F.; Tang, X.; Zhao, X.; Wan, C.; Liu, S. Climatology of the planetary boundary layer height over China and its characteristics during periods of extremely temperature. *Atmos. Res.* **2023**, *294*, 106960. [CrossRef]
26. Zhu, Z.; Zhang, M.; Huang, Y.; Zhu, B.; Han, G.; Zhang, T.; Liu, B. Characteristics of the planetary boundary layer above Wuhan, China based on CALIPSO. *Atmos. Res.* **2018**, *214*, 204–212. [CrossRef]
27. Elgendy, D.; Tolba, O.; Kamel, T. The impact of increasing urban surface albedo on outdoor air and surface temperatures during summer in newly developed areas. *Sci. Rep.* **2025**, *15*, 25165. [CrossRef]
28. Liu, Y.; Chu, C.; Zhang, R.; Chen, S.; Xu, C.; Zhao, D.; Meng, C.; Ju, M.; Cao, Z. Impacts of high-albedo urban surfaces on outdoor thermal environment across morphological contexts: A case of Tianjin, China. *Sustain. Cities Soc.* **2024**, *100*, 105038. [CrossRef]
29. Diana Grecia, A.-M.; Sergio Arturo, T.-S.; Marlenne, G.-R. Review: Implications of Air Pollution on Trees Located in Urban Areas. *Earth* **2025**, *6*, 38. [CrossRef]
30. Yan, K.; Gao, S.; Yan, G.; Ma, X.; Chen, X.; Zhu, P.; Li, J.; Gao, S.; Gastellu-Etchegorry, J.-P.; Myneni, R.B.; et al. A global systematic review of the remote sensing vegetation indices. *Int. J. Appl. Earth Obs. Geoinf.* **2025**, *139*, 104560. [CrossRef]
31. Lai, P.; Marshall, M.; Darvishzadeh, R.; Tu, K.; Nelson, A. Characterizing crop productivity under heat stress using MODIS data. *Agric. For. Meteorol.* **2024**, *355*, 110116. [CrossRef]
32. Lee, J.K.; Woo, S.Y.; Kwak, M.J.; Park, S.H.; Kim, H.D.; Lim, Y.J.; Park, J.H.; Lee, K.A. Effects of elevated temperature and ozone in *Brassica juncea* L.: Growth, physiology, and ROS accumulation. *Forests* **2020**, *11*, 68. [CrossRef]

 https://doi.org/10.3390/atmos17010109

33. Kašpar, J.; Král, K.; Levanič, T.; Adamič, P.C.; Čater, M. Climate growth limitations of European beech and silver fir along the Carpathian arc—The recent state and future prospects. *Agric. For. Meteorol.* **2025**, *361*, 110323. [CrossRef]
34. Hu, S.; Ding, Y.; Zhu, C. Sensitivity and responses of chloroplasts to heat stress in plants. *Front. Plant Sci.* **2020**, *11*, 375. [CrossRef] [PubMed]
35. Ojasalo, A.; Lindqvist, H.; Willberg, E.; Toivonen, T.; Maeda, E.E. Vegetation phenology along urban-rural gradients across European capital cities. *Urban For. Urban Green.* **2025**, *112*, 129004. [CrossRef]
36. Percival, G. Heat tolerance of urban trees—A review. *Urban For. Urban Green.* **2023**, *86*, 128021. [CrossRef]
37. Al-Zahrani, H.S.; Alharby, H.F.; Fahad, S. Antioxidative defense system, hormones, and metabolite accumulation in different plant parts of two contrasting rice cultivars as influenced by plant growth regulators under heat stress. *Front. Plant Sci.* **2022**, *13*, 911846. [CrossRef]
38. De Boeck, H.J.; De Velde, H.; De Groote, T.; Nijs, I. Ideas and perspectives: Heat stress: More than hot air. *Biogeosciences* **2016**, *13*, 5821–5825. [CrossRef]
39. Hanley, P.A.; Arndt, S.K.; Livesley, S.J.; Szota, C. Relating the climate envelopes of urban tree species to their drought and thermal tolerance. *Sci. Total Environ.* **2021**, *753*, 142012. [CrossRef]
40. Santiago, J.-L.; Martilli, A.; Martin, F. On dry deposition modelling of atmospheric pollutants on vegetation at the microscale: Application to the impact of street vegetation on air quality. *Bound. Layer. Meteorol.* **2017**, *162*, 451–474. [CrossRef]
41. Buccolieri, R.; Santiago, J.-L.; Rivas, E.; Sáanchez, B. Reprint of: Review on urban treemodelling in CFD simulations: Aerodynamic, deposition and thermal effects. *Urban For. Urban Green.* **2019**, *37*, 56–64. [CrossRef]
42. Redondo-Bermúdez, M.d.C.; Gulenc, I.T.; Cameron, R.W.; Inkson, B.J. 'Green barriers' for air pollutant capture: Leaf micromorphology as a mechanism to explain plants capacity to capture particulate matter. *Environ. Pollut.* **2021**, *288*, 117809. [CrossRef]
43. Akarsu Varşak, N.; Ender Altay, E. Thermal Comfort and Green Spaces: The Role of Temperature-Regulating Elements in Neighborhood Parks. *Sustainability* **2024**, *16*, 9801. [CrossRef]
44. Lindén, J.; Gustafsson, M.; Uddling, J.; Watne, Å.; Pleijel, H. Air pollution removal through deposition on urban vegetation: The importance of vegetation characteristics. *Urban For. Urban Green.* **2023**, *81*, 127843. [CrossRef]
45. Chiam, Z.Y.; Song, X.P.; Lai, H.R.; Tan, H.T.W. Particulate matter mitigation via plants: Understanding complex relationships with leaf traits. *Sci. Total Environ.* **2019**, *688*, 398–408. [CrossRef] [PubMed]
46. Corada, K.; Woodward, H.; Alaraj, H.; Collins, C.M.; de Nazelle, A. A systematic review of the leaf traits considered to contribute to removal of airborne particulate matter pollution in urban areas. *Environ. Pollut.* **2021**, *269*, 116104. [CrossRef]
47. Delaria, E.R.; Place, B.K.; Liu, A.X.; Cohen, R.C. Laboratory measurements of stomatal NO_2 deposition to native California trees and the role of forests in the NO_x cycle. *Atmos. Chem. Phys.* **2020**, *20*, 14023–14041. [CrossRef]
48. McCarthy, M.J.; Herrero, H.V.; Insalaco, S.A.; Hinten, M.T.; Anyamba, A. Satellite remote sensing for environmental sustainable development goals: A review of applications for terrestrial and marine protected areas. *Remote Sens. Appl. Soc. Environ.* **2025**, *37*, 101450. [CrossRef]
49. Araminienė, V.; Sicard, P.; Černiauskas, V.; Coulibaly, F.; Varnagirytė-Kabašinskienė, I. Estimation of air pollution removal capacity by urban vegetation from very high-resolution satellite images in Lithuania. *Urban Clim.* **2023**, *51*, 101594. [CrossRef]
50. Yao, M.; Smith, M.; Peng, C. Modelling the effects of vegetation and urban form on air quality in real urban environments: A systematic review of measurements, methods, and predictions. *Urban For. Urban Green.* **2025**, *105*, 128693. [CrossRef]
51. Mărmureanu, L.; Leca, Ş.; Pitar, D.; Pascu, I.; De Marco, A.; Sicard, P.; Chivulescu, Ş.; Dobre, A.C.; Badea, O. Estimation of plant pollution removal capacity based on intensive air quality measurements. *Environ. Res.* **2024**, *261*, 119703. [CrossRef]
52. Ferrini, F.; Fini, A.; Mori, J.; Gori, A. Role of vegetation as a mitigating factor in the urban context. *Sustainability* **2020**, *12*, 4247. [CrossRef]
53. Gustafsson, M.; Linden, J.; Johansson, E.M.M.; Watne, Å.K.; Pleijel, H. Air pollution removal with urban greenery—Introducing the Vegetation Impact Dynamic Assessment model (VIDA). *Atmos. Environ.* **2024**, *323*, 120397. [CrossRef]
54. Smith, I.A.; Fabian, M.P.; Hutyra, L.R. Urban green space and albedo impacts on surface temperature across seven United States cities. *Sci. Total Environ.* **2023**, *857*, 159663. [CrossRef]
55. Yu, H.; Piao, Y. The Cooling Effect and Its Stability in Urban Green Space in the Context of Global Warming: A Case Study of Changchun, China. *Sustainability* **2025**, *17*, 2590. [CrossRef]
56. Wang, A.; Wang, J.; Zhang, R.; Cao, S.-J. Mitigating urban heat and air pollution, considering green and transportation infrastructure. *Transp. Res. Part A Policy Pract.* **2024**, *184*, 104079. [CrossRef]
57. Zhou, W.; Yu, Y.; Zhang, S.; Xu, J.; Wu, T. Methods for quantifying the cooling effect of urban green spaces using remote sensing:A comparative study. *Landsc. Urban Plan.* **2025**, *256*, 105289. [CrossRef]

 https://doi.org/10.3390/atmos17010109

58. Sheng, Q.; Ji, Y.; Jia, C.; Jiang, L.; Li, C.; Huang, Z.; Ma, C.; Zhang, X.; Chen, H.; Wang, T.; et al. Differential effects of cooling and humidification in urban green spaces and thresholds of vegetation community structure parameters: A case study of the Yangtze River Delta region. *Cities* **2025**, *159*, 105765. [CrossRef]
59. Azizi, S.; Azizi, T. Urban Climate Dynamics: Analyzing the Impact of Green Cover and Air Pollution on Land Surface Temperature—A Comparative Study Across Chicago, San Francisco, and Phoenix, USA. *Atmosphere* **2024**, *15*, 917. [CrossRef]
60. Galieni, A.; D'Ascenzo, N.; Stagnari, F.; Pagnani, G.; Xie, Q.; Pisante, M. Past and future of plant stress detection: An overview from remote sensing to positron emission tomography. *Front. Plant Sci.* **2021**, *11*, 609155. [CrossRef]
61. Copernicus Climate Change Service & World Meteorological Organization. European State of the Climate: Report 2024. 2024. Available online: https://climate.copernicus.eu/esotc/2024 (accessed on 5 November 2025).
62. National Institute of Statistics. INS Baze de Date Statistice. 2025. Available online: http://statistici.insse.ro:8077/tempo-online/#/pages/tables/insse-table (accessed on 10 September 2025).
63. National Institute of Statistics. Green space surface per counties and localities (municipalities and towns). In *Statistics*; N.I.o., Ed.; GOS103A; National Institute of Statistics: Kolkata, India, 2025.
64. Copernicus Urban Atlas. Available online: https://land.copernicus.eu/local/urban-atlas/urban-atlas-2018 (accessed on 5 November 2025).
65. Dușcu, D.-M.; Rîșnoveanu, G. Understanding visitor preferences: Perceived importance of anthropogenic and natural forest features in supplying cultural ecosystem services. *For. Ecosyst.* **2025**, *13*, 100306. [CrossRef]
66. Popescu, L.; Popescu, R.; Catalina, T. Indoor particle's pollution in Bucharest, Romania. *Toxics* **2022**, *10*, 757. [CrossRef]
67. MERRA. 2025. Available online: http://www.soda-pro.com/web-services/meteo-data/merra (accessed on 10 September 2025).
68. Copernicus. 2025. Available online: https://cds.climate.copernicus.eu/ (accessed on 12 September 2025).
69. GIOVANNI. Geospatial Interactive Online Visualization and Analysis Infrastructure GIOVANNI. 2025. Available online: https://giovanni.gsfc.nasa.gov/giovanni (accessed on 15 September 2025).
70. NOAA Physical Sciences Laboratory. Available online: https://psl.noaa.gov/data/composites/day/ (accessed on 20 September 2025).
71. NASA MODIS. Available online: https://modis.ornl.gov/ (accessed on 20 February 2025).
72. Nour Eldeen, N.; Mao, K.; Yuan, Z.; Shen, X.; Xu, T.; Qin, Z. Analysis of the spatiotemporal change in land surface temperature for a long-term sequence in Africa (2003–2017). *Remote Sens.* **2020**, *12*, 488–511. [CrossRef]
73. Phan, T.N.; Kappas, M.; Nguyen, K.T.; Tran, T.P.; Tran, Q.V.; Emam, A.R. Evaluation of MODIS land surface temperature products for daily air surface temperature estimation in Northwest Vietnam. *Int. J. Remote Sens.* **2019**, *40*, 5544–5562. [CrossRef]
74. ORNL DAAC. *MODIS and VIIRS Land Products Global Subsetting and Visualization Tool*; ORNL DAAC: Oak Ridge, TN, USA, 2018. [CrossRef]
75. Yoo, C.; Im, J.; Cho, D.; Yokoya, N.; Xia, J.; Bechtel, B. Estimation of all-weather 1 km MODIS land surface temperature for humid summer days. *Remote Sens.* **2020**, *12*, 1398. [CrossRef]
76. Russo, E.; Domeisen, D.I.V. Increasing intensity of extreme heatwaves: The crucial role of metrics. *Geophys. Res. Lett.* **2023**, *50*, e2023GL103540. [CrossRef]
77. Perkins-Kirkpatrick, S.; Lewis, S. Increasing trends in regional heatwaves. *Nat. Commun.* **2020**, *11*, 3357. [CrossRef]
78. Domeisen, D.I.V.; Eltahir, E.A.B.; Fischer, E.M.; Knutti, R.; Perkins-Kirkpatrick, S.E.; Schär, C.; Seneviratne, S.I.; Weisheimer, A.; Wernli, H. Prediction and projection of heatwaves. *Nat. Rev. Earth Environ.* **2023**, *4*, 36–50. [CrossRef]
79. López-Bueno, J.A.; Navas-Martín, M.A.; Linares, C.; Mirón, I.J.; Luna, M.Y.; Sánchez-Martínez, G.; Culqui, D.; Díaz, J. Analysis of the impact of heatwaves on daily mortality in urban and rural areas in Madrid. *Environ. Res.* **2021**, *195*, 110892. [CrossRef]
80. Maggiotto, G.; Miani, A.; Rizzo, E.; Castellone, M.; Piscitelli, P. Heatwaves and adaptation strategies in a Mediterranean urban context. *Environ. Res.* **2021**, *197*, 111066. [CrossRef]
81. Costa, D.F.; Gomes, H.B.; Silva, M.C.L.; Zhou, L. The most extreme heatwaves in Amazonia happened under extreme dryness. *Clim. Dyn.* **2022**, *59*, 281–295. [CrossRef]
82. Eastman, J.R.; Sangermano, F.; Machado, E.A.; Rogan, J.; Anyamba, A. Global Trends in Seasonality of Normalized Difference Vegetation Index (NDVI), 1982–2011. *Remote Sens.* **2013**, *5*, 4799–4818. [CrossRef]
83. Huang, S.; Tang, L.; Hupy, J.P.; Wang, Y.; Shao, G. A commentary review on the use of normalized difference vegetation index (NDVI) in the era of popular remote sensing. *J. For. Res.* **2021**, *32*, 2719. [CrossRef]
84. Al-Hameed, A.A.K. Spearman's correlation coefficient in statistical analysis. *Int. J. Nonlinear Anal. Appl.* **2022**, *13*, 3249–3255.
85. Shi, Z.; Jian, D.G.; Wang, Y. Spatiotemporal dependence of compound drought—Heatwave and fire activity in China. *Weather. Clim. Extrem.* **2024**, *45*, 100695. [CrossRef]

86. de la Iglesia Martinez, A.; Labib, S.M. Demystifying normalized difference vegetation index (NDVI) for greenness exposure assessments and policy interventions in urban greening. *Environ. Res.* **2023**, *220*, 115155. [CrossRef]
87. Eisfelder, C.; Asam, S.; Hirner, A.; Reiners, P.; Holzwarth, S.; Bachmann, M.; Gessner, U.; Dietz, A.; Huth, J.; Bachofer, F.; et al. Seasonal Vegetation Trends for Europe over 30 Years from a Novel Normalised Difference Vegetation Index (NDVI) Time-Series—The TIMELINE NDVI Product. *Remote Sens.* **2023**, *15*, 3616. [CrossRef]
88. Klimavičius, L.; Rimkus, E.; Stonevičius, E.; Mačiulytė, V. Seasonality and long-term trends of NDVI values in different land use types in the eastern part of the Baltic Sea basin. *Oceanologia* **2023**, *65*, 171–181. [CrossRef]
89. Prăvălie, R.; Sîrodoev, I.; Nita, I.-A.; Patriche, C.; Dumitraşcu, M.; Roşca, B.; Tişcovschi, A.; Bandoc, G.; Săvulescu, I.; Mănoiu, V.; et al. NDVI-based ecological dynamics of forest vegetation and its relationship to climate change in Romania during 1987–2018. *Ecol. Indic.* **2022**, *136*, 108629. [CrossRef]
90. Copernicus Climate Change Service. Copernicus: Record-Breaking Heat Stress in Southeastern Europe During Summer 2024. 2024. Available online: https://climate.copernicus.eu/copernicus-record-breaking-heat-stress-southeastern-europe-during-summer-2024 (accessed on 5 November 2025).
91. Ionita, M.; Nagavciuc, V. 2024: The year with too much summer in the eastern part of Europe. *Weather* **2025**, *80*, 288–295. [CrossRef]
92. Xin, J.; Zhang, Y.; Bai, W.; Song, Z. Response of urban heat island effects within the planetary boundary layer to heatwaves and impact of horizontal advection over Shanghai. *Atmos. Res.* **2024**, *311*, 107721. [CrossRef]
93. Alinasab, N.; Mohammadzadeh, N.; Karimi, A.; Mohammadzadeh, R.; Gál, T. A measurement-based framework integrating machine learning and morphological dynamics for outdoor thermal regulation. *Int. J. Biometeorol.* **2025**, *69*, 1645–1662. [CrossRef]
94. Bachofen, C.; Peillon, M.; Meili, N.; Bourgeois, I.; Grossiord, C. High transpirational cooling by urban trees despite extreme summer heatwaves. *Urban For. Urban Green.* **2025**, *107*, 128819. [CrossRef]
95. Wang, N.; Shen, L.; Fei, W.; Liu, Y.; Zhao, H.; Liu, L.; Wang, A.; He, B.-J. Seasonal Effects of Urban Morphology on the Thermal Environment Based on Automated Machine Learning: A Case Study of Beijing. *Remote Sens.* **2025**, *17*, 3150. [CrossRef]
96. Ma, W.; Yu, Z.; Chen, J.; Yang, W.; Zhang, Y.; Hu, Y.; Shao, M.; Hu, J.; Zhang, Y.; Zhang, H.; et al. What drives the cooling dynamics of urban vegetation via evapotranspiration and shading under extreme heat? *Sustain. Cities Soc.* **2025**, *130*, 106659. [CrossRef]
97. Pardo, S.K.; Paredes-Fortuny, L. Uneven evolution of regional European summer heatwaves under climate change. *Weather. Clim. Extrem.* **2024**, *43*, 100648. [CrossRef]
98. Dong, Y.; Wang, J.; Sargai. Multi-day dynamics and mechanisms of urban vegetation cooling during heatwaves: An engineering resilience perspective from Beijing, China. *Urban Clim.* **2025**, *64*, 102653. [CrossRef]
99. Wang, L.; Hu, P.; Zheng, H.; Bai, J.; Liu, Y.; Cao, X.; Hellwich, O.; Liu, T.; Bao, A.; Chen, X. A physics-constrained deep learning framework for actual evapotranspiration estimation using ground station observations and remote sensing data. *Environ. Model. Softw.* **2025**, *192*, 106585. [CrossRef]
100. Wai, K.-M. Is Integrating Tree-Planting Strategies with Building Array Sufficient to Mitigate Heat Risks in a Sub-Tropical Future City? *Buildings* **2025**, *15*, 1913. [CrossRef]
101. Afzal, M.T.; Tripathi, N.K.; Ninsawat, S.; Pal, I. Urban heatwave vulnerability assessment and ecological evaluation of land use/land cover in Karachi, Pakistan: A developing megacity perspective. *Environ. Sustain. Indic.* **2025**, *28*, 100965. [CrossRef]
102. Zoran, M.; Savastru, D.; Tautan, M.; Tenciu, D.; Stanciu, A. Spatiotemporal Analysis of Air Pollution and Climate Change Effects on Urban Green Spaces in Bucharest Metropolis. *Atmosphere* **2025**, *16*, 553. [CrossRef]
103. Muradyan, V.; Tepanosyan, G.; Asmaryan, S.; Saghatelyan, A.; Dell'Acqua, F. Relationships between NDVI and climatic factors in mountain ecosystems: A case study of Armenia. *Remote Sens. Appl. Soc. Environ.* **2019**, *14*, 158–169. [CrossRef]
104. Vicente-Serrano, S.M.; Martín-Hernández, N.; Reig, F.; Azorin-Molina, C.; Zabalza, J.; Beguería, S.; Domínguez-Castro, F.; El Kenawy, A.; Peña-Gallardo, M.; Noguera, I.; et al. Vegetation greening in Spain detected from long term data (1981–2015). *Int. J. Remote Sens.* **2019**, *41*, 1709–1740. [CrossRef]
105. Giuliani, G.; Egger, E.; Italiano, J.; Poussin, C.; Richard, J.-P.; Chatenoux, B. Essential Variables for Environmental Monitoring: What Are the Possible Contributions of Earth Observation Data Cubes? *Data* **2020**, *5*, 100. [CrossRef]
106. Ward, K.; Lauf, S.; Kleinschmit, B.; Endlicher, W. Heatwaves and urban heat islands in Europe: A review of relevant drivers. *Sci. Total Environ.* **2016**, *569–570*, 527–539. [CrossRef] [PubMed]
107. Lai, H.-C.; Dai, Y.-T.; Mkasimongwa, S.W.; Hsiao, M.-C.; Lai, L.-W. The Impact of Atmospheric Synoptic Weather Condition and Long-Range Transportation of Air Mass on Extreme PM10 Concentration Events. *Atmosphere* **2023**, *14*, 406. [CrossRef]
108. Li, R.; Li, B.; Yuan, Y.; Liu, W.; Zhu, J.; Qi, J.; Liu, H.; Ma, G.; Jiang, Y.; Li, Y.; et al. Improvement of FAPAR Estimation Under the Presence of Non-Green Vegetation Considering Fractional Vegetation Coverage. *Remote Sens.* **2025**, *17*, 603. [CrossRef]
109. Liang, S.; Ma, W.; Sui, X.; Wang, M.; Li, H. An Assessment of Relations between Vegetation Green FPAR and Vegetation Indices through a Radiative Transfer Model. *Plants* **2023**, *12*, 1927. [CrossRef]

110. Alibakhshi, S.; Naimi, B.; Hovi, A.; Crowther, T.W.; Rautiainen, M. Quantitative analysis of the links between forest structure and land surface albedo on a global scale. *Remote Sens. Environ.* **2020**, *246*, 111854. [CrossRef]
111. Dong, X.; Ye, Y.; Yang, Y.; Li, X. Planning for green infrastructure based on integration of multi-driving factors: A case study in pilot site of sponge city. *Sustain. Cities Soc.* **2023**, *93*, 104549. [CrossRef]
112. Chang, H.-S.; Lin, Z.-H.; Hsu, Y.-Y. Planning for green infrastructure and mapping synergies and trade-offs: A case study in the Yanshuei River Basin, Taiwan. *Urban For. Urban Green.* **2021**, *65*, 127325. [CrossRef]
113. Motlagh, O.R.K.; Darand, M. Identification of the Driving factors impacts of Land Surface Albedo over Iran: An analysis with the MODIS data. *J. Atmos. Sol. Terr. Phys.* **2024**, *265*, 106378. [CrossRef]
114. Badugu, A.; Arunab, K.S.; Mathew, A. Predicting land surface temperature using data-driven approaches for urban heat island studies: A comparative analysis of correlation with environmental parameters. *Model. Earth Syst. Environ.* **2024**, *10*, 1043–1076. [CrossRef]
115. Yu, C.; Xu, D.; Peng, Y.; Zhou, S.; Lu, D. Analysis of land surface temperature drivers and marginal effect on the Yangtze River Delta urban agglomeration. *Ecol. Indic.* **2025**, *178*, 114123. [CrossRef]
116. Pang, G.; Chen, D.; Wang, X.; Lai, H.-W. Spatiotemporal variations of land surface albedo and associated influencing factors on the Tibetan Plateau. *Sci. Total Environ.* **2022**, *804*, 150100. [CrossRef] [PubMed]
117. Moncada-Morales, G.A.; Verichev, K.; López-Guerrero, R.E.; Carpio, M. A Global Review of Vegetation's Interaction Effect on Urban Heat Mitigation Across Different Climates. *Urban Sci.* **2025**, *9*, 361. [CrossRef]
118. Lin, J.; Zhang, H.; Chen, M.; Wang, Q. Socioeconomic disparities in cooling and warming efficiencies of urban vegetation and impervious surfaces. *Sustain. Cities Soc.* **2023**, *92*, 104464. [CrossRef]
119. Bai, X.; Yu, Z.; Wang, B.; Zhang, Y.; Zhou, S.; Sha, X.; Li, S.; Yao, X.; Geng, X. Quantifying threshold and scale response of urban air and surface temperature to surrounding landscapes under extreme heat. *Build. Environ.* **2023**, *247*, 111029. [CrossRef]
120. Yu, Z.; Chen, J.; Chen, J.; Zhan, W.; Wang, C.; Ma, W.; Yao, X.; Zhou, S.; Zhu, K.; Sun, R. Enhanced observations from an optimized soil-canopy-photosynthesis and energy flux model revealed evapotranspiration-shading cooling dynamics of urban vegetation during extreme heat. *Remote Sens. Environ.* **2024**, *305*, 114098. [CrossRef]
121. Seo, J.; Park, J. Spatiotemporal analysis of interactions between building configuration and urban green space on outdoor air temperature during heatwaves: IoT sensor-based insights from Seoul, South Korea. *Urban For. Urban Green.* **2025**, *113*, 129019. [CrossRef]
122. Shi, X.; She, D.; Xia, J.; Zhang, Y.; Tang, R.; Kong, D. A modified PML model considering anthropogenic heat to estimate urban evapotranspiration. *Urban Clim.* **2025**, *64*, 102632. [CrossRef]
123. Augusto, B.; Roebeling, P.; Rafael, S.; Ferreira, J.; Ascenso, A.; Bodilis, C. Short and medium- to long-term impacts of nature-based solutions on urban heat. *Sustain. Cities Soc.* **2020**, *57*, 102122. [CrossRef]
124. Vulova, S.; Rocha, A.D.; Meier, F.; Nouri, H.; Schulz, C.; Soulsby, C.; Tetzlaff, D.; Kleinschmit, B. City-wide, high-resolution mapping of evapotranspiration to guide climate-resilient planning. *Remote Sens. Environ.* **2023**, *287*, 113487. [CrossRef]
125. Húdoková, H.; Petrik, P.; Petek-Petrik, A.; Konôpková, A.; Leštianska, A.; Střelcová, K.; Kmet, J.; Kurjak, D. Heat-stress response of photosystem II in five ecologically important tree species of European temperate forests. *Biologia* **2022**, *77*, 671–680. [CrossRef]
126. Esperon-Rodriguez, M.; Power, S.A.; Tjoelker, M.G.; Marchin, R.M.; Rymer, P.D. Contrasting heat tolerance of urban trees to extreme temperatures during heatwaves. *Urban For. Urban Green.* **2021**, *66*, 127387. [CrossRef]
127. Kisvarga, S.; Horotán, K.; Neményi, A.; Jana, T.; Istvánfi, Z.; Orlóci, L. The ecology of urban Vegetation: Trade-Offs, stressors, and adaptive strategies. *Environ. Sustain. Indic.* **2025**, *28*, 100887. [CrossRef]
128. Roy, A.; Bhattacharya, T.; Kumari, M. Air pollution tolerance, metal accumulation and dust capturing capacity of common tropical trees in commercial and industrial sites. *Sci. Total Environ.* **2020**, *722*, 137622. [CrossRef]
129. Fan, P.; Wang, H.; Imbroglini, C. Hourly impact of urban forests on land surface temperature based on machine learning. *Results Eng.* **2025**, *28*, 107584. [CrossRef]
130. Du, M.; Li, N.; Hu, T.; Yang, Q.; Chakraborty, T.; Venter, Z.; Yao, R. Daytime cooling efficiencies of urban trees derived from land surface temperature are much higher than those for air temperature. *Environ. Res. Lett.* **2024**, *19*, 044037. [CrossRef]
131. Toscan, P.C.; Seong, K.; Jiao, J.; Ribeiro, C.A.L.R.; Carvalho, F.A.C.; Oliveira, M.L.; Pereira, E.B. Impact of nature-based solutions (NBS) on urban surface temperatures and land cover changes using remote sensing and machine learning. *Remote Sens. Appl. Soc. Environ.* **2025**, *39*, 101721. [CrossRef]
132. Sharmin, M.; EsperonRodriguez, M.; Clackson, L.; Pfautsch, S.; Power, S.A. Heat Impact of Urban Sprawl: How the Spatial Composition of Residential Suburbs Impacts Summer Air Temperatures and Thermal Comfort. *Atmosphere* **2025**, *16*, 899. [CrossRef]
133. Yılmaz, M.; Çulpan, H.C.; Can, G.; Toros, H.; Tayanç, M. Environment, development, and health: The vulnerability of sub-populations in İstanbul to heat extremes. *Environ. Dev. Sustain.* **2024**. [CrossRef]

134. Kovács, L.N.; Jordan, G.; Berglund, F.; Holden, B.; Niehoff, E.; Pohl, F.; Younssi, M.; Zevallos, I.; Ágoston, C.; Varga, A.; et al. Acting as we feel: Which emotional responses to the climate crisis motivate climate action. *J. Environ. Psychol.* **2024**, *96*, 102327. [CrossRef]
135. Nishioka, S. A challenge for sustainability science: Can we halt climate change? *Sustain. Sci.* **2024**, *19*, 7–18. [CrossRef]
136. Wu, W.; Xu, Y. Factors affecting climate adaptation behavior among grain farmers in China. *Environ. Dev. Sustain.* **2025**, *27*, 13933–13946. [CrossRef]
137. Oliveira, M.C.Q.D.; de Miranda, R.M.; de Fátima Andrade, M.; Kumar, P. Impact of urban green areas on air quality: An integrated analysis in the metropolitan area of São Paulo. *Environ. Pollut.* **2025**, *372*, 126082. [CrossRef]
138. Degirmenci, T.; Aydin, M. A path towards the sustainable environment from environmental governance: Testing the LCC and EKC hypotheses with the environmental policy stringency and government effectiveness in France. *J. Knowl. Econ.* **2025**, 1–27. [CrossRef]
139. Liu, L.; Zhao, J.; Li, Z. Sustainable Urban Development: A Comprehensive Model for Assessing Health Risks from Compounded Pollution in Xi'an. *Sustainability* **2024**, *16*, 9799. [CrossRef]
140. Hojabri, J.; Nguyen-Huy, T. Spatio-temporal dynamics of vegetation and water bodies in the Lake Urmia Basin (1984–2022): An analysis using MODIS and Landsat data with new perspective. *J. Arid. Environ.* **2026**, *232*, 105500. [CrossRef]

https://doi.org/10.3390/atmos17010109

atmosphere

MDPI

Article

Bridging Heritage Conservation and Energy Efficiency: Retrofitting Historic Social Housing in Mediterranean Cities

Zoe Kanetaki [1,*], Evgenia Tousi [2], Athina Mela [2], Eleni Kanetaki [3], Gianluca Pappaccogli [4] and Emmanouel Proestakis [1]

1 Faculty of Engineering, Department of Mechanical Engineering, University of West Attica, 12241 Athens, Greece; eproest@uniwa.gr
2 Faculty of Engineering, Department of Civil Engineering, University of West Attica, 12241 Egaleo, Greece; etousi@uniwa.gr (E.T.); amela@uniwa.gr (A.M.)
3 School of Applied Arts & Culture, Department of Conservation of Antiquities and Works of Art, University of West Attica, 12241 Egaleo, Greece; elkanet@uniwa.gr
4 Department of Biological and Environmental Sciences and Technologies, University of Salento, 73100 Lecce, Italy; gianluca.pappaccogli@unisalento.it
* Correspondence: zoekanet@uniwa.gr

Abstract: Historic social housing in Mediterranean cities faces the dual challenges of energy inefficiency and cultural preservation. This study presents a pilot methodological framework for energy retrofitting of historical residence buildings, using the Kaisariani Asia Minor refugee housing complex in Athens as a case study. A bibliometric analysis revealed a research gap, as clusters concerning heritage retrofitting and social housing remain weakly connected, highlighting limited interdisciplinary integration between cultural conservation and energy-efficient design. The proposed framework combines historical analysis, energy consumption assessment, and technical evaluation to examine three retrofit scenarios that integrate thermal insulation, upgraded HVAC systems, renewable domestic hot water, and photovoltaic installations. Results demonstrate that substantial performance improvements can be achieved without compromising architectural authenticity. The most comprehensive scenario achieved a 97% reduction in primary energy demand, a 63–76% decrease in heating and cooling loads, and significant CO_2 emission reductions, maintaining economic feasibility with a payback period of approximately ten years. The findings emphasize that conservation-compatible retrofitting can transform obsolete housing into low-energy buildings, fostering environmental, social, and cultural sustainability. Beyond quantitative energy gains, the study underlines the importance of integrating heritage values and community identity into urban regeneration strategies, offering a transferable model for Mediterranean municipalities seeking to align climate action with cultural continuity.

Keywords: urban microclimate; thermal comfort; sustainable urban space; thermal performance; retrofit; historical social housing; Mediterranean areas; sustainability; cultural sustainability; aging building stock

1. Introduction

In the era of sustainable urban transitions, cities worldwide face mounting pressure to modernize infrastructure, reduce environmental impacts, and promote inclusive growth. However, this transformation often presents significant challenges to the preservation and integration of cultural heritage. As urban landscapes evolve to meet sustainability targets, historic sites, traditional practices, and community identities are frequently at risk of marginalization or erasure. From this point of view, cultural sustainability plays a critical role in ensuring that progress does not come at the expense of cultural identity.

European cities are undergoing a major shift from greenfield development to the regeneration and in situ retrofitting of existing urban housing, a process underscored by the urgent need to meet ambitious climate and sustainability targets at scale [1–3]. A central challenge within this transition is the deep energy renovation of traditional social housing—often protected for its architectural or cultural value—to achieve near-zero energy standards while safeguarding historical identity and social fabric [1,4]. The technical complexities are considerable, including the integration of internal insulation, moisture, and thermal bridge mitigation, and the selection of minimally invasive, reversible interventions that respect conservation principles [5,6]. At the same time, these projects must navigate economic feasibility and long-term affordability, especially under non-profit and public housing constraints [1,5,7].

Increasingly, scholars and practitioners are advocating for the explicit recognition of cultural sustainability as an essential fourth pillar in urban development, alongside the more established environmental, economic, and social dimensions [4]. However, operationalizing this concept remains a challenge; while some research proposes frameworks for balancing energy performance and heritage value, the development of quantitative indicators and participatory tools to capture intangible assets—such as authenticity, place attachment, and community rituals—lags behind technical innovation [6].

Recent years have seen the growth of methodological approaches that combine technical assessment methods (such as life-cycle analysis and energy simulation) with participatory, user-centered design processes, particularly in the context of social housing pilots [2,6,8,9]. Decision-support frameworks based on European standards, such as EN 16883, are increasingly being adopted to guide the selection of "conservation-compatible" retrofit strategies that maximize energy, comfort, and cultural value [6,10]. Yet, empirical studies demonstrating the alignment of environmental and cultural goals at the block or district scale remain limited, and lessons from innovative models—such as the block-scale, community-oriented retrofits of Vienna—have only partially influenced mainstream practice [1,11].

Positioned at the intersection of environmental and cultural sustainability, this study investigates the energy retrofit of a former social housing complex in Kaisariani, within the metropolitan area of Athens, Greece. To elaborate further, this study investigates the integration of heritage retrofitting within social housing, recognizing that many social housing complexes—particularly those constructed in the early twentieth century—face the dual challenge of improving energy performance while preserving architectural character. Although not all social housing developments possess formal heritage status, the proposed framework is designed to accommodate both historically significant buildings and more contemporary examples. The primary emphasis, however, is placed on social housing complexes that exhibit notable architectural value, in order to acknowledge their unique contribution to cultural sustainability. Architecture, as a tangible form of heritage, conveys layered narratives across time and shapes the evolving character of urban landscapes. By foregrounding both technical and cultural dimensions, this approach provides a flexible and adaptable model suitable for a wide range of housing typologies.

The selected case study, rich in historical and social significance, serves as a compelling case study for examining how energy-efficient interventions can be harmonized with heritage preservation, offering insights into sustainable urban transitions that respect both ecological imperatives and cultural values [12,13]. The Kaisariani neighborhood is deeply rooted in the events following the 1922 Asia Minor Catastrophe, which led to the forced displacement of over 1.5 million refugees. In response to these historic pressures, urban areas across Greece were rapidly transformed into refugee settlements, with Kaisariani becoming one of the most emblematic examples [14,15]. The selected social housing complex reflects design principles inspired by the Viennese superblocks and the iconic Karl Marx-Hof, yet it currently suffers from severe deterioration [16,17].

As a pilot initiative, this research aims to establish a methodological framework for the energy retrofit of similar social housing developments that share comparable architectural and historical attributes. It is important to note that, despite the fact that several English-language studies have investigated energy retrofitting of Greek residential and social housing, these works mainly focus on technical performance, cost effectiveness, or workflow integration rather than on heritage-sensitive methodological frameworks. Studies such as those by [18–20] provide valuable insights into energy-efficient renovation practices in Greece, but do not address the specific challenges of historic or refugee housing typologies. The Tavros Project, implemented by the Hellenic Passive House Institute and Passivistas, represents an important pilot initiative in applying Passive House principles to social housing in Athens; however, it does not extend to the older interwar housing stock that forms part of Greece's refugee housing heritage. Therefore, there remains a significant gap in the literature regarding comprehensive, conservation-compatible frameworks for the energy retrofitting of historic social housing, which this study aims to fill through the examination of the Kaisariani case. The study not only seeks to safeguard the layered cultural heritage of the area—shaped by displacement, resettlement, and community resilience—but also to propose contemporary technical strategies for the large-scale energy upgrading of aging social housing stock. Three core research questions frame the scope of this study, as presented below:

RQ1: How can historic social housing in Mediterranean climates be retrofitted to improve energy efficiency while preserving cultural and architectural heritage?

RQ2: What type of methodological framework may be effective in cases of historic housing complexes, with the mission to achieve contemporary energy efficiency goals?

RQ3: Which retrofit scenario provides the best balance between cost effectiveness, CO_2 reduction, and preservation values?

This study opens a crucial dialog on the often-neglected issue of integrating energy efficiency with heritage preservation, underscoring the significance of the early 20th-century Modernist social housing complexes in Europe. By focusing on the retrofitting of historic social housing complexes, such as the Kaisariani complex in Athens, it sets a solid foundation for further application and development in this area. The proposed methodological framework combines technical retrofitting strategies with cultural sustainability issues, addressing both environmental, cultural and social dimensions, which have been largely overlooked in existing research. This work paves the way for future studies and practical applications in similar Mediterranean contexts, offering a novel approach that can be adapted and expanded upon in the field of heritage-compatible energy retrofitting.

2. Literature Review

In order to focus on the most recent scientific developments in the research field, a bibliometric analysis of keywords has been performed to define the research gap.

2.1. Bibliometric Analysis of Keywords Methodology

An initial step before reviewing the related works consists of an advanced document search in the Scopus database, based on a set of targeted keywords. The specific query includes keywords derived from the manuscript title, combined with additional keywords and limitations in the time period that this search is limited to.

The query returning 126 related articles has been launched as follows:

(TITLE-ABS-KEY ("energy efficiency" OR "building retrofit" OR "energy retrofit" OR "thermal performance") AND TITLE-ABS-KEY ("historic building*" OR "heritage housing" OR "social housing" OR "residential building*" OR "old building*" OR "aged building*" OR "aging building*") AND TITLE-ABS-KEY ("Mediterranean climate*" OR "southern Europe" OR "Mediterranean region" OR "Mediterranean countr*") AND TITLE-ABS-KEY ("framework" OR "methodology" OR "pilot study" OR "case study" OR "model")) AND PUBYEAR > 2009 AND PUBYEAR < 2026.

The time period has been limited after the year 2010 in order to refine the results and focus on the most recent advances in the field. An RIS file has been exported from the Scopus database, which was inserted into the VOSviewer (2023) software in order to analyze bibliographic data of various scientific publications and journals, as well as the link strength and the occurrence [21–23]. A keyword co-occurrence analysis has been conducted for the specific time period [24–26]. Network maps have been generated as shown in Figure 1, indicating the frequency or co-occurrence, identifying five research clusters based on the weight of the keywords. The links and weights of the clustered items in the mapping represent all core components. It is important to note that the inclusion of the term "social housing" within the bibliometric search string was a deliberate methodological choice. While social housing represents a specific subcategory of residential buildings, it was incorporated to capture studies that engage with both energy efficiency and social dimensions in the context of historically or culturally significant housing stock. This decision broadens the scope of analysis beyond purely architectural or technical retrofits, encompassing research that addresses the intersection of energy performance, heritage preservation, and social sustainability. In Mediterranean cities, many early-to-mid-twentieth-century social housing complexes have now acquired historical or cultural value, reflecting collective memory and urban identity. Therefore, including "social housing" in the search strategy provides a complementary rather than alternative perspective to other typologies such as "historic" or "heritage" buildings, enriching the bibliometric mapping and ensuring a more holistic understanding of the research field.

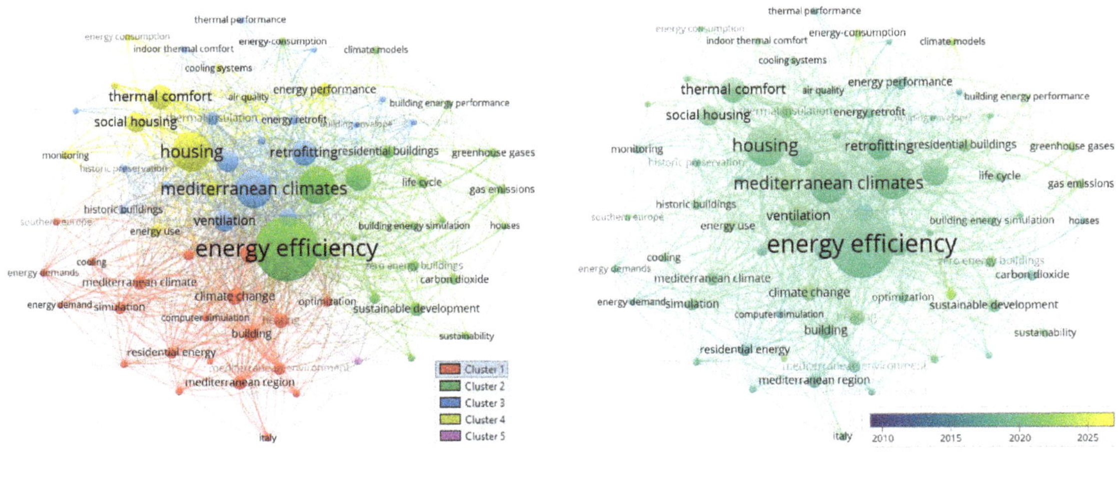

(a) (b)

Figure 1. (**a**) Network visualization map of keywords indicating five clusters and (**b**) overlay visualization map of keywords from the time period 2010–2026.

2.2. *Bibliometric Analysis: Networks Maps of Keywords*

The network visualization map reveals a keyword co-occurrence network where nodes represent keywords and their size represents their co-occurrence. The size of the nodes reveals the strength of the keywords. The size of each circle (node) shows its frequency of occurrence within the dataset. The color coding (as shown in Figure 1a) represents distinct clusters of related terms, identified through co-occurrence analysis.

An overlay visualization of the imported data is presented in Figure 1b, including similar information as shown in Figure 1a, with the addition of the color bar indicating the color mapping in the selected time period of publication. The legend bar at the bottom of Figure 1b indicates the time period within which the query has been launched in the Scopus database (from 2010 to 2026). Keywords (nodes) in the colors yellow and vivid green are derived from the most recent publications (2020–2026), therefore representing the most recent trends. Those keywords are "climate models", "energy consumptions", "climate models", "greenhouse gases", and "sustainability". Dark green and purple nodes indicate keywords from older publications (2010–2015). Larger nodes with darker or mid-green tones represent continued research over a long period of time. The keywords such as "energy efficiency", "housing", and "thermal comfort" are represented in large green nodes, indicating a continued research trend over a longer period.

Consequently, a cluster analysis has been performed, identifying five clusters bringing to light five scientific themes, research themes or subject areas. The keywords are represented as "items". A total of 72 items have been grouped in the five clusters, as shown in Appendix A, Table A1.

Based on the network analysis of keywords performed for the time period 2010–2026, the research gap identified underlines the necessity for integrating heritage conservation, social equity, and climate adaptation within a unified framework for Mediterranean housing.

2.3. *Related Works*

2.3.1. Historical Background

In September 1922, the persecution of Greeks in Asia Minor began and was completed in 1924–1925 under the supervision of the League of Nations (LON). At that time, Greece had 5 million inhabitants, limited natural resources, political turmoil, and was economically devastated [15]. During that difficult period, it accepted 1.5 million refugees, dramatically increasing the country's population [27]. Of the refugees who arrived, 47% settled in rural areas and 53% in urban areas [14]. The sudden arrival of such a large population created serious problems in terms of housing the refugees. Initially, there was no free space left in the city. Schools, churches, warehouses, and theaters were flooded with people who then found shelter by any means possible in the suburbs [28].

In 1922, Athens had approximately 200,000 inhabitants, which meant that there was one house for every ten inhabitants. In November of the same year, the law "On the requisition of real estate for the settlement of refugees" was passed [29]. Three agencies were created with the aim of housing refugees: the Refugee Care Fund (RCF) 1922–1925, followed by the Refugee Rehabilitation Committee (EAP) 1924–1930 and the Ministry of Welfare 1922–1940 [30]. Between 1925 and 1930, 12 main and 34 smaller settlements were created 1–4 km beyond the boundaries of the already built-up area of Athens, one of which is Kaisariani [31].

The Asia Minor refugee settlements established in Greece during the 1920s represent a distinct architectural typology that embodies both the urgency of mass resettlement and the enduring social identity of displaced communities [32,33]. In the early phase, Asia Minor refugee housing was characterized by ground-floor, load-bearing masonry construction and modest floor plan dimensions, mirroring traditional building practices. In contrast, later projects evolved into two-story or three-story apartment buildings that employed mixed construction techniques, demonstrating a gradual refinement in structural design and adaptation to urban housing needs [34]. Early-phase constructions drew inspiration from vernacular architectural traditions, whereas later designs were strongly influenced by the principles of European Modernism. These neighborhoods, such as Kaisariani, Nea Ionia, and Kokkinia, were characterized by modest, repetitive housing blocks constructed with locally available materials and organized around shared courtyards that fostered communal life [34]. The modest geometry and minimal ornamentation of early refugee housing reflected both economic hardship and a collective spirit of adaptation. Today, these settlements represent valuable living heritage, embodying early social housing policies and memories of displacement and integration. Preserving their material and spatial integrity is essential when improving energy performance or comfort. Despite growing European interest in conservation-compatible retrofitting, Greece lacks comprehensive research on the energy rehabilitation of such historic housing, with few examples—like the Tavros Project—addressing more recent buildings. Earlier refugee complexes, including Kaisariani, remain underexplored, highlighting the need for adaptable, preservation-oriented retrofit frameworks.

2.3.2. Social Housing and Heritage Value in Mediterranean Europe: Challenges and Opportunities for Retrofitting

Across the recent literature, social housing is increasingly and explicitly framed as legitimate residential built heritage through mainstream, values-based approaches that foreground social, urban, technological, and architectural significance rather than formal listing alone, positioning estates as "ordinary," "modern," or "urban" heritage and calling for evaluations that extend beyond fabric to encompass resident profiles and urban context [35]. For modern mass-housing ensembles, authenticity and integrity are being recalibrated at the attribute level—façade rhythms, balcony/loggia systems, stair/landing cores, and open-space structures—with layered change accepted as part of significance, thereby refining thresholds for permissible adaptation [36]. Operationalization is advanced by typological and model/type frameworks that classify post-war stocks and distinguish character-defining from modifiable elements, yielding archetypes and reference models suitable for comparative evaluation and retrofit planning [37,38].

Italian case studies from Parma show how carefully planned, low-cost retrofits can significantly improve the performance of aging social housing without disrupting residents or altering the historic character of the buildings. These projects highlight how phased renovation, clear assessment methods, and close collaboration with local authorities can deliver meaningful energy savings while supporting vulnerable households [39].

Due to the heat waves and periods of extreme air temperatures, sustainable social retrofitting strategies are emerging as a high priority for many countries worldwide. These extraordinary circumstances will impact an increasing number of regions in the future, intensifying their effects on the social, economic, cultural and environmental aspects [40,41].

From this point of view, international experience indicates that sustainability in social housing extends well beyond the objective of improving energy efficiency. It encompasses a broader framework of environmental performance, community well-being, and long-term affordability [42]. Contemporary research underscores a significant shift from traditional, maintenance-based approaches toward integrated retrofit frameworks that simultaneously address environmental, economic, and social dimensions [43]. Within this paradigm, sustainability interventions are understood as both technical transformations—including the enhancement of thermal comfort, the reduction of carbon emissions, and the improvement of building envelopes—and as social processes, grounded in participatory planning, tenant empowerment, and inclusive design practices.

The renewal of social housing, with an emphasis on historic buildings, across Europe provides a rich source of precedents relevant to the Kaisariani context. Gaining insights from the European experience, studies have increasingly focused on the heritage significance of post-war social housing estates, which are often undervalued despite their architectural and historical importance. Research on estates like the Western Garden Cities in Amsterdam reveals how elements such as parcellation, balconies, and green spaces are highly valued in sustainable refurbishments, while modifications like external insulation and PVC-U windows are seen as detrimental to their heritage value. These findings reflect broader trends across Europe, where the preservation of post-war housing is becoming a key issue in balancing sustainability and cultural heritage [44].

Siza's SAAL housing in Porto represents one of the most politically engaged examples of Portuguese social housing, integrating resident participation with a sensitive response to the existing urban fabric. The project challenged tabula rasa redevelopment practices, showing how limited resources can generate culturally and socially meaningful architecture. Key SAAL projects such as the S. Victor housing (1974–1977) and the Bouça complex illustrate the embedding of modernist architectural principles into dense historic neighborhoods. These cases remain significant precedents for socially informed and contextually responsive housing of how collaborative design processes addressed the post-revolution housing shortage [45].

In the Netherlands, the "Energiesprong" model has introduced prefabricated retrofit systems capable of transforming existing housing blocks into net-zero energy dwellings, utilizing standardized façade modules, integrated photovoltaic panels, and high-performance insulation [46]. Similarly, in the United Kingdom, programs such as "Retrofit for the Future" have demonstrated the feasibility of achieving substantial energy reductions through a combination of passive and active retrofit measures, emphasizing the role of occupant behavior and community co-management in ensuring long-term effectiveness [47]. In France, national initiatives under the "Grenelle Environnement" framework have promoted the large-scale renovation of social housing estates, coupling improvements in energy performance with goals of social inclusion and urban integration [48]. These interventions highlight the potential of sustainability to act as both an environmental and a social catalyst in urban regeneration. Furthermore, Scandinavian examples reveal the value of cooperative governance structures and circular renovation practices, where sustainability principles are embedded not only in physical design but also in the long-term maintenance and management of housing [49].

Focusing on the Mediterranean region, relevant studies [50] explore how differences in apartment location, insulation, and heating behavior affect energy use and comfort in multi-unit buildings. Using a social housing block in a Mediterranean climate as a case study, the researchers simulated various heating scenarios and validated them with real consumption data. They found that apartments in advantaged positions (e.g., centrally located or south-facing units) can "borrow" heat from neighboring units, benefiting even when heated for fewer hours, while disadvantaged apartments (e.g., top or north-facing units) lose more heat and require higher energy input to maintain comfort. This unequal heat exchange—termed "heat theft"—creates fairness issues when heating costs are shared via sub-metering systems.

Other relevant studies develop methods to analyze how the thermal envelope dynamics of Mediterranean low-income housing affect energy flows in changing climatic conditions. They employ building stock models to simulate moderate energy flow regimes under various future climate scenarios, assessing how the envelopes' capacity to buffer thermal variations influences heating and cooling demand in low-income housing. They argue that taking into account these dynamic envelope behaviors is crucial when projecting energy performance under climate change, especially in regions with mild climates like the Mediterranean [8,51]. There are also studies that apply the MedZEB cost-optimal methodology to public housing in Spain's Comunitat Valenciana region, showing that targeted energy retrofits, especially to the building envelope, can achieve major energy savings and CO_2 reductions at reasonable cost. It emphasizes that climate-specific, cost-effective renovation strategies are key to improving comfort, advancing the decarbonization of Mediterranean public housing [52]. Valuable knowledge also comes from studies that examine the gap between theoretical and actual energy performance of Portuguese dwellings, showing that occupant behavior, building maintenance, and local conditions cause significant discrepancies. These performance gaps affect thermal comfort and highlight the need for targeted, high-resolution interventions to improve both energy efficiency and living conditions [53]. Other studies explore hybrid heating systems (heat pumps plus renewables) to retrofit historic Mediterranean homes in Seville, showing that they can cut CO_2 emissions and fossil fuel use while maintaining comfort and heritage features [54]. Focusing on the case of Spain, there are studies that underscore the significance of simplified models to determine the energy demand of existing buildings, with a particular emphasis on social housing. For example, the case study of social housing in Zaragoza, Spain, presents a model that predicts heating and cooling energy demands in buildings without insulation, offering a practical approach for assessing energy needs in older residential stock. These models are essential tools for guiding urban regeneration efforts and optimizing energy efficiency interventions in vulnerable housing sectors [55]. Like other studies in Mediterranean Europe, research in Zaragoza compares two energy retrofit strategies for social housing with cultural value. The findings indicate that the approach involving coordinated exterior modifications, including external insulation, is more effective in reducing energy consumption and emissions, ultimately enabling the transformation of these buildings into nearly zero energy buildings (nZEBs) [56]. Decision-making for renovating Mediterranean social housing requires practical, evidence-based approaches that can guide interventions in vulnerable households. While many studies address building decarbonization in broad terms, other studies have focused on providing interactive, open-access tools that optimize passive retrofit strategies for southern Spain. These tools support informed decision-making by demonstrating the effectiveness of phased interventions and quantifying incremental improvements in thermal comfort and cost efficiency, offering a practical framework to enhance the energy performance, health, and comfort of social housing stock [57].

Similar studies in Cyprus have explored the energy performance and thermal comfort of social housing, focusing on how climatic conditions, building orientation, and environmental factors influence occupants' comfort levels and energy efficiency, with findings suggesting that a neutral indoor temperature of 28.5 °C is optimal for residents in subtropical and semi-arid climates [58].

There are also studies to show that Italian historic buildings can be refurbished in a cost-effective way to improve energy efficiency while preserving their architectural and cultural value. Research demonstrates that using advanced simulation tools, such as EMAR, allows for accurate modeling of building performance, helping to identify retrofit strategies that optimize energy savings. These studies highlight the importance of balancing technical, economic, and conservation considerations, ensuring that interventions reduce energy consumption and CO_2 emissions without compromising the historical integrity of the buildings [59]. In [60], the authors examined a Thermo-Reflective Multi-Layer System (TRMS) developed for lightweight emergency shelters, demonstrating high thermal efficiency with a U-value of 0.85 $W/(m^2 \cdot °C)$ and substantially reduced heating demand compared to standard tents. Although focused on temporary structures, the results highlight the potential of thin multilayer insulation as a conservation-compatible retrofit solution for historic and social housing, particularly in Mediterranean contexts where façade alterations are restricted.

Among the selected relevant cases, Corviale in Rome, Italy, holds a prominent position because of its ambitious approach to transforming a large-scale, post-war Brutalist structure into a more vibrant, sustainable, and socially cohesive community. The project stands out for its focus on improving the quality of life for residents through the incorporation of green spaces, energy-efficient technologies, and the creation of flexible public areas that foster social interaction. Its success lies in balancing the preservation of architectural identity with the integration of contemporary design solutions, making it a benchmark for modern urban renewal [61,62].

When the Bouça complex was finally completed and upgraded many years later, it became clear that expectations about comfort and daily living had changed. The process brought residents back into the conversation, and the new metro station and other improvements gradually opened the area to a wider mix of people, including students and young professionals who were not part of the original community [63].

Collectively, these European cases emphasize the necessity of multi-scalar and interdisciplinary approaches, integrating technical innovation with institutional reform and community engagement—principles that can inform the sustainable regeneration of the Kaisariani social housing complex. However, despite the efforts made, major challenges—such as financial constraints, informational gaps, and social vulnerabilities—continue to slow the progress of public housing retrofits. Evidence from pilot initiatives shows that integrated approaches—combining technical upgrades, tenant engagement, and effective governance—can help overcome these obstacles. Such interventions not only reduce energy consumption and emissions but also improve residents' quality of life, foster social inclusion, and strengthen the long-term socio-economic resilience of communities, highlighting the strategic value of investing in public housing renovations [64].

Early-to-mid-20th-century social housing complexes across Mediterranean Europe, such as those in Italy, Spain, Greece, and Cyprus, represent a significant socio-cultural and architectural legacy, often blending modernist design with communal spatial arrangements [65–67]. While these buildings are historically important, their age and widespread distribution present challenges in energy efficiency and comfort. Retrofitting these structures through sensitive interventions, like the Kaisariani complex in Athens, allows for the preservation of architectural value while improving performance and living conditions.

3. Materials and Methods

The methodological structure of this study has been designed to ensure a transparent and replicable process that logically connects the research objectives, the analytical tools employed, and the implementation tasks undertaken. The workflow integrates three main components, analyzed in six distinct phases: (i) the theoretical and bibliometric framing of the research gap, (ii) the development of a methodological framework combining heritage conservation principles with energy assessment tools, and (iii) the practical application of this framework to a representative case study in Kaisariani, Athens. Each stage builds upon the previous one, forming a coherent chain that links conceptual foundations to empirical validation. Figure 2 presents the overall workflow diagram.

Figure 2. Workflow Chart of the Study.

3.1. Location and Historical Background of the Selected Case Study

Kaisariani was established between 1923 and 1924, following the Asia Minor Catastrophe, as part of the large-scale refugee housing program organized by the Refugee Rehabilitation Commission (RRC) [31]. Following the development of the methodological framework, the process proceeds to its empirical implementation through a case study analysis. The Kaisariani housing complex was selected due to its representative typology, construction period, and cultural significance as an interwar refugee settlement. The case study serves as an applicative demonstration of the proposed framework, allowing the verification of methodological coherence and the assessment of real-world applicability in terms of energy performance, conservation compatibility, and replicability potential.

Built on the then-outskirts of Athens, the settlement provided homes for thousands of displaced families from Asia Minor who had fled to Greece. Kaisariani was developed in three phases: the initial refugee settlement (1922–1935), expansion eastward around the Skopeftirio and National Gymnasium (1935–1960), and continued growth toward Mount Hymettus (1960–present) [68]. The municipality of Kaisariani is home to different architectural types of social housing, starting from interwar ground-floor and two-story houses to Modernist social housing complexes (Figure 3). The selected case study is one of the unique examples of housing that derives elements from the famous Viennese superblocks, organized around a communal open space (Figure 4). All types of former social housing in the area are subject to deterioration, providing poor living conditions when evaluated against contemporary standards (Figure 3). The urban layout of the area follows a typical Hippodamian grid, being densely built so as to cover the Asia Minor refugee housing needs, as presented in Figure 3.

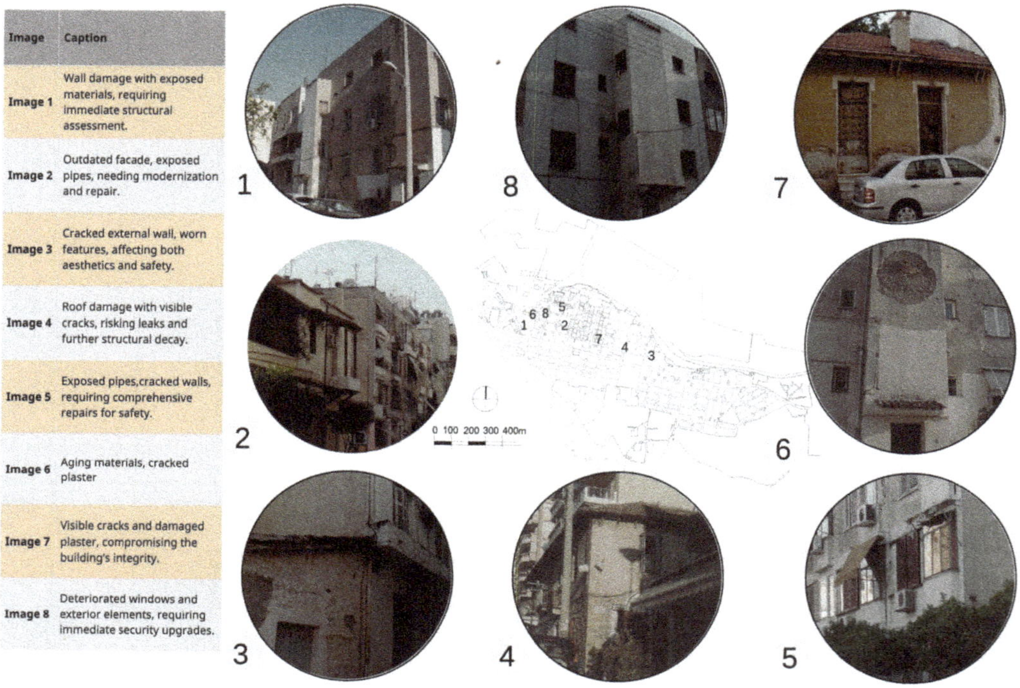

Figure 3. Types of Asia Minor refugee housing rehabilitations in Kaisariani—authors' field work.

Figure 4. Location of the selected social housing complex in Kaisariani, within the grey circle.

Today, in the municipality of Kaisariani, there are many social residences dating from 1940 to 1970, with enormous energy consumption and CO_2 emissions. In view of the above and following actions taken by the Directorate of Special Projects for Area Upgrading (DEEAP) for the regeneration of the area, there is a need for energy upgrading of the buildings.

Based on the existing literature, early 20th-century social housing in Greece—particularly the refugee settlements established after the Asia Minor Catastrophe—underwent a gradual transformation in ownership status. Initially constructed and managed by public authorities such as the Refugee Rehabilitation Commission and later the Ministry of Welfare, these dwellings were eventually transferred to their occupants [12]. Once specific financial and administrative requirements were met, such as the completion of installment payments or proof of long-term residence, full ownership rights were granted to the beneficiaries, effectively converting the social housing stock into privately owned property [32]. This policy shift reflected both the socio-political priorities of post-war reconstruction and the broader national objective of promoting social integration and stability among refugee populations.

3.2. Description of the Social Housing Complex

In 1936–1939, apartment buildings were constructed in Kaisariani, applying the principles of the Modern Movement of the interwar period. In terms of urban planning, they are arranged in blocks, usually in parallel rows. The apartment blocks of Kaisariani are arranged in Π- or Γ-shaped formations, adhering to a rudimentary urban framework inspired by the Hippodamian plan [69–71], as shown in Figure 4. Single-level, two-level or even three-level apartment buildings are constructed, which, due to a low budget, a mixed construction with horizontal elements of reinforced concrete has been selected, combined with the use of domestic materials: The walls are made of stone, either plastered or unplastered. Each residence floor is designed to accommodate 6, 8, 10, or 12 families. The three-level buildings have no lifts, with staircases serving between two and twelve apartments. They include one- and two-room dwellings, a kitchen and a toilet, with an average area of 30 square meters, as well as communal laundry rooms on rooftops or in basements, which are often converted into residences, as there are not enough apartments. Water and electricity installations were included in the building design; however, at the time of construction, the surrounding urban infrastructure had not yet been fully developed.

In terms of morphology, the Kaisariani complex features prominent staircases, entrance areas, and laundry rooms on the terraces, which are covered with tiles. Most of these are still preserved, although they show signs of wear, mainly on the plaster, staircases, and balconies, due to damage to the reinforced concrete and window frames. The complex also features a large courtyard with a fountain, although several architectural elements—particularly plaster surfaces, staircases, and balconies—show visible deterioration due to material decay.

According to the building plans received from the Directorate of Special Projects for Area Upgrading (hereinafter referred to as DEEAP) and a visit to the site, the building complex has a load-bearing structure made of stone masonry. The roofs are covered by 1/3 with tiles, and the rest consists of a flat roof. The entire complex shows significant damage and deficiencies in its load-bearing structure. Due to its age, it lacks insulation. All buildings have old wooden window frames with single glazing and visible damage and broken glass.

Over the years, the building has suffered serious damage and deterioration. Due to its historical value, this complex is worth studying, proposing interventions to improve its energy efficiency.

3.3. Current Status of Kaisariani Social Housing Building Complex

Figure 4 presents a comprehensive visual documentation of the selected social housing complex in Kaisariani, Athens, illustrating a clear state of physical degradation and long-term neglect. The photographs, taken in September 2025, capture multiple forms of structural and surface damage, collectively reflecting the poor quality of housing and the absence of adequate maintenance interventions over time. Several recurring issues are evident across the depicted units. In points 1, 4, and 7 (Figure 5), extensive plaster detachment and exposure of underlying masonry indicate significant deterioration of the façade surfaces. These damages suggest prolonged exposure to moisture, temperature fluctuations, and possibly insufficient protective coatings, all of which accelerate material decay. The compromised façade integrity not only diminishes the visual coherence of the complex but also poses safety hazards due to the risk of falling debris. Points 2, 3, 5, and 6 (Figure 5) reveal evidence of poor upkeep in communal and private exterior areas. Corroded metal elements, patched surfaces, and inconsistent window and balcony treatments point out uncoordinated repair attempts, rather than systematic rehabilitation. The presence of makeshift structures and temporary coverings further reflects socio-economic constraints and the lack of institutional maintenance frameworks. In point 8 (Figure 5), the broken window exemplifies the neglect of basic infrastructure components, such as openings and joinery, which are essential for habitability, energy efficiency, and user comfort. Such conditions are indicative of a broader pattern of housing obsolescence, where physical decline parallels social marginalization.

Image	Caption
point 1	Cracked wall, peeling plaster, moisture and aging materials.
point 2	Exposed pipes, aging balconies, neglect and lack of modernization.
point 3	Cracked plaster, requiring immediate repair to prevent further damage.
point 4	Dilapidated facade with wear,outdated features, needing restoration.
point 5	Broken windows and exposed elements, affecting safety and energy efficiency.
point 6	Damaged stucco and aging paint, reducing both aesthetics and energy efficiency.
point 7	Crumbled plaster and exposed masonry, risking structural integrity.
point 8	Worn-out window frames and shutters, posing security concerns.

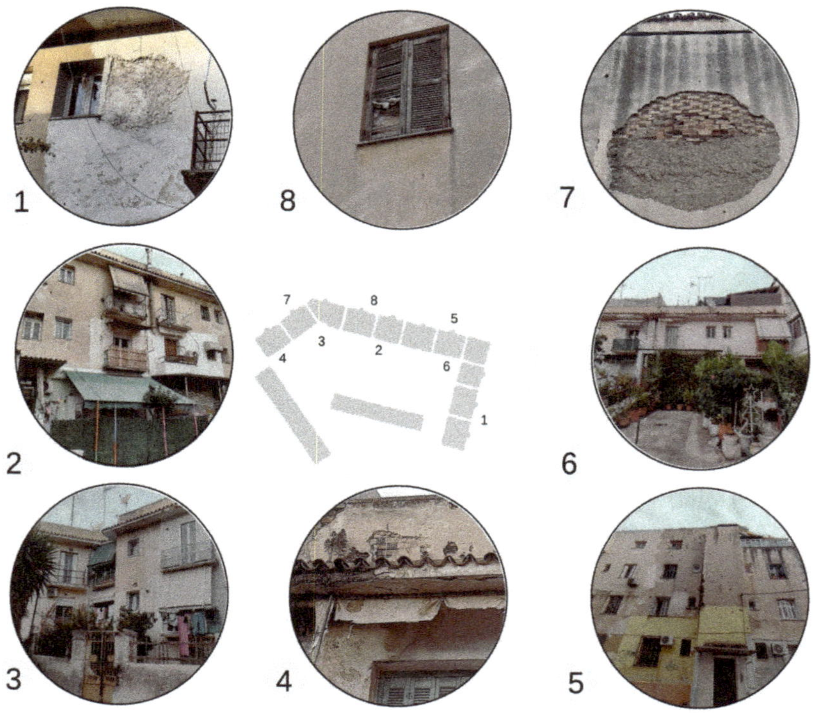

Figure 5. Current status of Kaisariani social housing building complex.

The current condition of the Kaisariani social housing complex reflects not only physical deterioration but also symptoms of energy poverty, a growing issue in aging urban housing across Greece. The combination of non-insulated masonry structures, obsolete mechanical systems, and deteriorated openings results in extremely high energy demand for heating and cooling, as evidenced by annual consumption values exceeding 469 kWh/m^2 for heating alone. Residents, many of whom belong to lower-income or elderly populations, face disproportionate energy costs relative to household income, often leading to under-heating in winter and overheating in summer. This situation undermines indoor comfort, health, and social well-being, while exacerbating inequality within historically significant neighborhoods. The case of Kaisariani illustrates how energy poverty intersects with heritage conservation challenges, highlighting the urgent need for targeted, socially inclusive retrofit strategies that enhance both energy performance and living quality in historic social housing.

Overall, the complex exhibits symptoms of advanced material fatigue, inadequate upkeep, and insufficient public investment. The visible damages are not merely aesthetic but signify deeper structural and environmental vulnerabilities. This case exemplifies the broader challenges that social housing in Kaisariani faces, being marked by limited maintenance and funding. The gradual erosion of living standards within historically significant yet under-resourced urban communities raises skepticism for the future of these neighborhoods.

3.4. Energy Performance Evaluation Tools

In the Kaisariani case study, the Energy Performance of Buildings Regulation (KENAK) serves as a key framework for evaluating and comparing energy retrofit scenarios [8,9]. To be more specific, KENAK is the Greek Regulation on the Energy Performance of Buildings, aligned with the European Standards [72]. Introduced in 2010, it establishes the framework for designing, constructing, and renovating energy-efficient buildings in compliance with EU directives [73].

- EN ISO 52016-1:2017—Energy Performance of Buildings—Energy needs for heating and cooling, internal temperatures, and sensible and latent heat loads [74].
- EN 16883:2017—Guidelines for improving the energy performance of historic buildings [75,76].
- EPBD (Directive 2018/844/EU)—Energy Performance of Buildings Directive [77].

KENAK sets standards for insulation, heating, cooling, lighting, and renewable energy use, ensuring that buildings meet specific energy performance criteria. It also mandates the issuance of an Energy Performance Certificate, classifying buildings from A+ (most efficient) to G (least efficient), with the ultimate goal of achieving nearly zero-energy buildings across Greece.

In addition, KENAK provides standardized benchmarks—known as reference buildings—which mirror the studied structures in geometry, orientation, function, and operation, but meet only the minimum national energy efficiency requirements [72]. By using this model, the study was able to assess actual buildings against a regulated baseline across critical performance indicators such as heating, cooling, lighting, total energy demand, and operational cost. All evaluations were performed according to Greek national codes and EN ISO standards, incorporating Mediterranean climate parameters to ensure accuracy. To be more specific, the assessment of the energy performance of the Kaisariani social housing complex was carried out using TEE-KENAK 1.31_March2018, the official software tool of the Greek Regulation on the Energy Performance of Buildings (KENAK). The program applies standardized calculation procedures to evaluate energy demand and classify buildings according to national benchmarks, rather than dynamic simulation methods. The analysis followed the official KENAK methodology, assuming residential use with 18 h of daily occupancy, seven days per week, and continuous operation throughout the year. Internal heat gains from occupants, lighting, and appliances were included following the default KENAK parameters, while air infiltration was set at 0.5 air changes per hour, consistent with the characteristics of pre-1979 uninsulated masonry buildings. The efficiency of heating, cooling, and domestic hot water systems was defined according to the relevant KENAK reference values. For each of the three retrofit scenarios, the boundary conditions—including thermal transmittance (U-values), insulation thickness, and the introduction of renewable systems such as solar water heaters and photovoltaic panels—were defined and compared with the existing and reference building models. In parallel, 3D CAD models of the complex were used to support the geometrical description of thermal zones and to verify building envelope characteristics. The results were then evaluated against the reference building (RB) defined by KENAK, allowing a consistent comparison of energy demand, efficiency class, and operational cost under each scenario.

Data were processed and visualized through analytical tools like Microsoft Excel, enabling a systematic analysis of energy inefficiencies and the economic feasibility of different upgrade options within Kaisariani's dense urban context.

It is important to note that thermal insulation plays a central role in these comparisons. Greek building legislation distinguishes four categories based on insulation standards:

- Pre-1979 buildings: Non-insulated, built before thermal standards were enforced.
- 1979–2010 buildings: Comply with the first Thermal Insulation Regulation and are considered "basically insulated."
- 2010–2017 buildings: Constructed under the first phase of KENAK (2010).
- Post-2017 buildings: Meet the revised and stricter KENAK standards.

The selected case study falls within the first category, the pre-1979 built housing reserve, suffering from numerous inefficiencies in terms of energy performance.

A key innovation in this study lies in the application of the Greek KENAK energy performance evaluation system. Traditionally, separate files are required for each building or thermal zone within a complex. In contrast, we developed a novel approach by consolidating all ten buildings of the Kaisariani social housing complex into a single KENAK file, with each building represented as a distinct thermal zone. This adaptation allowed for a more efficient and streamlined analysis, reducing the complexity typically associated with handling multiple files. By using one integrated file, we were able to perform detailed energy simulations across all buildings in the complex simultaneously, ensuring consistency and improving the overall accuracy of the results. This methodology not only simplifies the workflow but also provides a scalable model for energy assessments in multi-building housing complexes, offering a more cohesive and efficient approach to evaluating energy retrofitting scenarios.

3.5. Data Collection

The following diagram (Figure 6) illustrates the data collection process. These parameters were essential for evaluating thermal dynamics and energy performance, and they served as baseline values for assessing the retrofit scenarios.

The social housing complex is composed of ten interconnected buildings, with adjoining side walls, as shown in Figure 7. All buildings consist of three levels, with the exception of building D, which is two levels high. Each building consists of six apartments, two per floor, with the exception of building C, which consists of twelve apartments, four per level. As presented in Figure 7, in the Kaisariani social housing complex, surfaces in contact with external air include the building façades and roof areas directly exposed to outdoor conditions, where most thermal losses occur due to the absence of insulation.

Adiabatic surfaces refer to the shared walls between adjacent apartments or buildings within the same block, where no significant heat transfer takes place because both sides are similarly conditioned (Figure 8). Surfaces in contact with unheated space correspond to structural elements bordering non-conditioned areas, such as stairwells, which experience moderate heat exchange and contribute to overall energy inefficiency compared to fully external walls.

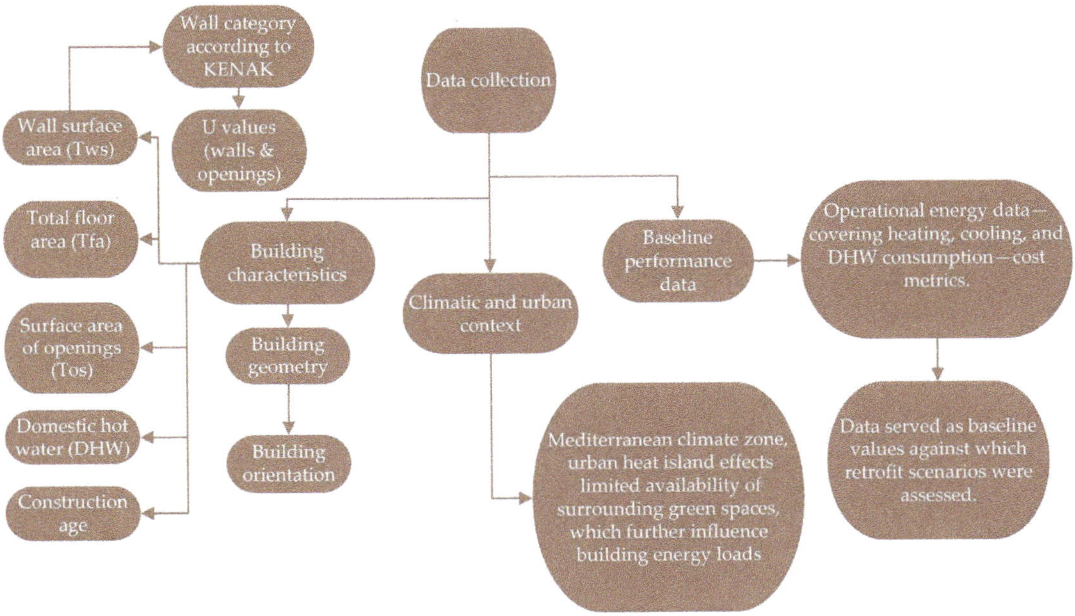

Figure 6. Data collection process flowchart.

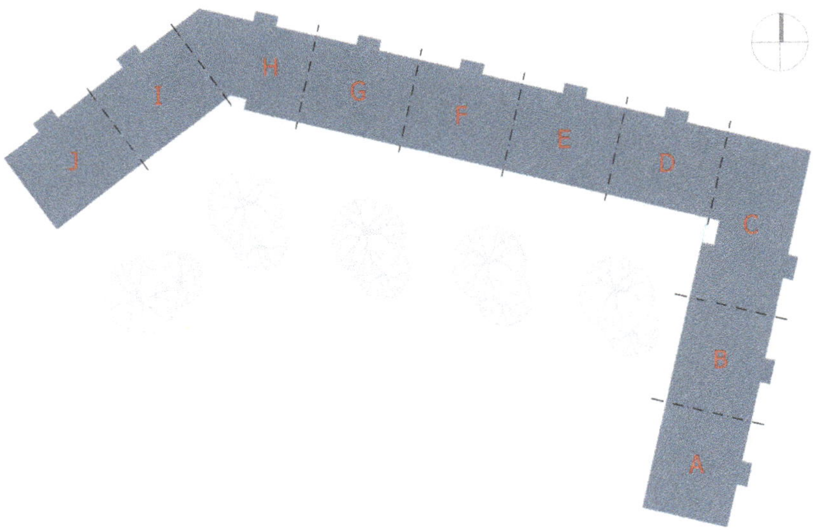

Figure 7. Outlines and names of the ten buildings (A–J) corresponding to ten thermal zones within the housing complex (drafted using the CAD file provided by DEEAP).

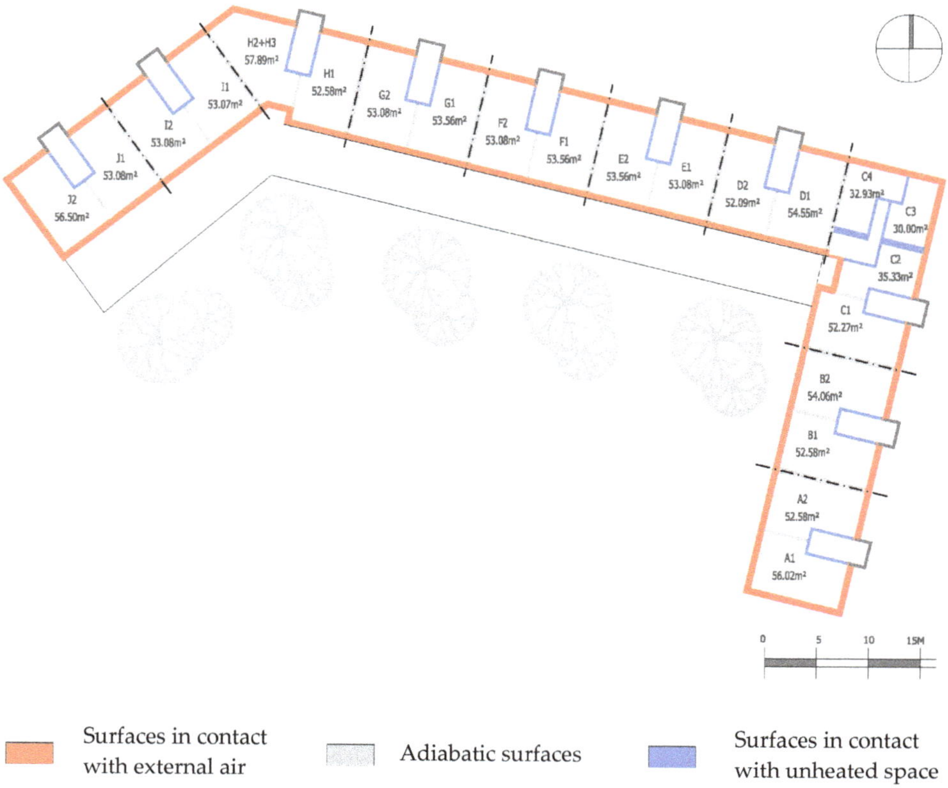

Figure 8. Typical floor layout EB with designated properties and wall categories according to KENAK (drafted using the CAD file provided by DEEAP).

The standard operating hours for domestic use are 18 h/day, 7 h/week and 12 months/year, as elaborated in Appendix E.

Table 1 presents the main geometric, functional, and energy-related characteristics of the ten buildings comprising the Kaisariani social housing complex.

Table 1. Area characteristics and thermal demands of the ten existing buildings (EBs).

Building Name	Number of Levels	Total Floor Area * (m^2)	Number of Apartments	Automation Category	Energy Efficiency Category	Heating Energy Demand (Year)	Cooling Energy Demand (Year)	DHW Energy Demand
A	3	316.80	6	D	G			
B	3	311.11	6	D	G			
C	3	469.71	12	D	G			
D	2	207.40	4	D	G			
E	3	310.98	6	D	G	469.4	82.6	31.3
F	3	310.98	6	D	G			
G	3	310.98	6	D	G			
H	3	322.5	6	D	G			
I	3	309.51	6	D	G			
J	3	319.80	6	D	G			

* The total floor area includes the sum of the surface of each level measured externally (including external walls) and the flat roof surface, as shown in Appendix B. The surface area of windows (openings) is deducted.

All buildings may be described as low-rise structures of two to three levels, consistent with the interwar typology of social housing. The total floor area ranges from 207.40 m^2 in Building D to 469.71 m^2 in Building C, which accommodates twelve apartments. Across the complex, all buildings are classified under automation category D and energy efficiency category G, indicating outdated systems and very poor energy performance according to current standards. Typical values of energy demand may be found in Table 1, referring to building E: an annual heating energy demand of 469.4 kWh/m^2, cooling demand of 82.6 kWh/m^2, and domestic hot water (DHW) demand of 31.3 kWh/m^2. These values reflect extremely high energy requirements typical of uninsulated masonry structures with obsolete installations.

Table 2 provides detailed thermal characteristics of the ten buildings within the Kaisariani social housing complex, emphasizing parameters directly related to envelope performance and energy demand. The heated floor area ranges from 181.30 m^2 in Building D to 376.77 m^2 in Building C, reflecting variations in building size and occupancy capacity. The wall surfaces in contact with unheated spaces (WCUSs) vary between 158.15 m^2 and 258.70 m^2, indicating significant exposure to non-conditioned zones such as stairwells and basements. U-values have been estimated for examining the thermal transmittance of materials [9,78]. Consequently, it is important to better understand the role that the thermal transmittance of the building envelope elements has on air-conditioning consumption. This paper analyzes the effect of different U-values on building design in the Mediterranean region [79]. In total, 192,000 residential buildings were randomly generated for sixteen distinct locations, and the energy consumption was assessed for each. It was found that in northern Mediterranean locations, as U-values decreased, energy consumption also decreased. However, in warmer climates, low thermal transmittances tended to significantly increase energy consumption. Hence, the lower the latitude, the higher the U-values should be, in order to prevent increasing the cooling demands. Additionally, geometry-based indexes were correlated with the building's energy performance. For high U-values, it was found that bigger buildings worsen the energy performance, and larger windows tend to improve it. For low U-values, bigger north-facing windows were beneficial. There is an adequate interval of values for which the geometry has a lower impact, which is wider and higher for lower latitudes, thus meaning that not only does the building performance improve, but architects are also freer to explore alternatives [79].

Table 2. Thermal characteristics of the ten buildings within the Kaisariani social housing complex, emphasizing parameters directly related to envelope performance and energy demand.

Building Name	Heated Floor Area (m^2)	Walls Surface in Contact with Unheated Spaces (WCUSs) * (m^2)	Average U ** Value of (WCUS) (W/m^{2}K)	Ground-Contact Surfaces (m^2)	Total Openings' Surface (Tos) (m^2)	DHW *** System Capacity (Kw)
A	277.65	158.58	3.58	105.60	35.10	24
B	271.95	165.58	3.58	103.70	34.98	24
C	376.77	258.7	3.58	156.57	61.89	48
D	181.30	164.79	3.58	103.70	22.76	16
E	271.83	158.15	3.58	103.66	37.29	24
F	271.83	158.15	3.58	103.66	37.29	24
G	271.83	158.15	3.58	103.66	37.29	24
H	283.45	164.79	3.58	107.50	33.59	24
I	270.36	158.58	3.58	103.17	36.09	24
J	280.65	158.15	3.58	106.60	37.29	24

* Unheated spaces refer to common spaces within the buildings, such as corridors, building stairs and storage areas. ** U value expresses the amount of heat that passes through 1 m^2 of wall for every 1 K temperature. Better thermal insulation material properties are expressed with lower U-values. *** Consumptions are calculated according to Table A4 in Appendix F (see Abbreviations table).

The calculation of heat transfer in windows has a direct impact on the thermal transmittance of highly insulated glazing components. A series of experimental tests was carried out in order to calculate the U-value of active insulated windows using the Heat Flux Meter method (ISO 9869-1 [80]); this was to compare the insulation properties of traditional single glass, double glazing with different aerogel fillings and vacuum glazing windows. The use of this heat flux method utilized an environmental chamber to provide a temperature difference of 15 °C, reporting the U-values as follows: traditional double glazing 3.09 W/m^2·K, vacuum glazing 1.12 W/m^2·K, and double glazing with aerogel pillars 2.52 W/m^2·K. On the other hand, double glazing with KGM wheat starch reported 3.40 W/m^2·K, double glazing with granulated aerogel 2.07 W/m^2·K, and heat insulation solar glass 1.84 W/m^2·K. Vacuum glazing recorded optimal results under these experimental conditions, describing a U-value 78% lower when compared to traditional single glazing window units (5.15 W/m^2·K). Installation of windows with lower thermal transmittance is expected to increase in the global market to meet the current construction codes, aimed at achieving net-zero carbon buildings [78,81,82].

The retrofit of the most energy-intensive buildings represents an opportunity to improve their energy efficiency or to reduce their energy demand. This paper proposes combining computer-aided design (CAD) modeling and the use of TEE-KENAK. 1.31_March2018 energy efficiency software to build a methodology for calculating, visualizing, and analyzing building parameters in order to provide retrofit scenarios. Five retrofit scenarios were implemented using the energy software, including the initial operating cost, capital cost, and payback period to be evaluated. At the same time, a three-dimensional CAD model was created to perform daylighting and shading simulations to visualize and design the role of building orientation under actual use conditions. These retrofit scenarios were evaluated individually and then combined to examine their performance in terms of cost effectiveness and energy efficiency. The simulation results show the importance of the building's orientation, as this directly affects the thermal properties of the walls and openings, as well as the daylighting areas. The simulation results were also used to define the parameters that affect the interoperability of the retrofit solutions. Finally, in addition to the significant reduction in calculation time, the coupling of the Autocad 2021 (educational version) software with the energy efficiency software allowed access to information that was not available at the outset [9].

All buildings share a consistently high average U-value of 3.58 W/m^{2}K, confirming the absence of thermal insulation and the poor heat retention capacity of the existing masonry structure. Ground-contact surfaces range from 103.17 m^2 to 156.57 m^2, while total opening surfaces (Tos) span from 22.76 m^2 to 61.89 m^2, showing proportional relationships between window areas and building size. The domestic hot water (DHW) system capacities vary from 16 kW in the smallest units to 48 kW in the largest, aligning with the differences in total dwelling area.

For all studied buildings, the automation category is D, ranking from A (best automation category) to D (poor automation category), as elaborated in Appendix C.

3.6. The Three Energy Upgrade Scenarios

The proposed energy upgrade scenarios for the Kaisariani social housing complex are designed to address the significant inefficiencies highlighted in the current state of the buildings. They follow a stepwise approach, beginning with essential envelope and system improvements, and progressing toward the integration of renewable energy technologies (Table 3). The selection of technologies and intervention strategies for the retrofit scenarios was guided by four principal criteria to ensure both technical and cultural suitability. First, all proposed measures were evaluated for compatibility with conservation constraints, prioritizing reversibility, minimal visual impact, and respect for the architectural integrity of the historic housing fabric. Second, the solutions were designed for adaptability to Mediterranean climatic conditions, considering the specific thermal and environmental characteristics of the region. Third, the methodology ensured full compliance with established regulatory frameworks, including EN 16883:2017 [83], which provides guidelines for improving energy performance in historic buildings, and the KENAK, governing building energy performance in Greece. Finally, each intervention was assessed for technical feasibility and cost effectiveness, particularly in relation to large-scale applications in social housing contexts. This multi-criteria approach provided a robust and replicable basis for defining the three retrofit scenarios tested in the study.

Table 3. Retrofit solutions/scenario.

Scenario	Retrofit Solutions	Detailed Interventions
1	Thermal insulation of the building envelope—upgraded heating and upgraded cooling	Addition of a central heat pump for each building Installation of new air-conditioning units
2	Scenario 1 + replacement of window frames and installation of a solar water system for DHW	Average U value of new windows = 2 instead of 5 in the EB (W/m^{3}K)
3	Scenario 3 + Photovoltaics	Contribution of RES–CHP

Scenario 1 employs thermal insulation of the building envelope, in combination with upgrades to both the heating and cooling systems. A central heat pump is suggested for each building, ensuring more efficient and uniform heating distribution. In parallel, new air-conditioning units are installed to modernize cooling provision. This scenario represents the baseline intervention strategy, focusing on the most critical aspects of energy efficiency—thermal protection of the envelope and modernization of mechanical systems.

Scenario 2 builds on the first intervention by incorporating additional measures to improve both envelope performance and hot water supply. Alongside the thermal insulation and system upgrades of scenario 1, this scenario suggests the replacement of window frames with more efficient alternatives, reducing heat transfer through the openings, and mitigating air infiltration. Furthermore, a solar water heating system is introduced to cover a significant share of domestic hot water (DHW) demand through renewable energy, reducing reliance on conventional systems, with the aim of lowering operational costs.

Scenario 3 represents the most comprehensive intervention package. It includes all measures from the first two scenarios—thermal insulation, efficient heating and cooling, improved window performance, and solar water heating—while also introducing energy-efficient glazing and a photovoltaic (PV) system. The adoption of high-performance windows further enhances the building envelope, minimizing energy losses and improving indoor comfort. The integration of PV panels marks a transition toward active energy production, enabling partial on-site generation of electricity. This scenario thus offers the greatest potential for reducing both energy consumption and greenhouse gas emissions, positioning the buildings closer to contemporary standards of nearly zero-energy performance.

Together, these scenarios illustrate a progressive strategy for upgrading the historic refugee housing complex, ranging from essential insulation and system modernization to the integration of renewable technologies and self-generation of electricity. The three proposed retrofit scenarios were carefully designed to be minimally invasive, ensuring full respect for the historical morphology and architectural character of the Kaisariani social housing complex. Aligned with EN 16883:2017—"Guidelines for Improving the Energy Performance of Historic Buildings" [75], developed by CEN/TC 346, the proposed retrofit approach for the Kaisariani social housing complex adopts a conservation-compatible framework that enhances energy performance while safeguarding heritage values. Following the principles of compatibility, reversibility, and minimal intervention, all three scenarios integrate energy-efficient measures—such as internal insulation, discreet renewable systems, and morphologically consistent window replacements—without altering the original façades or spatial typology [83]. This approach ensures that energy upgrades meet KENAK standards while fully respecting the complex's architectural integrity and cultural significance.

4. Results

4.1. Existing Building (EB)

The interwar social housing buildings in Kaisariani, Athens, constructed to accommodate Asia Minor refugees, represent a modest but historically significant part of the city's built environment. The housing stock is largely uniform in form and size: most blocks are three-story buildings containing six apartments each. Only two stand out—Building C, which has twelve apartments on three levels, and Building D, which is smaller, with two storeys and four apartments. The total floor areas of these buildings vary between 207.40 m^2 (Building D) and 469.71 m^2 (Building C), while the corresponding heated floor areas are consistently lower, from 181.30 m^2 to 376.77 m^2. Envelope characteristics reflect early 20th-century construction practices, with the total surface of openings ranging from 22.76 m^2 to 61.89 m^2. Existing openings contribute to significant air infiltration rates of 475–887 m^3/h, a clear indicator of poor airtightness, functioning as a major source of thermal losses. Domestic hot water (DHW) demand is also notable, ranging from 109.20 m^3/year in the smaller buildings to 327.60 m^3/year in the larger ones. System capacities scale accordingly, from 16 kW to 48 kW. Despite differences in building size and consumption levels, all buildings fall into the same categories: automation class D and energy efficiency class G, which are low rating categories under current standards.

4.2. Heating and Cooling Energy Demand

Comparison Among EB, RB, and the Three Scenarios

Figure 9 illustrates the heating and cooling energy demand of the Kaisariani social housing buildings under different energy upgrade scenarios compared with the existing building (EB) and the reference building (RB). The existing building demonstrates very high heating demand, at 469.4 kWh/m^2, reflecting the poor thermal performance of the current envelope and systems. Scenario 1—introducing insulation and upgraded heating and cooling—reduces demand to 134.3 kWh/m^2, corresponding to a 71% reduction compared to the existing condition. Scenarios 2 and 3, which add further measures such as improved window systems, solar water heating, and photovoltaics, achieve an even greater improvement, lowering demand to 114 kWh/m^2, or a 76% reduction relative to the existing building. Although these values remain significantly higher than the benchmark of the reference building (41.7 kWh/m^2), the results underscore the substantial energy-saving potential of the proposed interventions, with scenarios 2 and 3 approaching three-quarters of the total reduction needed to align with modern standards.

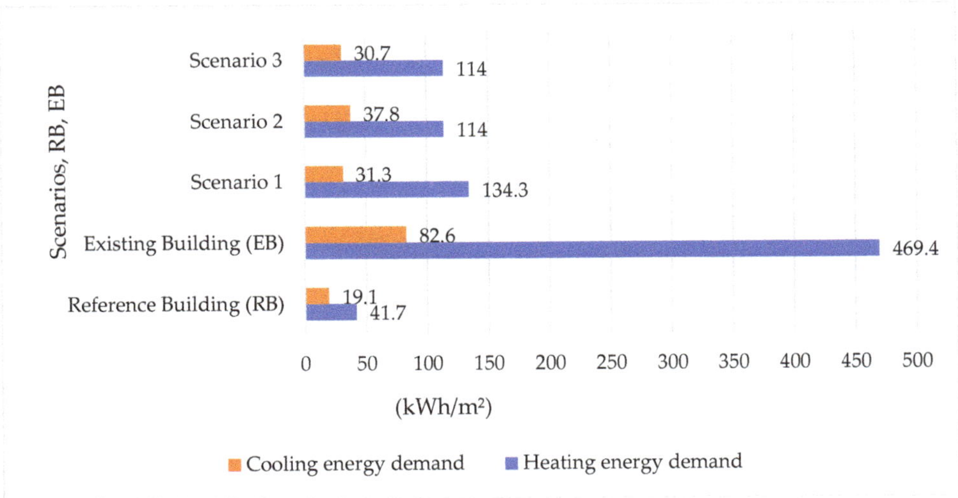

Figure 9. Heating and cooling energy demand (kWh/m^2).

4.3. Cooling Energy Demand

Comparison Among EB, RB, and the Three Scenarios

The cooling energy demand of the Kaisariani social housing buildings in their existing state (EB), under the proposed upgrade scenarios, and against the benchmark of the reference building (RB) are also shown in Figure 9. The existing building exhibits very high cooling demand at 82.6 kWh/m^2, which reflects both inadequate envelope performance and outdated mechanical systems. Scenario 1, incorporating insulation and upgraded heating/cooling, reduces demand to 31.3 kWh/m^2, corresponding to a 62% reduction compared to the existing building. Scenario 2, which adds new window frames and a solar water system, results in a slightly higher demand of 37.8 kWh/m^2, equating to a 54% reduction, suggesting that while this intervention improves heating efficiency, it may slightly compromise summer performance due to altered solar gains. Scenario 3, which introduces energy-efficient glazing and photovoltaics in addition to the previous measures, achieves the lowest cooling demand at 30.7 kWh/m^2, representing a 63% reduction relative to the existing condition. Despite these substantial improvements, all scenarios remain above the reference building level of 19.1 kWh/m^2, underscoring both the progress achieved and the remaining gap toward optimal energy performance.

4.4. Domestic Hot Water Demand

Comparison Among EB, RB, and the Three Scenarios

Figure 10 illustrates the domestic hot water (DHW) energy demand for the existing Kaisariani buildings, the three upgrade scenarios, and the reference building (RB). In their current state, the buildings exhibit a DHW demand of 31.3 kWh/m^2, which remains unchanged under scenario 1, as no interventions related to water heating are included. By contrast, scenarios 2 and 3 integrate solar water heating systems, dramatically reducing demand to just 1.4 kWh/m^2. This represents a 95% reduction compared to both the existing building and scenario 1, and it performs significantly better than the reference building benchmark of 14.6 kWh/m^2. These results highlight the pivotal role of renewable-based DHW systems in achieving substantial efficiency gains, demonstrating that relatively targeted interventions in water heating can achieve deeper reductions than broader envelope or system upgrades alone.

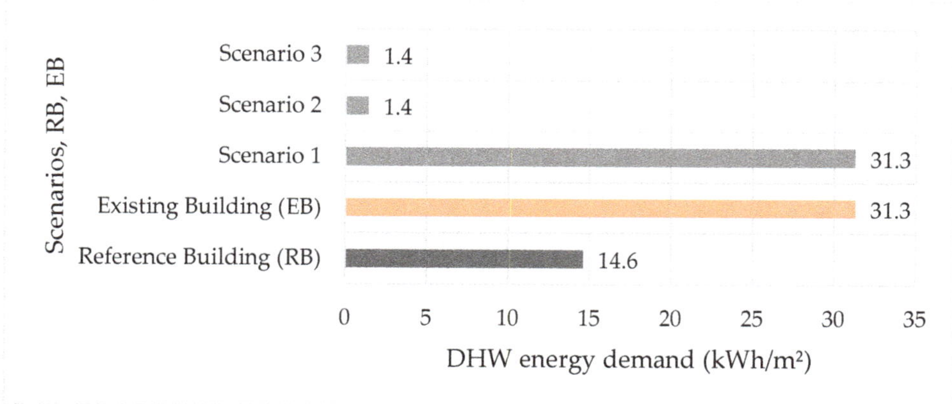

Figure 10. DHW energy demand: EB, RB, and the three scenarios.

4.5. Total Energy Demand

Comparison Among EB, RB, and the Three Scenarios

The existing building (EB) has the highest energy demand at 583.2 units, followed by scenario 1 at 196.9 units, scenario 2 at 153.2 units, and the reference building (RB) at 75.3 units. Scenario 3 shows the lowest demand at 17.5 units, indicating significant energy reduction potential in comparison to the existing building (Figure 11).

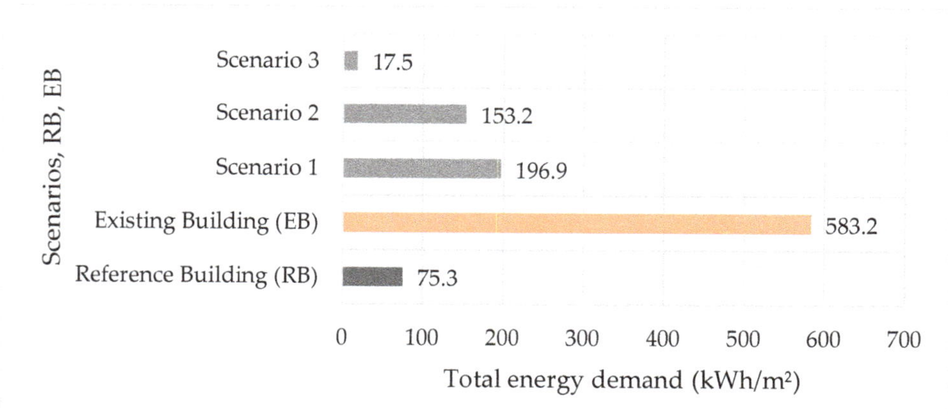

Figure 11. Total energy demand: EB, RB (kWh/m^2), and the three scenarios.

4.6. Automation Category

According to the KENAK framework, the automation category reflects the degree of control and monitoring of a building's energy systems. In all three retrofit scenarios for the Kaisariani social housing complex, the automation category improves from D to C, indicating a shift from minimal or manual operation to a more coordinated, semi-automated management of energy functions. This upgrade results from the installation of modern HVAC systems, central heat pumps, and renewable energy technologies that enable better regulation of indoor comfort conditions and energy use. Under KENAK classification, category C represents buildings equipped with automated control of heating, cooling, and domestic hot water systems, allowing for improved energy performance, reduced operational losses, and enhanced occupant comfort—marking a crucial step toward compliance with contemporary energy efficiency standards.

4.7. Reduction in Primary Energy Demand (% and kWh/m^2)

Comparison Among the Three Scenarios

As presented in Figure 12, the percentage of the reduction in primary energy demand varies from 66.2% in scenario 1 to 97% observed in scenario 3. In between, scenario 2 is found with a significant reduction percentage of 73.7%. These percentages correspond to kWh/m^2 reductions at 386.3 in scenario 1, 439 in scenario 2, and 565.7 in scenario 3 (see Figure 12). Undoubtedly, scenario 3 is regarded as the most efficient in terms of reduction in primary energy demand as it achieves the best possible outcome.

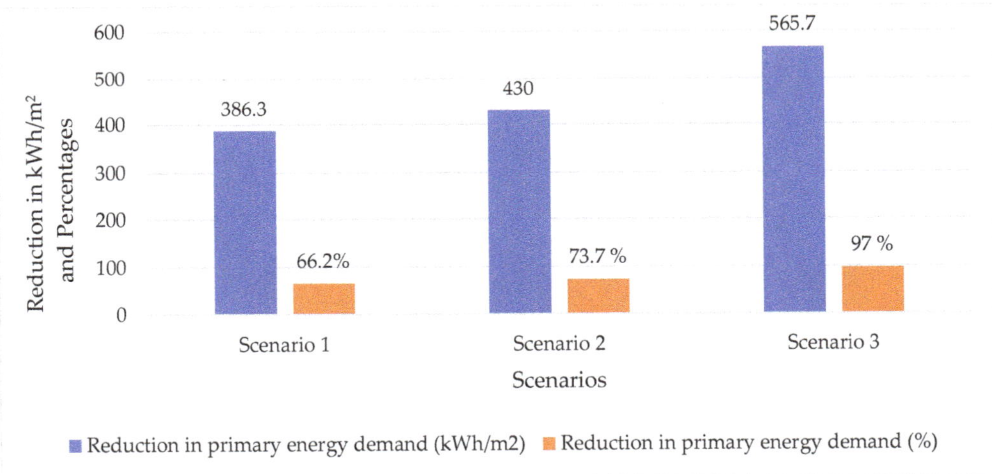

Figure 12. Reduction in primary energy demand in (kWh/m^2), and reduction percentages.

4.8. Reduction of CO_2 Emissions

Comparison Among the Three Scenarios

Figure 13 illustrates the reduction of CO_2 emissions under the three scenarios, each reflecting a different combination of energy efficiency measures. Scenario 1, which includes thermal insulation of the building envelope, upgraded heating with central heat pumps, and upgraded cooling with new air-conditioning units, achieves a reduction of 131.9 units. Scenario 2 adds the replacement of window frames and the installation of a solar water system for domestic hot water to the measures of scenario 1, increasing the reduction to 155.7 units. Scenario 3 further enhances the package by incorporating energy-efficient windows, a solar water system, and photovoltaics, resulting in the highest reduction of 184.3 units. The results demonstrate that more comprehensive interventions, particularly those integrating renewable energy technologies, yield greater reductions in CO_2 emissions. However, the progressive improvements also suggest that the most significant gains may come with higher upfront costs and technical complexity, raising questions about the balance between feasibility, affordability, and environmental benefit.

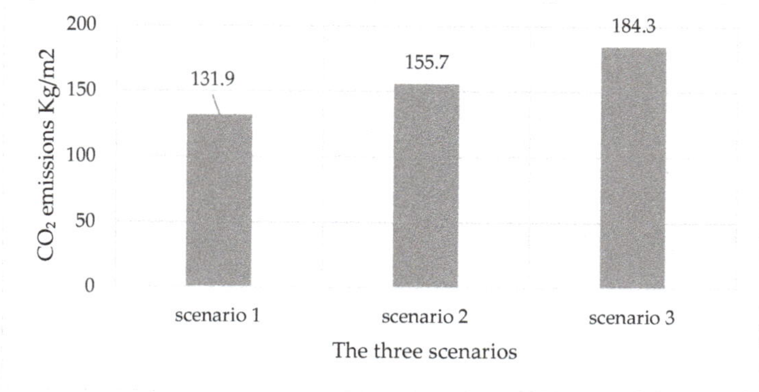

Figure 13. Reduction of CO_2 emissions Kg/m^2.

4.9. Economic Dimensions of the Suggested Scenarios

The three scenarios differ not only in their capacity to reduce CO_2 emissions but also in their economic implications. Scenario 1, focused on thermal insulation and the upgrading of heating and cooling systems, represents the least costly intervention, with an initial investment of EUR 520,568 (Figure 14). While it delivers the lowest emission reduction (131.9 units), its lower financial requirement makes it a relatively accessible option. Scenario 2, which adds window replacement and a solar water system for domestic hot water, increases the investment to EUR 721,276 but achieves a higher emission reduction of 155.7 units, suggesting an improved balance between cost and environmental performance (Figure 13). Scenario 3, incorporating energy-efficient windows and photovoltaics, demands the highest capital expenditure of EUR 869,276 yet results in the largest emission reduction of 184.3 units. Overall, the results highlight a clear trade-off: higher upfront investments are associated with greater reductions in CO_2 emissions, underscoring the importance of considering payback periods, long-term energy savings, and supportive financial mechanisms when evaluating the feasibility of each scenario. To be more specific, the repayment period for scenario 1 is 8.3 years, for scenario 2 it is 9.8 years, and for scenario 3 three the repayment period is 10 years (Figure 14).

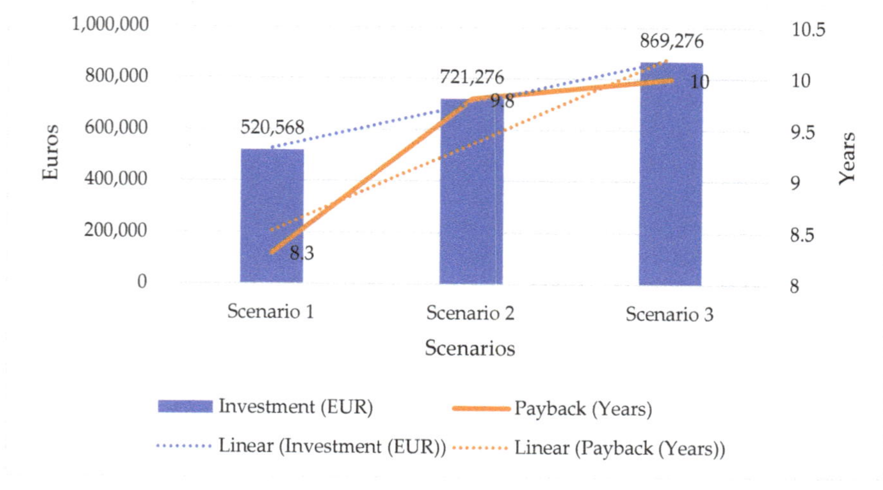

Figure 14. The initial investment cost in euros, repayment period in years, and trendlines.

The linear trendline for investment represented by the blue dotted line in Figure 14 shows how the initial investment cost increases between the three retrofit scenarios, indicating a steady and proportional rise in investment. Accordingly, the linear trendline for payback period in years, represented by the orange dotted line, indicates that more expensive scenarios (2 and 3) take more time to recover the investment through cost savings. The payback period also follows a positive linear trend, though less steeply. Even though scenario 3 is the most expensive, its financial performance remains relatively stable, with only a modest increase in payback time.

The three scenarios show progressive financial scaling: higher investment yields improved or enhanced system performance, while maintaining a manageable and nearly proportional repayment period, indicating economically reasonable progression between the three retrofit scenarios.

Regarding the operational costs (Figure 15), the existing building reaches the highest value of 98,634.1 euros, which is far more expensive compared to scenario 1 (36,220.8 euros). Scenarios 2 and three perform better in terms of operational costs, with values ranging from 24,939.8 euros to 11,412.2 euros, showing the long-term benefits of a targeted energy upgrade in social housing accommodation. It is important to note that scenario 3 outperforms the reference building, whose operational costs are estimated at 20,162.8 euros, a value similar to the costs of scenario 2.

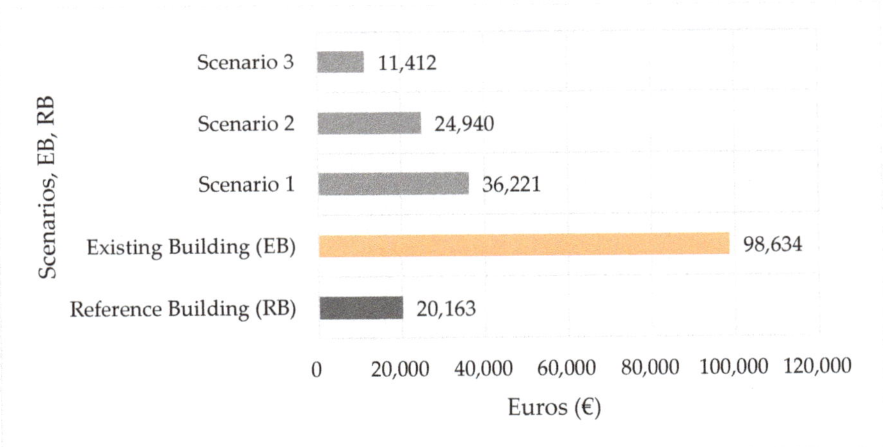

Figure 15. Operational costs in euros, EB, RB and the three scenarios in EUR.

5. Discussion

5.1. Discussion of Findings from the Bibliometric Analysis

The bibliometric analysis of keywords revealed dominant terms represented as nodes of bigger diameter, which include "energy efficiency", "mediterranean climate", "housing", "thermal comfort" and "retrofitting". Those red-colored keywords belong in the primary thematic area. The network visualization map shows the prominence of Spain, Cyprus, Italy and southern European countries as pioneers in the literature activity. Emerging research trends are also revealed: The keywords "monitoring", "occupant behavior", and "thermal comfort" seem to indicate a transition towards data-driven retrofit frameworks.

Recent research on social housing across Europe shows that many older buildings still struggle with poor indoor conditions, high energy demand and rising vulnerability to energy poverty. Studies from Spain and Italy reveal that even modest retrofit measures can noticeably improve comfort and reduce heating needs. At the same time, the lack of reliable data on how residents use their homes remains a major barrier, making continuous monitoring essential for shaping more realistic energy strategies [84].

Recent work highlights how important it is to monitor real indoor conditions and everyday user behavior before planning renovations in social housing. In southern European cities, simple passive measures and small renewable additions can significantly lower energy needs and ease energy poverty [85].

Green circles (nodes) indicate specialized topics such as "building energy simulation", "sustainable development", "life cycle", and "residential buildings". Those green nodes (circles of smaller diameter) are linked to the dominant nodes but represent subfields of research. Lighter-colored and smaller-sized circles represent distinct subfields which are also connected to each other and to the core (red) nodes, such as "zero energy buildings", "thermal insulation", and "building envelopes".

Keywords positioned around the dominant ones demonstrate their limited integration with the core keywords (red nodes), which may lead us to conclude their limited integration in the red nodes and, consequently, to a research gap: the clusters addressing heritage building retrofits and those concentrating on social housing and occupant behavior are clearly separated, with limited direct interaction. The above limitation indicates a research gap in the recent literature activity and more specifically in the studies relating to historical housing preservation and energy-efficient retrofitting within the Mediterranean area.

An overall pattern may be distinguished for the five clusters generated by the cluster analysis: Clusters 1–3 include technical and design-focused keywords, such as simulation, optimization, retrofitting, and thermal performance. Cluster 4 includes keywords related to policy, housing, and monitoring aspects. Finally, Cluster 5 includes keywords focused on strategy and decision-making.

5.2. Discussion of Findings from the Buildings Energy Performance Analysis and Retrofit Scenarios

The findings of this study demonstrate that historic social housing in Mediterranean climates can achieve substantial energy efficiency improvements while preserving architectural and cultural heritage through carefully designed, minimally invasive and conservation-compatible interventions (RQ1). The proposed methodological framework, which integrates historical analysis, technical assessment, and scenario-based energy evaluation, proves effective for systematically assessing retrofit options, aligning them with both KENAK standards and heritage preservation principles (RQ2). Applied to the Kaisariani social housing complex, the framework highlights the feasibility of achieving up to a 97% reduction in primary energy demand and a 63–76% reduction in heating and cooling loads, even in deteriorated interwar masonry structures. In addition to the technical assessment, the analysis takes into account the specific climatic context of Athens, which plays a decisive role in the performance of retrofit measures. The city is classified as Climate Zone B under the Greek KENAK, characterized by hot–dry summers, mild winters, and high solar radiation levels typical of Mediterranean environments. These climatic parameters directly influence both the heating and cooling demands of the Kaisariani housing complex. The relatively mild winter conditions contribute to moderate heating loads, while intense summer solar gains increase the importance of shading, glazing performance, and natural ventilation. For instance, in scenario 2, the introduction of new window frames with higher solar transmittance slightly increased cooling loads despite improving overall insulation performance—demonstrating the need to balance energy efficiency interventions with climatic responsiveness.

Among the three retrofit scenarios examined, the third scenario—combining thermal insulation, efficient HVAC systems, renewable domestic hot water, and photovoltaic installations—emerges as the optimal solution, offering the best balance between cost effectiveness, CO_2 reduction, and preservation of cultural identity (RQ3). While deeper interventions entail higher initial investments and slightly longer payback periods (8.3–10 years), their long-term operational savings and environmental benefits justify the cost, particularly when paired with supportive funding schemes.

This study aligns with contemporary approaches that balance heritage conservation with energy performance, cost efficiency, and user comfort. At the neighborhood and district scales, such methods group buildings by typology and protection level, test retrofit packages, and deploy representative strategies consistent with established conservation frameworks [86]. Stock-level archetypes, combined with scenario-based restriction levels, enable the translation of qualitative heritage constraints into quantitative optimization and life-cycle cost analyses. Life-cycle assessments further reveal embodied–operational trade-offs, often supporting preservation through selective, low-footprint insulation and microclimate measures rather than demolition—particularly relevant in cooling-dominated contexts [87]. Procedural tools and governance instruments are evolving in parallel, as relevant studies indicate that reforms to energy performance certification processes in multi-owner historic buildings can mitigate heritage risks and better align retrofit recommendations with occupant needs and objectives [88]. As with the Spanish experience, where municipal "image vulnerability" classifications have guided energy retrofits in post-war housing without precluding aesthetic or heritage upgrades, the case of Kaisariani, Athens, illustrates how façade and envelope transformations can be regulated to balance energy efficiency with the recovery of character-defining features. This approach enables coordinated interventions across neighborhood, building, and dwelling scales, ensuring that energy performance improvements also reinforce local identity and social value [89]. Meta-reviews of sustainability rating systems highlight the need for indicator sets that balance environmental, economic, and social dimensions while adapting assessment criteria to heritage-specific constraints—an agenda that supports greater standardization and comparability across cities and building archetypes. The Kaisariani case contributes to this discourse by illustrating how such balanced frameworks can be contextualized within urban fabrics, aligning sustainability goals with cultural and social values rooted in collective housing heritage [90]. Together, these strands reinforce both the conceptual legitimacy and the practical capacity to integrate historic social housing into comparative studies of urban quality, regeneration, and climate adaptation—particularly in Mediterranean contexts, where selective envelope upgrades, shading interventions, ventilation preservation, and open-space microclimate strategies emerge as priority, with heritage-compatible levers for sustainable transformation [91].

Building on previous research into the energy retrofit of historic and social housing, this study proposes an integrated methodological framework that connects energy performance evaluation, cultural heritage assessment, and socio-spatial context analysis. The approach combines bibliometric evidence with scenario-based modeling aligned with both EN 16883:2017 and KENAK, offering a transferable decision-support tool for Mediterranean municipalities.

Applied to the Kaisariani Asia Minor refugee complex—a historically and socially significant ensemble—the framework demonstrates how energy retrofitting can extend beyond technical and economic goals to serve as a catalyst for cultural sustainability. The Kaisariani pilot shows that conservation-aligned retrofit strategies can enhance living conditions, reduce energy poverty, and strengthen community resilience without compromising authenticity. In this way, it provides a transferable model for Mediterranean cities, illustrating how aging social housing can evolve into sustainable, nearly zero-energy environments while preserving historical and social values. More broadly, the study contributes to expanding the discourse on energy-efficient retrofitting from isolated building optimization toward the urban, cultural, and policy dimensions of sustainability, bridging the gap between heritage conservation, social equity, and climate adaptation. The representativeness of the Kaisariani case and the scalability of the proposed framework are underscored by its alignment with broader Mediterranean patterns of historic social housing. The complex reflects common features of early twentieth-century refugee and social housing—masonry construction, limited thermal protection, and culturally significant architectural elements—found across Spain, Italy, Portugal, and other southern European contexts. Evidence from comparable retrofit initiatives, such as the MedZEB project in Lisbon and the upgrading of post-war estates in Barcelona, further supports the transferability of this study's approach. These parallels demonstrate that the methodological structure developed in this study is well suited for application to similar housing typologies with analogous climatic conditions, construction systems, and heritage constraints, offering a strong basis for regional adaptation and wider implementation [90–92].

6. Conclusions

This study has explored how historic social housing in Mediterranean climates can be effectively upgraded to meet contemporary energy performance standards while maintaining their architectural integrity. The research confirms that deep energy retrofitting and heritage preservation are not mutually exclusive, but can be successfully integrated through a structured, conservation-compatible framework (RQ1). The results demonstrate that substantial energy savings and emission reductions can be achieved in aging, historically significant housing stock through targeted, minimally invasive interventions (RQ2). By combining thermal insulation, high-efficiency HVAC systems, renewable domestic hot water systems, and photovoltaic installations, the most comprehensive retrofit scenario achieved a 97% reduction in primary energy demand, alongside significant decreases in heating and cooling loads (63–76%) and CO_2 emissions (RQ3). The comparative analysis of retrofit scenarios highlights that while more extensive interventions require higher initial investment, their long-term benefits—reflected in energy savings, operational cost reductions, and emission cuts—justify their adoption. The payback period of around ten years underlines their practical feasibility, especially when integrated into broader funding frameworks for social or public housing regeneration.

A limitation of the present study is the absence of a detailed analysis of thermal comfort conditions. Although the research concentrated on improving the energy performance of historic social housing through conservation-compatible interventions, the evaluation of indoor comfort parameters was beyond its scope. Given that thermal comfort directly influences occupants' health, well-being, and the overall effectiveness of retrofit measures, future research should incorporate both simulated and user-based comfort assessments. In particular, the integration of indoor thermal comfort simulations would enhance the current findings by providing a more complete understanding of how retrofit scenarios affect indoor environmental quality. Such analyses would contribute to the development of more holistic, socially responsive, and user-centered retrofit strategies for Mediterranean historic housing.

Beyond technical outcomes, the research developed and validated a methodological framework that integrates historical, technical, and environmental assessment tools to guide decision-making in similar contexts. This framework offers a practical pathway for public authorities, engineers, and conservation professionals to balance environmental performance, cultural values, and economic feasibility in the renovation of historic urban housing. The approach ensures compliance with both the Greek KENAK and the European EN 16883:2017 standard [75], bridging the gap between climate policy goals and heritage conservation principles.

The broader significance and purpose of this work extend beyond a single case study. The Kaisariani pilot demonstrates that energy retrofitting can serve as a strategic tool for sustainable urban regeneration, simultaneously addressing environmental, social, and cultural dimensions. It promotes improved living conditions, mitigates energy poverty, and reinforces the collective identity of historically significant communities. Furthermore, the proposed framework is applicable to similar housing typologies across Greece and southern Europe, where interwar and post-war social housing estates share comparable climatic, architectural, and socio-economic conditions. Similar challenges related to aging social housing, poor energy performance, and heritage preservation are widely observed across Mediterranean Europe, underscoring the broader relevance of the issues addressed in this study. Evidence from retrofit initiatives in countries such as Spain, Italy, and Portugal confirms that these concerns are not unique to Kaisariani, reinforcing the applicability of the proposed framework to comparable contexts across the region. Applying this model could support the design of context-sensitive retrofit strategies that align climate neutrality targets with architectural preservation.

The proposed retrofit framework demonstrated outstanding quantitative results, achieving up to a 97% reduction in primary energy demand, 63–76% decreases in heating and cooling loads, and an estimated payback period of approximately ten years, confirming both its technical and economic feasibility. Beyond the Kaisariani case, the methodology shows strong replicability potential for other Mediterranean historic or social housing complexes, particularly those sharing comparable climatic conditions, masonry construction typologies, and heritage preservation constraints. From a policy perspective, the framework provides a strategic tool for local authorities and urban planners, enabling the integration of energy-efficient retrofitting within municipal regeneration programs and EU-funded renovation initiatives. In a broader sense, the study contributes to climate neutrality objectives while reinforcing cultural continuity and social cohesion, demonstrating that sustainable energy transitions can coexist harmoniously with the preservation of architectural identity and collective memory in Mediterranean cities.

Author Contributions: Conceptualization, Z.K. and E.T.; methodology, Z.K.; software, E.P. and Z.K.; validation, E.T., E.P. and Z.K.; formal analysis, E.T. and G.P.; investigation, A.M. and E.K.; resources, A.M. and E.K.; data curation, Z.K., G.P. and E.P.; writing—original draft preparation, Z.K., E.K., G.P., E.P. and E.T.; writing—review and editing, Z.K.; E.K. and E.T.; visualization, A.M., E.T. and E.K.; supervision, Z.K. and E.T.; project administration, Z.K. and E.T.; funding acquisition, Z.K. All authors have read and agreed to the published version of the manuscript.

Funding: This research received no external funding.

Institutional Review Board Statement: Not applicable.

Informed Consent Statement: Not applicable.

Data Availability Statement: The original contributions presented in this study are included in the article. Further inquiries can be directed to the corresponding author.

Acknowledgments: The authors would like to acknowledge the Directorate of Special Area Upgrade Projects (DEEAP) of Athens for providing CAD plans and necessary information. The authors would also like to acknowledge Kipriani Liakopoulou and Sofia Lebesi for their preliminary research on the studied buildings. The authors also would like to acknowledge the Editor of "Atmosphere"journal for issuing an APC waiver for this article.

Conflicts of Interest: The authors declare no conflicts of interest.

Abbreviations—Nomenclature Section

The following abbreviations are used in this manuscript:

RB	Reference building
EB	Existent building
LED	Light-emitting diode
PV	Photovoltaics
HVAC	Heating, ventilation, and air conditioning
TIBE	Thermal insulation of the building envelope
Tws	Total wall surface
Tos	Total openings' surface
Tfa	Total floor area
WCUS	Unheated spaces refer to common spaces within the buildings, such as corridors, building stairs, and storage areas
ROI	Return-on-investment
EPBD	Energy Performance of Buildings Directive
VOS	Visualization of similarities
TCG	Technical Chamber of Greece
DHW	Domestic hot water
RESs	Renewable energy sources
CHP	Combined heat and power
U	Thermal transmittance (W/m^{2}K), ACH (air changes per hour) U-values express the amount of heat that passes through 1 m^2 of wall for every 1 K temperature; better thermal insulation material properties are expressed with lower U-values
COP	Coefficient of performance
EER	Energy efficiency ratio

Appendix A. Keywords Included in Each One of the Five Clusters

Table A1. Table of keywords by cluster.

Cluster 1	Cluster 2	Cluster 3	Cluster 4	Cluster 5
building, buildings climate change computer simulation cooling energy demand, energy demands energy management energy savings heating historic building Italy Mediterranean climate, Mediterranean environment, Mediterranean region numerical model optimization performance assessment residential energy simulation southern Europe	architectural design building energy performance building energy simulation building energy simulation carbon dioxide climate models energy energy simulation energy consumption environmental impact gas emissions greenhouse gases house HVAC life cycle residential building residential buildings sustainable development zero energy building	air conditioning building envelope building retrofits climate cost effectiveness energy conservation energy retrofit energy saving historic buildings indoor thermal comfort insulation residential clean times occupant behavior retrofitting thermal insulation thermal performance ventilation	air quality, cooling systems, energy consumption, energy performance, energy policy, energy use, housing, housing stock, monitoring, social housing, thermal comfort	Decision-making

Appendix B. Area Calculation-Floor Area

For the accurate calculation of the areas of the structural elements, the building facades were redesigned in CAD software for the most accurate and direct calculation of the areas. The total floor area includes the surface of each level in external measurements (external walls included). The surface of the windows (openings are deducted).

The area occupied by the building's load-bearing structure (conventional method of calculating the area occupied by the building's load-bearing structure as a percentage of its facade area in cases where it is not possible to record the load-bearing structure) was calculated according to T.O.T.E.E. 20701-1/2017. A total of 20% was selected as the percentage of reinforced concrete for the buildings, for which a permit was issued before 1980, and there are up to two floors and 18% for buildings constructed from 1980–1999.

Table A2. Surface area calculation by percentage in façade area.

Number of Floors	Up to 2	2 < Floors < 5	5 or More
Before 1980	15%	20%	23%
1980 to 1999	18%	23%	28%

Appendix C. Automatic Control Devices: Impact of Automatic Control Devices on Energy Consumption

The use of automatic control and regulation devices significantly reduces energy consumption across various end uses such as heating, cooling, air conditioning, and lighting. When systems are equipped with either central or local automatic controls, the energy required to meet specific functional loads is notably decreased.

The EN 15232:2007 [93] standard categorizes building automation and control systems into four efficiency classes—A, B, C, and D [2]:

- Class D (lowest efficiency): Manual control of terminal units and distribution networks, and absence of room thermostats.
- Class A (highest efficiency): Automatic, independent control of terminal unit operation per zone or functional space, which includes the use of thermostats and/or thermostatic valves in each space, as well as ON/OFF control per zone.

In order to assess the energy efficiency of the building and issue an energy certificate, the calculations take into account the photovoltaic (PV) systems installed to cover all or part of its electricity needs (with or without offsetting the energy produced by the PV with the electricity consumed—NET METERING) and not for the production of electricity fed into the grid (sold to the electricity supplier), which is not offset against own consumption.

Appendix D

Table A3. Typical U-Values according to TOTEE 20701-2/2010.

Construction Type	Wall Exposure	With no Thermal Insulation (W/m^2K)	With Insufficient Thermal Insulation (W/m^2K)	With Adequate Insulation (W/m^2K)
Reinforced concrete (unplastered)	Exposed to air	3.65	2.75	0.60–0.80
Reinforced concrete with plaster (both sides)	Exposed to air	3.40	2.60	0.50–0.70

Table A3. *Cont.*

Construction Type	Wall Exposure	With no Thermal Insulation (W/m^{2}K)	With Insufficient Thermal Insulation (W/m^{2}K)	With Adequate Insulation (W/m^{2}K)
Brick masonry (25 cm + plaster both sides)	Exposed to air	1.70	1.20	0.45–0.60
Double brick wall with air gap and insulation	Exposed to air	1.20	0.90	0.40–0.50
Wall adjacent to unheated space	—	Multiply by 0.50–0.70 correction factor	—	—

Appendix E. Hours of Operation—Building Operating Period

For the section on hours of operation and building operating period, T.O.T.E.E. 20701-1/2017 was used (standard building operating hours per use), with the basic building category being housing and the use of buildings in thermal zones being primary residence building (more than one apartment). With the two options from above, we have the following:

- Hours of operation: 18;
- Days of operation per week: 7;
- Period of operation in months: 12.

Appendix F

Table A4. DHW consumption for residential buildings.

Building Use or Thermal Zone	Daily DHW Consumption [ℓ/Person/Day]	Annual DHW Consumption Per Bedroom [m^3/Bedroom/Year]
apartment building	50	27.38

References

1. Kovacic, I.; Summer, M.; Achammer, C. Life-Cycle Oriented Renovation Strategies for Social Housing Stock. *Organ. Technol. Manag. Constr. Int. J.* **2013**, *5*, 881–891. [CrossRef]
2. Lucchi, E.; Delera, A.C. Enhancing the Historic Public Social Housing through a User-Centered Design-Driven Approach. *Buildings* **2020**, *10*, 159. [CrossRef]
3. Georgopoulou, E.; Mirasgedis, S.; Sarafidis, Y.; Giannakopoulos, C.; Varotsos, K.V.; Gakis, N. Climate Change Impacts on the Energy System of a Climate-Vulnerable Mediterranean Country (Greece). *Atmosphere* **2024**, *15*, 286. [CrossRef]
4. Oorschot, L.; Spoormans, L.; El Messlaki, S.; Konstantinou, T.; de Jonge, T.; van Oel, C.; Asselbergs, T.; Gruis, V.; de Jonge, W. Flagships of the Dutch Welfare State in Transformation: A Transformation Framework for Balancing Sustainability and Cultural Values in Energy-Efficient Renovation of Postwar Walk-up Apartment Buildings. *Sustainability* **2018**, *10*, 2562. [CrossRef]
5. Sugar, V.; Kita, M.; Ballo, Z. Energetic Retrofit of Historical Downtown Buildings: Cost Effectiveness and Financing Options. *Int. J. Energy Prod. Manag.* **2019**, *4*, 273–286. [CrossRef]
6. Buda, A.; Hansen, E.J.; Rieser, A.; Giancola, E.; Pracchi, V.N.; Mauri, S.; Marincioni, V.; Gori, V.; Fouseki, K.; López, C.S.P.; et al. Conservation-Compatible Retrofit Solutions in Historic Buildings: An Integrated Approach. *Sustainability* **2021**, *13*, 2927. [CrossRef]
7. Massimo, D.E.; Del Giudice, V.; Malerba, A.; Bernardo, C.; Musolino, M.; De Paola, P. Valuation of Ecological Retrofitting Technology in Existing Buildings: A Real-World Case Study. *Sustainability* **2021**, *13*, 7001. [CrossRef]
8. Jacques, S.; Stergiou, C.; Kanetaki, Z. A Novel BIM-Based Process Workflow for Building Retrofit. In *Cost-Effective Energy-Efficient Methods for Refurbishment and Retrofitting of Buildings*; Woodhead Publishing Series in Civil and Structural Engineering; Woodhead Publishing: Cambridge, UK; Elsevier: Amsterdam, The Netherlands, 2025; pp. 189–216, ISBN 978-0-443-23975-5.
9. Sofias, K.; Kanetaki, Z.; Stergiou, C.; Jacques, S. Combining CAD Modeling and Simulation of Energy Performance Data for the Retrofit of Public Buildings. *Sustainability* **2023**, *15*, 2211. [CrossRef]

10. Trachte, S.; Stiernon, D. P-Renewal Project: A Reflexive Contribution to the Evolution of Energy Performance Standards for the Renovation of Historic Buildings. *Heritage* **2024**, *7*, 1539–1568. [CrossRef]
11. Boeri, A.; Longo, D. Eco-Technologies for Retrofitting Social Housing Stock: Strategies for an Italian Case Study. *WIT Trans. Ecol. Environ.* **2012**, *165*, 375–386. [CrossRef]
12. Tousi, E.; Karadoulama, K.; Papaioannou, I.; Patsea, A.; Skrepi, A.; Spentza, E.; Voskos, T.; Zafeiropoulos, G. Issues of Urban Conservation and Collective Memory. The Case of the Asia Minor Post-Refugee Urban Neighbourhood Germanika at Nikea, Piraeus, Greece. *J. Sustain. Archit. Civ. Eng.* **2023**, *33*, 30–44. [CrossRef]
13. Tousi, E. Challenges on Urban Socio-Spatial Cohesion. The Case of Social Housing Complexes in the Regional Administrative Area of Piraeus in Greece. *J. Sustain. Archit. Civ. Eng.* **2021**, *29*, 21–32. [CrossRef]
14. Hill, M. The League of Nations and the Work of Refugee Settlement and Financial Reconstruction in Greece, 1922–1930. *Weltwirtschaftliches Arch.* **1931**, *34*, 265–283.
15. Pentzopoulos, D. *The Balkan Exchange of Minorities and Its Impact Upon Greece*; Walter de Gruyter GmbH & Co KG: Berlin, Germany, 2021; ISBN 978-3-11-241586-3.
16. Brooks, M. Karl Marx-Hof. Available online: https://www.designingbuildings.co.uk/wiki/Karl_Marx-Hof (accessed on 23 September 2025).
17. Brenner, A.-K.; Haas, W.; Rudloff, C.; Lorenz, F.; Wieser, G.; Haberl, H.; Wiedenhofer, D.; Pichler, M. How Experiments with Superblocks in Vienna Shape Climate and Health Outcomes and Interact with the Urban Planning Regime. *J. Transp. Geogr.* **2024**, *116*, 103862. [CrossRef]
18. Karakosta, C.; Vryzidis, I. Cost-Effective Energy Retrofit Pathways for Buildings: A Case Study in Greece. *Energies* **2025**, *18*, 4014. [CrossRef]
19. Tzortzopoulos, P.; Ma, L.; Soliman Junior, J.; Koskela, L. Evaluating Social Housing Retrofit Options to Support Clients' Decision Making—SIMPLER BIM Protocol. *Sustainability* **2019**, *11*, 2507. [CrossRef]
20. Synnefa, A.; Vasilakopoulou, K.; Masi, R.F.D.; Kyriakodis, G.-E.; Londorfos, V.; Mastrapostoli, E.; Karlessi, T.; Santamouris, M. Transformation through Renovation: An Energy Efficient Retrofit of an Apartment Building in Athens. *Procedia Eng.* **2017**, *180*, 1003–1014. [CrossRef]
21. Bukar, U.A.; Sayeed, M.S.; Razak, S.F.A.; Yogarayan, S.; Amodu, O.A.; Mahmood, R.A.R. A Method for Analyzing Text Using VOSviewer. *MethodsX* **2023**, *11*, 102339. [CrossRef]
22. van Eck, N.J.; Waltman, L. *VOSviewer Manual*; Leiden University: Leiden, The Netherlands, 2017.
23. Chakraborty, S.; Gonzalez-Triana, Y.; Mendoza, J.; Galatro, D. Insights on Mapping Industry 4.0 and Education 4.0. *Front. Educ.* **2023**, *8*, 1150190. [CrossRef]
24. Zaino, R.; Ahmed, V.; Alhammadi, A.M.; Alghoush, M. Electric Vehicle Adoption: A Comprehensive Systematic Review of Technological, Environmental, Organizational and Policy Impacts. *World Electr. Veh. J.* **2024**, *15*, 375. [CrossRef]
25. Ahad, M.T.; Bhuiyan, M.M.H.; Sakib, A.N.; Becerril Corral, A.; Siddique, Z. An Overview of Challenges for the Future of Hydrogen. *Materials* **2023**, *16*, 6680. [CrossRef]
26. Lepsch-Cunha, N.; Muraro, V.; Nascimento, H.E.M.; Mazoni, A.; Nunez, C.V.; Bonacelli, M.B.M. Technical-Scientific Production and Knowledge Networks about Medicinal Plants and Herbal Medicines in the Amazon. *Front. Res. Metr. Anal.* **2024**, *9*, 1396472. [CrossRef]
27. Chatziiosif, C. The refugee shock, the constants and transformations of the Greek economy. In *History of Greece in the Twentieth Century, 1922–1940: The Interwar*; Chatziiosif, C., Ed.; Vivliorama: Athens, Greece, 2002; Volume 2.1, pp. 9–57.
28. Heirs of the Greek Catastrophe: The Social Life of Asia Minor Refugees in Piraeus. Available online: https://www.berghahnbooks.com/title/HirschonHeirsNew (accessed on 18 October 2025).
29. Gemenetzi, G.; Papageorgiou, M. Spatial and social aspects of the housing policies for refugees and immigrants in Greece: A critical overview. *Greek Rev. Soc. Res.* **2017**, *148*, 39–74. [CrossRef]
30. Europe, A.P. The Refugees of 1922: Archival Material from the General State Archives. Available online: https://www.archivesportaleurope.net/blog/the-refugees-of-1922-archival-material-from-the-general-state-archives/ (accessed on 23 September 2025).
31. Athens Social Atlas. Available online: https://www.athenssocialatlas.gr/en/homepage/ (accessed on 18 October 2025).
32. Tousi, E. Changing Socio-Spatial Identities. The Case of the Asia Minor Refugee Urban Settlements in the Greater Athens-Piraeus Region in Greece. *J. Sustain. Archit. Civ. Eng.* **2024**, *35*, 153–167. [CrossRef]
33. Tousi, E. A Retrospect on Social Housing in Greece: The Case of Refugee Dwellings. *Des. Princ. Pract. Int. J. Annu. Rev.* **2016**, *10*, 1–11. [CrossRef]
34. Evgenia, T. Initial planning and current situation of refugee housing and outdoor public spaces in the post-refugee urban settlement of Nikea in Attica. *Athens Soc. Atlas.* **2024**. [CrossRef]

35. La Arquitectura Ordinaria Del Siglo Xx Como Patrimonio Cultural: Tres Barrios de Promoción Oficial de Madrid. Available online: https://www.scielo.cl/scielo.php?script=sci_arttext&pid=S0250-71612017000300269&lng=en&nrm=iso&tlng=en (accessed on 14 November 2025).
36. Özmen, A. The Adaptation of Modern Housing Heritage: Authenticity and Integrity in the Case of Cité Modèle. *Vitr.-Int. J. Archit. Technol. Sustain.* **2024**, *9*, 22495. [CrossRef]
37. Gilles, C.T.; Silva, N.J. Técnicas de refuerzo sísmico para la recuperación estructural del patrimonio arquitectónico chileno construido en adobe. *Inf. Constr.* **2018**, *70*, e252. [CrossRef]
38. Bernardo, G. Critical and Typological Analysis of the Habitat in Spain after Developmentalism. *Vitr.-Int. J. Archit. Technol. Sustain.* **2022**, *7*, 18–29. [CrossRef]
39. Gherri, B.; Cavagliano, C.; Orsi, S. Social Housing Policies and Best Practice Review for Retrofit Action—Case Studies from Parma (IT). *IOP Conf. Ser. Mater. Sci. Eng.* **2017**, *245*, 082038. [CrossRef]
40. Esposito, A.; Pappaccogli, G.; Donateo, A.; Salizzoni, P.; Maffeis, G.; Semeraro, T.; Santiago, J.L.; Buccolieri, R. Urban Morphology and Surface Urban Heat Island Relationship During Heat Waves: A Study of Milan and Lecce (Italy). *Remote Sens.* **2024**, *16*, 4496. [CrossRef]
41. Pappaccogli, G.; Giangrande, F.; Esposito, A.; Donateo, A.; Lionello, P.; Buccolieri, R. Dynamics of Urban Heat Island Intensity in Lecce, Italy: Seasonal, Diurnal and Heat Wave Influence. *Bull. Atmos. Sci. Technol.* **2024**, *5*, 8. [CrossRef]
42. Calama-González, C.M.; Escandón, R.; Suárez, R.; Alonso, A.; León-Rodríguez, Á.L. Household Energy Vulnerability Evaluation in Southern Spain through Parametric Energy Simulation Models and Socio-Economic Data. *Sustain. Cities Soc.* **2024**, *103*, 105276. [CrossRef]
43. Silva, L.P.P.; Najjar, M.K.; da Costa, B.B.F.; Amario, M.; Vasco, D.A.; Haddad, A.N. Sustainable Affordable Housing: State-of-the-Art and Future Perspectives. *Sustainability* **2024**, *16*, 4187. [CrossRef]
44. Havinga, L.; Colenbrander, B.; Schellen, H. Heritage Attributes of Post-War Housing in Amsterdam. *Front. Archit. Res.* **2020**, *9*, 1–19. [CrossRef]
45. Álvaro, S.; SAAL, S. Victor Social Housing, Porto, Portugal. 1974–1977. MoMA. Available online: https://www.moma.org/collection/works/161471 (accessed on 18 November 2025).
46. Micelli, E.; Giliberto, G.; Righetto, E. Is the Energy Transition of Housing Financially Viable? Unlocking the Potential of Deep Retrofits with New Business Models. *Buildings* **2025**, *15*, 1175. [CrossRef]
47. Gupta, R.; Gregg, M.; Passmore, S.; Stevens, G. Intent and Outcomes from the Retrofit for the Future Programme: Key Lessons. *Build. Res. Inf.* **2015**, *43*, 435–451. [CrossRef]
48. Whiteside, K.H.; Boy, D.; Bourg, D. France's 'Grenelle de l'environnement': Openings and Closures in Ecological Democracy. *Environ. Polit.* **2010**, *19*, 449–467. [CrossRef]
49. Mata, É.; Sasic Kalagasidis, A.; Johnsson, F. Cost-Effective Retrofitting of Swedish Residential Buildings: Effects of Energy Price Developments and Discount Rates. *Energy Effic.* **2015**, *8*, 223–237. [CrossRef]
50. Canale, L.; Battaglia, V.; Ficco, G.; Puglisi, G.; Dell'Isola, M. Dynamic Evaluation of Heat Thefts Due to Different Thermal Performances and Operations between Adjacent Dwellings. *Appl. Sci.* **2020**, *10*, 2436. [CrossRef]
51. Echarri-Iribarren, V. Conditioning Using Ceramic Floor Panels with Capillary Tube Mats and Solar Thermal Panels on the Mediterranean Coast: Energy Savings and Investment Amortisation. *Energy Build.* **2019**, *202*, 109334. [CrossRef]
52. Jareño Escudero, C.I.; Navarro Escudero, M.; Mifsut García, C.D.; Flores Fillol, M.; Salmerón Lissen, J.M. Potential of Energy Savings in the Public Housing Stock of Comunitat Valenciana Region by Applying the MedZEB Cost-Optimal Methodology. *Appl. Sci.* **2022**, *12*, 138. [CrossRef]
53. Calama-González, C.M.; Suárez, R.; León-Rodríguez, Á.L. Thermal Comfort Prediction of the Existing Housing Stock in Southern Spain through Calibrated and Validated Parameterized Simulation Models. *Energy Build.* **2022**, *254*, 111562. [CrossRef]
54. Caro, R.; Sendra, J.J.; González, C.M.M.; Caro, R. The Role of Hybrid Systems in the Decarbonization of Residential Heritage Buildings in Mediterranean Climate. a Case Study in Seville, Spain. *Energy Build.* **2021**, *250*, 111302. [CrossRef]
55. Marta, M.; Belinda, L.-M. Simplified Model to Determine the Energy Demand of Existing Buildings. Case Study of Social Housing in Zaragoza, Spain. *Energy Build.* **2017**, *149*, 483–493. [CrossRef]
56. López-Mesa, B.; Monzón-Chavarrías, M.; Espinosa-Fernández, A. Energy Retrofit of Social Housing with Cultural Value in Spain: Analysis of Strategies Conserving the Original Image vs. Coordinating Its Modification. *Sustainability* **2020**, *12*, 5579. [CrossRef]
57. Calama-González, C.M.; Escandón, R.; Suárez, R.; Ascione, F. Decision-Making for Renovating the Mediterranean Social Housing: A Practical Approach through an Interactive Open Access Tool. *Energy Build.* **2025**, *336*, 115629. [CrossRef]
58. Muhy Al-Din, S.S.; Saltik, B. Regulating Indoor Comfortable Temperature Limits for Sustainable Architectural Design in Mediterranean Climates. *Buildings* **2025**, *15*, 899. [CrossRef]

59. Stasi, R.; Ruggiero, F.; Berardi, U. Assessing the Energy Saving Potential of Historical Buildings Heritage in Mediterranean Climate. In Proceedings of the 2024 IEEE International Humanitarian Technologies Conference (IHTC), Bari, Italy, 27–30 November 2024; pp. 1–7.
60. Salvalai, G.; Imperadori, M.; Scaccabarozzi, D.; Pusceddu, C. Thermal Performance Measurement and Application of a Multilayer Insulator for Emergency Architecture. *Appl. Therm. Eng.* **2015**, *82*, 110–119. [CrossRef]
61. Moreira, S. Corviale, a One-Kilometer Residential Complex in Rome. Available online: https://www.archdaily.com/956906/corviale-a-one-kilometer-residential-complex-in-rome (accessed on 18 November 2025).
62. Caudo, G. Regenerate Corviale. In *Future Housing*; IntechOpen: London, UK, 2022; ISBN 978-1-83768-177-8.
63. Social Housing SAAL Bouca. Available online: https://architectuul.com/architecture/social-housing-saal-bouca (accessed on 18 November 2025).
64. Kumar, G.K.; Saboor, S.; Babu, T.P.A. Study of Various Glass Window and Building Wall Materials in Different Climatic Zones of India for Energy Efficient Building Construction. *Energy Procedia* **2017**, *138*, 580–585. [CrossRef]
65. Vittorini, R. Reloaded Corviale, a City with a Single Building (1973–1984). Mario Fiorentino Architect, Rome. *Docomomo J.* **2016**, *54*, 44–51. [CrossRef]
66. Navas-Carrillo, D.; Ostos-Prieto, J.; Rodríguez-Lora, J.-A. Housing Policy in Spain between 1939 and 1976: The Residential Phenomenon as a Catalyst for Urban Growth. *Hum. Rev. Int. Humanit. Rev. Rev. Int. Humanidades* **2023**, *18*, 1–17. [CrossRef]
67. Archnet > Publication > The Significance of User Participation in Architectural Design: The Case of Nicosia Social Housing Complex. Available online: https://www.archnet.org/publications/6181 (accessed on 18 November 2025).
68. Mela, A.; Tousi, E.; Varelidis, G. Assessing Urban Public Space Quality: A Short Questionnaire Approach. *Urban Sci.* **2025**, *9*, 56. [CrossRef]
69. Kaimaris, D.; Kalyva, D. Aerial Remote Sensing and Urban Planning Study of Ancient Hippodamian System. *Urban Sci.* **2025**, *9*, 183. [CrossRef]
70. GRAMMENOS, F.; CRAIG, B.; POLLARD, D.; GUERRERA, C. Hippodamus Rides to Radburn: A New Model for the 21st Century. *J. Urban Des.* **2008**, *13*, 163–176. [CrossRef]
71. 10 Modern Cities Influenced by Ancient Hippodamian Urban Planning. Available online: https://www.thearchaeologist.org/blog/10-modern-cities-influenced-by-ancient-hippodamian-urban-planning (accessed on 23 September 2025).
72. Technical Chamber of Greece. *TEE KENAK 1.29.1.19: Software Inspection and Certification of Energy Buildings, Study of Energy Efficiency Boiler Inspection/Heating Installations and Facilities Air Conditioning-Manual*; Technical Chamber of Greece Publications: Athens, Greece, 2012.
73. Energy Performance of Buildings Directive. Available online: https://energy.ec.europa.eu/topics/energy-efficiency/energy-efficient-buildings/energy-performance-buildings-directive_en (accessed on 2 June 2025).
74. *ISO 52016-1:2017(En)*; Energy Performance of Buildings—Energy Needs for Heating and Cooling, Internal Temperatures and Sensible and Latent Heat Loads—Part 1: Calculation Procedures. ISO: Geneva, Switzerland, 2017.
75. *EN 16883:2017*; Conservation of Cultural Heritage—Guidelines for Improving the Energy Performance of Historic Buildings. CEN: Brussels, Belgium, 2017.
76. Leijonhufvud, G.L.; Buda, A.; Broström, T. Assessing and Enhancing EN 16883:2017. *IOP Conf. Ser. Earth Environ. Sci.* **2021**, *863*, 012033. [CrossRef]
77. *ISO 9869-1:2014*; Thermal Insulation—Building Elements—In-Situ Measurement of Thermal Resistance and Thermal Transmittance—Part 1: Heat Flow Meter Method. ISO: Geneva, Switzerland, 2014.
78. Thermal Transmittance (U-Value) Evaluation of Innovative Window Technologies. *Future Cities Environ.* **2020**, *6*, 1–13. [CrossRef]
79. Fernandes, M.S.; Rodrigues, E.; Gaspar, A.R.; Costa, J.J.; Gomes, Á. The Impact of Thermal Transmittance Variation on Building Design in the Mediterranean Region. *Appl. Energy* **2019**, *239*, 581–597. [CrossRef]
80. Carbonari, A. Retrofit of Massive Buildings in Different Mediterranean Climates. Interactions Between Mass, Additional Insulations and Solar Control Strategies. *Future Cities Environ.* **2023**, *9*, 1–6. [CrossRef]
81. Standards, E. BS EN 16883:2017 Conservation of Cultural Heritage. Guidelines for Improving the Energy Performance of Historic Buildings. Available online: https://www.en-standard.eu/bs-en-16883-2017-conservation-of-cultural-heritage-guidelines-for-improving-the-energy-performance-of-historic-buildings/ (accessed on 18 October 2025).
82. Ascione, F.; de Rossi, F.; Iovane, T.; Manniti, G.; Mastellone, M. Energy Demand and Air Quality in Social Housing Buildings: A Novel Critical Review. *Energy Build.* **2024**, *319*, 114542. [CrossRef]
83. Seabra, B.; Pereira, P.F.; Corvacho, H.; Pires, C.; Ramos, N.M.M. Low Energy Renovation of Social Housing: Recommendations on Monitoring and Renewable Energies Use. *Sustainability* **2021**, *13*, 2718. [CrossRef]

84. Karoglou, M.; Kyvelou, S.S.; Boukouvalas, C.; Theofani, C.; Bakolas, A.; Krokida, M.; Moropoulou, A. Towards a Preservation–Sustainability Nexus: Applying LCA to Reduce the Environmental Footprint of Modern Built Heritage. *Sustainability* **2019**, *11*, 6147. [CrossRef]
85. Eriksson, P.; Milić, V.; Brostrom, T. Balancing Preservation and Energy Efficiency in Building Stocks. *Int. J. Build. Pathol. Adapt.* **2019**, *38*, 356–373. [CrossRef]
86. Berg, F.; Donarelli, A. Energy Performance Certificates and Historic Apartment Buildings: A Method to Encourage User Participation and Sustainability in the Refurbishment Process. *Hist. Environ. Policy Pract.* **2019**, *10*, 224–240. [CrossRef]
87. Lizundia, I.; Uranga, E.J.; Azcona, L. A Methodology to Regulate Transformation of a City's Appearance Due to Energy Efficiency Building Renovations: A Case Study: Errenteria (Spain). *Heritage* **2023**, *6*, 6112–6131. [CrossRef]
88. Karimi, F.; Valibeig, N.; Memarian, G.; Kamari, A. Sustainability Rating Systems for Historic Buildings: A Systematic Review. *Sustainability* **2022**, *14*, 12448. [CrossRef]
89. Ulu, M.; Durmuş Arsan, Z. Retrofit Strategies for Energy Efficiency of Historic Urban Fabric in Mediterranean Climate. *Atmosphere* **2020**, *11*, 742. [CrossRef]
90. Padula, M.; Picenni, F.; Malvezzi, R.; Laghi, L.; Lissén, J.M.S.; Flor, F.J.S.; Mateo-Cecilia, C.; Soto-Francés, L.; Assimakopoulos, M.-N.; Karlessi, T. MedZEB: A New Holistic Approach for the Deep Energy Retrofitting of Residential Buildings. *TECHNE-J. Technol. Archit. Environ.* **2018**, *1*, 127–133. [CrossRef]
91. Gomez, C.D.; Mira, P.J.R.; Almirall, P.G.; Bardon, C.C.; Grau, S.V. Refurbishment in Large Housing Estates: A Review on Restructuring and Upgrade. *Procedia Eng.* **2016**, *161*, 1932–1938. [CrossRef]
92. Museu d'historia de Bascelona (MUHBA). Barcelona 20th Cent, *Housing Newspaper*. 2022. Available online: https://www.barcelona.cat/museuhistoria/en/formats-and-activities/llibrets-de-sala/housing-newspaper-20th-century-barcelona (accessed on 18 October 2025).
93. *EN 15232:2007*; Energy Performance of Buildings—Impact of Building Automation, Controls and Building Management. CEN: Brussels, Belgium, 2007.

atmosphere

MDPI

Article

Energy Sustainability of Urban Areas by Green Systems: Applied Thermodynamic Entropy and Strategic Modeling Means

Carla Balocco [1,*], Giacomo Pierucci [1], Michele Baia [1], Costanza Borghi [2], Saverio Francini [3], Gherardo Chirici [2] and Stefano Mancuso [2]

1 Department of Architecture, University of Florence, Via della Mattonaia 8, 50121 Florence, Italy; giacomo.pierucci@unifi.it (G.P.); michele.baia@unifi.it (M.B.)
2 Department of Agricultural, Food, Environmental and Forestry Sciences and Technologies, University of Florence, Piazzale delle Cascine 18, 50144 Florence, Italy; costanza.borghi@unifi.it (C.B.); gherardo.chirici@unifi.it (G.C.); stefano.mancuso@unifi.it (S.M.)
3 Department of Agricultural and Food Sciences, University of Bologna, Viale Fanin 50, 40127 Bologna, Italy; saverio.francini@unifi.it
* Correspondence: carla.balocco@unifi.it

Abstract: Global warming, anthropogenic pressure, and urban expansion at the expense of green spaces are leading to an increase in the incidence of urban heat islands, creating discomfort and health issue for citizens. This present research aimed at quantifying the impact of nature-based solutions to support decision-making processes in sustainable energy action plans. A simple method is provided, linking applied thermodynamics to physics-informed modeling of urban built-up and green areas, high-resolution climate models at urban scale, greenery modeling, spatial georeferencing techniques for energy, and entropy exchanges evaluation in urban built-up areas, with and without vegetation. This allows the outdoor climate conditions and thermo-hygrometric well-being to improve, reducing the workload of cooling plant-systems in buildings and entropy flux to the environment. The finalization and post-processing of obtained results allows the definition of entropy footprints. The main findings show a decrease in greenery's contribution for different scenarios, referring to a different climatological dataset, but an increase in entropy that becomes higher for the scenario with higher emissions. The comparison between the entropy footprint values for different urban zones can be a useful support to public administrations, stakeholders, and local governments for planning proactive resilient cities and anthropogenic impact reduction and climate change mitigation.

Keywords: thermodynamics; entropy footprint; climate responsive urban areas; greenery thermodynamic modelling; urban heat islands mitigation; climate change adaptation

1. Introduction

Sustainability is a very broad and complex concept in the current climate change and global warming situation [1–8].

In the urban context, the concept of sustainability implies that urban areas should be considered open thermodynamic systems, thus requiring physical definition of entropy, which some of the literature has interpreted based on the connections between entropy and irreversibility [9].

From literature studies comparison, it is deduced that different applications of entropy to urban systems can generate misunderstanding, especially if applied to bioenergetics, ecology, and evolutionary biology [10–13]. Nevertheless, with the Second Law of Thermodynamics, strong physical approaches can be formulated, at large and small scales, and crucial insights into the sustainability and efficiency of urban processes can be drawn. In detail, statistical entropy, including Shannon's diversity, and informational entropy have been applied to generally describe diversity, heterogeneity, and spatial patterning of biological systems [14].

While some research studies are oriented towards the study of entropy flows as indicators of urban ecosystems' integrity [15], others aim to apply thermodynamics entropy to the environment [16], often combining GIS and remote sensing techniques [17]. For instance, there are several literature studies on entropy assessment at large and small scales [18–25], using spatial quantitative–qualitative analyses, implementing GIS, Spearman correlation models, and fractal models. The "low-entropy city" has been explored by considering the role of urban green infrastructure [26–29], also referring to Prigogine and Stengers [30] and, for the Italian context, to [31]. Recent research has used information entropy, highlighting that the higher the value of information entropy, the more diverse the land use functions [32].

Similarly, a recent study has highlighted that urban microclimate models can play an important role for sustainability, management of energy sources, urban design, and urban heat island (UHIs) control [33]. Here, the Shannon entropy coefficient has been used to study spatial and temporal entropy connected to the effective temperature index (ET) [34]. Furthermore, the urban entropy patterns [35] and Entropy Life Cycle Assessment have been applied as a screening tool for ecological footprint [36]. In this context, accurate urban-scale modeling, for energy and entropy analyses, requires high-resolution spatial data (on the order of kilometers) to incorporate orographic effects and reliably represent anthropogenic emission sources [37–39].

Moreover, comparisons of climatic historical data with high-resolution climate models (i.e., the so-called Convection Permitting regional Models, CPMs) to obtain results applicable to future scenarios and quantify impacts across different projections are necessary [40–45] to study entropy dynamics. In fact, CPMs, derived from downscaling regional climate models [37,38], allow more accurate simulations of energy balance, providing detailed information on the statistics and return periods of extreme events.

In this literature background, our method fills the literature's lack of physics-thermodynamics-informed modeling for the entropy assessment of urban areas. Starting from the physics of the built urban system, through the application of the Laws of Thermodynamics, develops a simple, easy-to-implement method useful for quantifying, via georeferenced modeling, entropy flows at transient conditions and entropic impacts of built-up urban areas, evaluating the effects of greenery to counteract climate change. This approach aligns with the IPCC (2025) [46] issues.

The method is based on a simplified and easy-to-implement model (without the need for large amounts of measured/detected data or obtained from the literature evidence and/or specific data sources). Entropy is the physical quantity that can be used to define the anthropic weight within the evolution of physical phenomena (i.e., heat exchanges at urban scale). The city is one of the contexts in which the environment is most forced by human action from morphology (relationship between built-up and green spaces) and energy use point of view. Therefore, the entropy variation (with and without greenery) can provide a measure of sustainable scenarios, considering the future trends of global warming.

The method provides fundamental indicators, i.e., the entropy footprints, and enables spatial assessment of energy use and entropy fluxes and environmental stress, distinguishing high-impact zones where an increase in greenery and climate adaptation strategies are most urgent. By applying entropy footprints analysis, it is possible to find out effective solutions for urban design, orienting policies and social action towards resilience and sustainability transitions and energy planning for mitigation of UHIs.

2. Materials and Methods

Energy and entropy analyses of the urban system are implemented by means of the First and Second Law of Thermodynamics application for two conditions, i.e., with and without greenery areas.

The proposed approach integrates simplified model of greenery and a building system to compute leaf surface temperatures, thermal energy exchanges, and entropy balance evaluations of any urban area. We use the by now common term "sustainability" to mean low environmental impact, energy quality, and durability [39,47]. So, the production of entropy is related to energy flows, thermal exchanges, and not to material waste or pollution. Therefore, the entropy "discharged" into the surrounding environment is read as waste heat only. Referring to [39,47], we start from the concept that real sustainability can only be obtained if total irreversible entropy flux, due to anthropogenic weight, is lower than the entropy flux from the Sun.

After an introduction to the general theoretical framework (Section 2.1), Section 2.2 briefly shows and discuss how to organize the climate database and the classification of urban green spaces, which are necessary inputs to identify and estimate all the model parameters, applied at the case study of a Florentine area in Italy (Section 2.3).

The process and method implementation are provided in the flowchart of Figure 1 and cited as a reference.

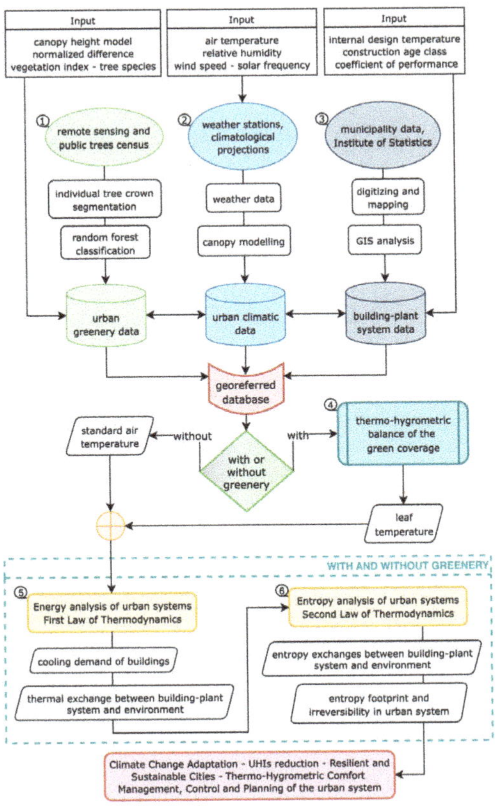

Figure 1. Flowchart of the process and method implementation.

2.1. Theoretical Framework

This study introduces a thermodynamics-informed entropy footprint method to quantify entropy exchanges in urban systems. The urban environment, conceptualized as an open thermodynamic system, is analyzed under two conditions—with and without greenery—through the application of the First and Second Laws of Thermodynamics. The energy balance is applied initially to buildings, and subsequently to the vegetation coverage, treating the leaf surface as continuous, homogeneous, and isotropic. This simplification enables the estimation of the leaf surface temperature and the resulting heat exchange with the environment—via latent and sensible heat—while neglecting physiological and soil interactions, i.e., disregarding the morphology, the physiological and water processes, of plant and leaves, with the soil and the atmosphere (point 4 of the flowchart in Figure 1).

The energy balance at steady-state conditions is described by governing equations applied to the composed system vegetation-air-building. The Second Law of Thermodynamics allows us to compute entropy variations resulting from the building's heat exchange and solar energy gains, as shown in the flowchart (branches 5–6 of Figure 1). Energy balance at the leaf surface is based on the method proposed in [48], which combine contributions of solar radiation, infrared emission, convection and evapotranspiration:

$$GHI \cdot \alpha_l = \varepsilon_l \cdot \sigma \cdot (T_l + 273)^4 + k_1 \left(\frac{v_w}{D_l} \right)^{0.5} (T_l - T_a) + L_l \left(\frac{\rho_{vl} - RH_a \rho_{va}}{r_l + k_2 \frac{D_l^{0.3} W_l^{0.2}}{v_w^{0.5}}} \right) \quad (1)$$

GHI is the global horizontal solar radiation on a specific site [W/m^2], T_l and T_a are, respectively, the surface temperature of the leaves and the ambient temperature [°C]; v_w is the wind velocity [m/s] and RH_a is the relative humidity of the air [-]. σ is the Stefan–Boltzmann constant [5.67×10^{-8} W/m^2 K^4]. Parameters such as absorption coefficient α_l [-], emissivity ε_l [-] and evaporation resistance of leaves r_l [s/m], latent heat of vaporization L_l [kg/m^3], vapor density at the surface leaves' temperature (ρ_{vl}) and at the ambient temperature (ρ_{va}) [kg/m^3], the characteristic dimension of leaves, in the direction transverse (W_l) and parallel (D_l) [m] to the wind, and empirical coefficients k_1 [J m^{-2} s$^{-1/2}$ °C^{-1})] and k_2 [s$^{1/2}$ m^{-1}] are taken from the literature experimental evidence [48].

Starting from the knowledge of the leaves' surface temperature, the green coverage is considered as an infinitely extended flat surface neglecting its thickness.

Therefore, the simplified energy balance in absence and presence of green coverage is expressed by the following equations (from Equations (2)–(5)). In particular, in Equations (2) and (3), the quantity of heat released into the air under the green coverage is considered. In Equations (4) and (5), the same quantity is described without greenery.

In the absence of green cover:

$$\alpha_a GHI + h_g(T_g - T_{a-g}) + \varepsilon_{effective}\sigma\left(T_g^4 - T_{a-g}^4\right) - mc_p(T_{a-g} - T_a) = 0 \quad (2)$$

$$(1 - \alpha_a)\alpha_g GHI - h_g(T_g - T_{a-g}) - \varepsilon_{effective}\sigma\left(T_g^4 - T_{a-g}^4\right) = 0 \quad (3)$$

In the presence of the green coverage:

$$\varepsilon_{effective}\sigma\left(T_l^4 - T_{a-g}^4\right) - h_g(T_{a-g} - T_g) + h_l(T_l - T_{a-g}) - mc_p(T_{a-g} - T_a) = 0 \quad (4)$$

$$\varepsilon_{effective}\sigma\left(T_l^4 - T_g^4\right) + h_s(T_{a-g} - T_g) + \varepsilon_{effective}\sigma(T_{a-g} - T_g) = 0 \quad (5)$$

with

$$m = v_w \rho_a \tag{6}$$

The meaning of the physical parameters used is α_a air adsorption coefficient, assumed 0.2 [48] in the visible range, while in the infrared wavelength range is taken equal to the emissivity ε_a 0.9; h_s is the heat transfer coefficient by convection between the ground and the air above [W/m^2 °C] (calculated through dimensionless numbers and their connected empirical correlations); T_g is the ground surface temperature [°C]; T_{a-g} is the air temperature above the ground [°C]; $\varepsilon_{effective}$ is the effective emissivity between two bodies/surfaces as a function of the emissivity of each: $\varepsilon_{effective} = \left(\frac{1}{\varepsilon_1} + \frac{1}{\varepsilon_2} - 1\right)^{-1}$ (e.g., for radiative heat exchange between the ground and the air above it, ε_1 and ε_2 are referred to them; $\varepsilon_{effective}$ is different for each term); m is the air flow rate on the considered surface as a function of wind velocity [kg/s m^2]; ρ_a is the air density at ambient temperature [kg/m^3]; T_a is the air temperature according to climate data [°C]; α_g is the ground adsorption coefficient [-] (e.g., 0.85 for asphalt, and the same value is used for emissivity ε_g [48]); and T_l is the leaves' surface temperature [°C] from Equation (1); h_l is the heat transfer coefficient by convection between the leaves and air below [W/m^{2}K] (for natural convection as a function of dimensionless numbers and connected correlations [49]).

The balances iteratively solved allow for determining the air temperature, affected by the presence of green cover, and the sunny ground without greenery, replacing, respectively, T_e and T_{e-G} of Equations (7) and (8). For this computation the overall average over summer hours of all the parameters must be considered. Due to the boundary conditions of central hours of some summer days, the leaf temperature can be lower than that one of the surrounding environment. Taking into account this phenomena, the outside air temperature values tend to decrease. The heat exchange of the building (branch 5 of the flowchart in Figure 1) in the absence (Q_b) and presence of greenery (Q_{b-G}), can be calculated as follows:

$$Q_b = U\,A_b(T_e - T_i) \tag{7}$$

$$Q_{b-G} = U\,A_b(T_{e-G} - T_i) \tag{8}$$

where T_e is the external air temperature without the green covering, T_{e-G} is the external air temperature with greenery and T_i the internal design temperature of the building (28 °C) in compliance with [50,51], A_b is the total dispersing surface and U the overall heat exchange coefficient that depends on the type of construction and average age of each building provided in [52]. In particular, the overall heat transfer coefficient U is related to the type of construction and materials used, which depend on the building's age. It significantly affects the amount of heat exchanged between the building and environment, mutual radiative heat exchange, increasing the overall entropy.

Considering the most significant climatic condition, i.e., summer, the total heat transfer occurs from the external environment to the internal one of building. Assuming a seasonal coefficient of performance of 2.7 (SCOP) the heat released to the external environment via a cooling system can be calculated in the absence and presence of greenery (Equations (9) and (10)), from the First Law of Thermodynamics:

$$Q_{b\_rel} = Q_b \frac{(1 + SCOP)}{SCOP} \tag{9}$$

$$Q_{b-G\_rel} = Q_{b-G} \frac{(1 + SCOP)}{SCOP} \tag{10}$$

Referring to the Second Law of Thermodynamics (branch 6 of the flowchart in Figure 1), the entropy variation [J/K] of a building at real environmental conditions, in the absence and presence of greenery, can be calculated with the following equations:

$$\Delta S_b = Q_{b\_rel}\left(\frac{1}{T_e}\right) \tag{11}$$

$$\Delta S_{b-G} = Q_{b-G\_rel}\left(\frac{1}{T_{e-G}}\right) \tag{12}$$

with T_e and T_{e-G} in Kelvin.

Using temperatures T_e, T_{e-G}, and Q_{Sun} (i.e., horizontal global solar radiation), A_{urban} is the total surface and α_m the mean absorption coefficient of the urban area (the latter is usually 0.6); the Sun entropy that naturally exists can be assessed by Equations (13) and (14):

$$\Delta S_{Sun,\,e} = \alpha_m A_{urban}\left(\frac{Q_{Sun}}{T_e}\right) \tag{13}$$

$$\Delta S_{Sun,\,e-G} = \alpha_m A_{urban}\left(\frac{Q_{Sun}}{T_{e-G}}\right) \tag{14}$$

Applying the Second Law of Thermodynamics, two entropy footprints (hereinafter referred to *ef* and *ef**) can be calculated. *ef* and *ef** are, respectively, the entropy weight of the urban built-up area in the absence and presence of greenery. They are expressed by the ratio between the entropy variation in buildings and the entropy variation due to the solar energy gain of the area. They provide a measure of the irreversibility produced in the environment and are calculated as follows:

$$ef = \frac{\Delta S_b}{\Delta S_{Sun,\,e}} \tag{15}$$

$$ef^* = \frac{\Delta S_{b-G}}{\Delta S_{Sun,\,e-G}} \tag{16}$$

Finally, a new irreversibility indicator can be defined by the ratio between *ef* and *ef**. This ratio expresses the building's entropy contribution when greenery is present compared to that generated in the absence of greenery.

2.2. Modeling Approach, Data Sources, and Pre-Processing

The proposed method, applied to a real urban area (Florence, Italy), is structured and follows the flowchart provided in Figure 1. The thermo-hygrometric modeling incorporates microclimatic influences and vegetation contributions inside the Urban Canopy Layer (UCL)—defined as the volume between the ground and the top of buildings and vegetation [53].

This region, where urban climate dynamics directly affect human comfort and energy demands [54], is characterized through high-resolution climate modeling and vegetation assessment. The modeling process includes remote sensing, meteorological data, and GIS-based building classification. Our research focuses on UCL-UHI because it is commonly characterized by field measurements and is the zone where thermal–hygrometric comfort, environmental quality, and people well-being are assessed [54]. Climatic inputs are derived from historical records and future projections using regional climate models. The climate change scenarios are based on the Very High-Resolution PROjections for Italy (VHR_PRO_IT) under RCP4.5 and RCP8.5 pathways [55,56]. These models provide hourly data at 2.2 km resolution for five datasets: historical (1981–2005), near-term, and long-term projections under both RCP4.5 and RCP8.5 forcing scenarios (Appendix A.1,). Starting from the modeling of past and future climate scenarios, five datasets are obtained: (i) historical years 1981–2005, (ii) RCP4.5 years 2006–2038, (iii) RCP4.5 years 2039–2070, (iv) RCP8.5 years 2006–2038, and (v) RCP8.5 years 2039–2070.

Each dataset includes air temperature, humidity, wind speed, and solar radiation that are collected with hourly frequency, for the entire time span over which they are defined. To apply entropy balances, a typical weather year (TWY) is generated for each dataset, condensing 30-year data into an annual sequence while maintaining representativeness (Appendix A.2). Details on climatological modeling and climate change scenarios construction are provided in Appendix A.1.

The characterization of greenery can be carried out by a dedicated modeling (branch 1 of the flowchart in Figure 1). Green cover surfaces and structural characteristics of urban vegetation (e.g., different species, total height, and diameter at breast height) can be collected either through the public tree census (which provides information on trees located in public areas) or by using optical and LiDAR remote sensing data, when census information is not available. This is the case of private green areas (i.e., private gardens), which still play a pivotal role in the total share of urban green surfaces. Moreover, each plant, with all its properties for specific species, can easily be referenced and connected to the urban area with GIS techniques.

In particular, to characterize green surface and tree species, the individual tree crown segmentation algorithm (ITC), based on the Canopy Height Model (CHM) have to be applied, using a variable moving window size [57]: treetops of trees in the upper canopy strata are identified through the local maxima algorithm [58] with a variable window filter [59]. The moving window [60] is used to scan the CHM so that any cell found to be the highest within the window can be classified as a treetop. The Marker-Controlled Watershed Segmentation algorithm [61] is then applied to delineate urban tree canopies.

In particular, the ITC algorithm allows for the quantification of the (i) number of urban trees (also including those not covered by the public tree census), and (ii) their canopy extent. Using available in situ data, or from Green-Trees Census and Municipal Green Plans, it is possible to classify each ITC-based canopy into broadleaved/conifer with very high-resolution imagery, by means of the application of machine learning models (e.g., Random Forest [62]).

The model's accuracy have to be assessed using out-of-bag (OOB) samples [63,64], which consist of data points that are not included in the bootstrap sample used to train a given tree [65]. These samples are used as a validation set for each decision tree in the Random Forest [66]. The OOB error is computed by aggregating the predictions made by each tree on its respective OOB samples. This procedure is usually applied for model's performance evaluation, without requiring a separate validation set, and provides the accuracy assessment [62].

Practical application to any area, as that one studied, allows the characterization of the existing vegetation and identification of basic species-specific parameters (e.g., the emissivity coefficient in the infrared and the evaporation resistance and/or vapor permeability of leaves). Usually, these parameters are obtained through direct measurement [67] or derived from dedicated studies carried out under local conditions [68]. For example, Tiralla et al. [69] collected information on ε for many different trees, shrubs, and herbaceous species, along with their standard deviation. Similarly, Mashabatu et al. [68] provided some useful information about the mean specific leaf resistance of five fruit tree orchard species. In Section 2.3, the above modeling method implementation is explained and adapted for the specific greenery of the urban area studied.

2.3. The Case Study: Typical Built-Up Urban Area of Florence

An urban area of the city of Florence (Figure 2), of Tuscany region (Italy), is the case study. The studied urban area belongs to the Municipal Environmental Energy Plan (PEAC) [70] (branch 3 of the flowchart in Figure 1). This built-up urban area is a representative of the city's urban development and includes buildings of different uses and construction periods (age classes of buildings) (Figure 2, in blue). In this area asphalt and green surfaces are present (Figure 2, in green).

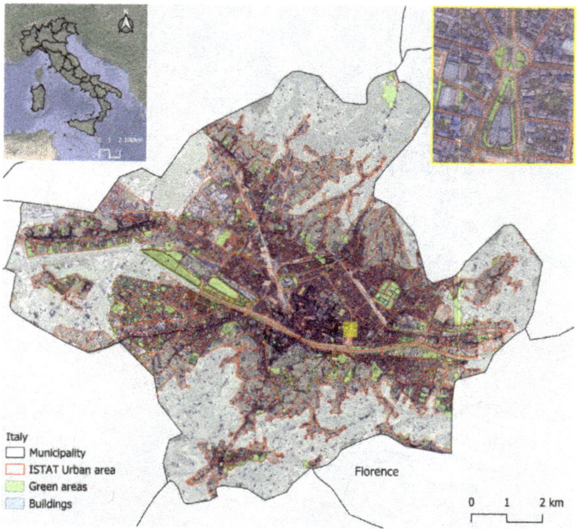

Figure 2. The area studied according to the boundaries defined by ISTAT for the urban area (in red). Green areas (in green) and buildings (in blue) refer to the Florence Municipality (in black).

The limits of this area were identified using the spatial boundary provided by the Italian National Institute of Statistics for the year 2011 [71], which includes the densely inhabited areas near recognized aggregation centers, i.e., squares and public complexes (Figure 2, in red). For this reason, the peripheral areas and those with low interaction between green system and built-up zones were excluded. All the necessary environmental parameters are derived from available databases/open-sources and census data from 2011 [71]. Georeferenced building data (i.e., three-dimensional geometry of buildings and their house number and construction age class, thermo-physical features, volume, height, base area, type of plant system, number of occupants, intended use, etc.) were extracted from PEAC [70,72]. By means of database interfacing and necessary information extraction, many useful data were derived from Tuscany Region GEOscopio (https://www.regione.toscana.it/-/geoscopio (accessed on 12 August 2025)) with QGIS software (version 3.40.1-Bratislava) post-processing [73].

The Florence Municipality covers 102.4 km^2, with a population of 352,622 in the urban area; the urban area case study is 6464 hectares (A_{urban}), with 115,000 buildings corresponding to 192 million m^2 total dispersing surface (A_b). The global heat exchange coefficient U [W/m^2 K] for all these buildings has an average value of 2.1 W/m^{2}K. The intended use of most buildings is "residential"; then follows "services" and "commercial".

The total dispersing surface of each building is connected with the global heat exchange coefficient as a function of the construction age class of each building, which takes into account the thermo-physical characteristics of the building envelope, as reported in the PEAC database [70,72]. Table 1 shows the value of the global heat exchange coefficient deduced from PEAC.

Table 1. Global heat exchange coefficient by building construction age class [W/m^{2}K].

Age Class	<1940	1940–1970	1971–1980	>1980
U [W/m^{2}K]	2.4	1.9	1.8	1.2

For green areas computation (branch 1 of the flowchart in Figure 1), the ITC algorithm was applied on the specific CHM, provided by the Florence Municipality for the year 2023, with 0.5 m spatial resolution. Two thresholds were applied at resulting segmentation for filtering out the non-tree features. Selecting trees, with heights between 3 and 35 m, the mean Normalized Difference Vegetation Index (NDVI)) was calculated for each identified tree crown (based on Sentinel-2 data composite for May–July 2023) using a minimum NDVI threshold of 0.5 [74]. Therefore, 273,528 trees, of public and private zones, were identified, for 9,858,840 m^2 total canopy cover, that is 15% of the total selected urban area.

This resulting total surface is used as the whole leaf-surface coverage (in Equation (1)) that is a square, with 3140 m side length, replacing the values for both the parameters D and W in Equation (1).

The specific trees classification was carried out: segmented tree crowns intersection with tree census points from the city of Florence, which provided a subset of 43,469 tree crowns for coniferous or broadleaf; five percentiles computation (10th, 30th, 50th, 70th, and 90th) based on four predictors and for each identified crowns [75] (i.e., the CHM and three bands from a high-resolution Florentine orthophoto; 0.2 m, for year 2023). Then, 20 predictors available for each tree crown were computed. Random Forest (RF) model [62] allowed for predicting tree species class (coniferous or broadleaf) using the above 20 predictors as input features. The Google Earth Engine RF classifier was exploited, setting the number of decision trees to 300. Modeling performance was evaluated by the OOB aggregating predictions and comparing reference data, and then constructing a confusion matrix. Due to the unbalance inside the dataset (i.e., 6269 coniferous trees vs. 37,200 broadleaves), the RF classification model performance was assessed by the Matthews Correlation Coefficient (MCC) [76,77] based on the confusion matrix. In detail, the MCC is a correlation measure between categorical predictions and observations: it equals 1 for perfect predictions, 0 for random agreement, and -1 when predictions and observations are entirely mismatched. The trained model was applied to the 273,528 tree crowns remaining, using the 20 predictors to classify all of them into coniferous and/or broadleaves, with a 0.83 MCC value. This classification provided a subset of 43,469 tree crowns (coniferous or broadleaf) according to the municipality's tree census.

Therefore, for the present study, the overall canopy coverage turned out 760,191 m^2 conifers and 9,098,648 m^2 broadleaves. Their mean emissivity values (inside the 8–14 μm interval) were derived from [78]: the emissivity was 0.982 ± 0.009 for conifers and 0.985 ± 0.010 for broadleaves; 0.985 emissivity mean value, weighted by the covering surface. Fundamental literature evidence [79] was used for the evaporation resistance of the oak leaves (i.e., 2000 s/m) that was applied to conifers and broadleaves trees. The used values of vapor diffusivity, despite not directly available for conifer species, and absorption coefficient of solar radiation 0.8, refer to [48].

Using to the aforementioned greenery parameters, and to the TWYs for each dataset, all the balances of Equations (1)–(5) were carried out for summer periods, i.e., from 01/04 to 31/10.

From the available hourly data, the hot hours (HHs) were selected, matching the following boundary conditions:

$$GHI > 200\frac{\mathrm{W}}{\mathrm{m}^2}\ 0.1\frac{\mathrm{m}}{\mathrm{s}} < v_w < 10.7\frac{\mathrm{m}}{\mathrm{s}}\ T_a > 28\ ^\circ\mathrm{C}$$

Those constraints identify the set of hot hour measurements, such that all the balances have physical meaning. The conditions on the solar radiation *GHI* and the exclusion of small values of v_w ensure thermodynamic balance validity [56]. Extreme wind conditions are avoided, taking into account the Beaufort wind force scale, i.e., an empirical scale that relates wind velocity to observed conditions at sea or on land: for the upper bound chosen, the scale indicates a condition of fresh breeze, characterized by the presence of wind that begins to move the large branches, but not the entire structure of the tree [79].

Filtering the summer data with the above boundary conditions, all the balances are solved for the two configurations (i.e., green presence and absence) in order to determine the hourly values of leaf temperature T_l, air volume temperature T_{a-g}, and ground temperature T_g. The average values of the external temperature with green coverage are

$$T_{e-G} = < T_{a-g}{}^{withG} >_{HH}$$

and without green coverage

$$T_e = < T_{a-g}{}^{withoutG} >_{HH}$$

where the $< >_{HH}$ represents the average over the selected hours (*HH*).

The mean values are used to solve Equations (7)–(14) and calculate the entropy footprint of Equations (15) and (16). The average value of *GHI* over the *HH* subset was used for the solar radiation Q_{Sun} in Equations (13) and (14).

3. Results

All the balances of Section 2 are solved under the hypothesis and boundary conditions extensively described above, obtaining the main following results: hourly summer temperature for the leaf coverage T_l; bottom mean air temperature T_{e-G}, i.e., between the leaf coverage and ground; mean air temperature above the ground without leaf coverage, T_e; cooling energy need of buildings in the presence and absence of greenery, respectively (Q_{b-G} and Q_b); heat released to the external environment, in the presence and absence of greenery (Q_{b-G_rel} and Q_{b_rel}); entropy increase, in the presence and absence of greenery, i.e., $\Delta S_{b\ -\ G}$ and ΔS_b; entropy footprints *ef* and *ef**, due to the aforementioned thermodynamic process; and entropy increases due to the Sun radiation hitting the ground $\Delta S_{Sun,\ e\ -\ G}$ and $\Delta S_{Sun,\ e}$.

In particular, a preliminary global sensitivity analysis was carried out, in order to evaluate how much greenery properties (i.e., emissivity, vapor resistance, and absorption coefficients) impact the balances and affect results. This sensitivity analysis showed a variation in leaf temperature of no more than one hundredth of a degree, and therefore, an absolute error lower than 2%, also showing the method robustness and its applicability to different contexts. Therefore, in the ranges of all the possible values for parameters, the leaf temperature variation obtained is negligible compared to the contribution of the parameters related to the boundary conditions (solar radiation, wind velocity, relative humidity).

Table 2 shows the main parameters for the five investigated scenarios, referring to [80]: Standard TWY, Historical TWY, "RCP4.5 years 2006–2038" TWY, "RCP4.5 years 2039–2070" TWY, "RCP8.5 years 2006–2038" TWY, and "RCP8.5 years 2039–2070" TWY.

Table 2. Studied area results: main temperatures and entropy footprints with and without greenery.

	Standard TWY	Historical TWY	RCP4.5 2006–2038 TWY	RCP4.5 2039–2070 TWY	RCP8.5 2006–2038 TWY	RCP8.5 2039–2070 TWY
T_i [°C]	28.0					
A_{urban} [m^2]	64,645,324					
A_b [m^2]	192,582,468					
U [W/m^{2}K]	2.1					
T_e [°C]	31.8	32.5	33.0	33.1	32.7	33.8
T_{e-G} [°C]	31.3	32.1	32.5	32.7	32.3	33.4
GHI [W/m^2]	641.3	610.6	606.4	589.6	597.0	581.3
Q_b [MWh]	1540	1831	2016	2085	1897	2337
Q_{b-G} [MWh]	1347	1670	1844	1921	1731	2174
Q_{b_rel} [MWh]	2110	2509	2763	2857	2600	3202
Q_{b-G_rel} [MWh]	1846	2289	2527	2633	2372	2979
$\Delta S_{Sun,e}$ [MJ/K]	293,658	278,941	276,589	268,791	272,575	264,473
$\Delta S_{Sun,e-G}$ [MJ/K]	294,117	279,303	276,973	269,146	272,942	264,819
ΔS_b [MJ/K]	24,925	29,566	32,510	33,597	30,620	35,578
ΔS_{b-G} [MJ/K]	21,836	27,008	29,777	31,001	27,967	35,006
ef [%]	8.5%	8.5%	9.7%	10.4%	9.1%	12.1%
ef^* [%]	7.4%	7.6%	8.7%	9.4%	8.1%	11.1%
ef/ef^* [%]	87.5%	89.1%	89.6%	90.6%	89.2%	91.8%

It must be noticed that the entropy footprints are already updated according to the weighted correction b found in Equation (A2), that is -0.021. The calibration effect is provided by Figure 3 (top graph): entropy footprints for the different datasets, highlighting the distance between the entropy footprints value calculated in respect to the same period (1981–2005) but with different sources (Standard TWY vs. Historical TWY).

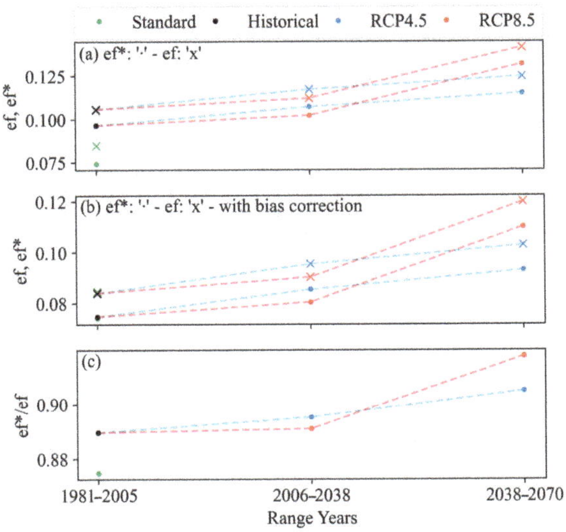

Figure 3. Entropy footprints [-] (**a**) without and (**b**) with bias correction. (**c**) Ratio between the entropy footprints for all the climatological datasets with bias correction. Standard refers to the dataset obtained from measured data from fixed station, while Historical refers to the dataset obtained from the historical model, and temporally overlapped to the Standard dataset. RCP4.5 and RCP8.5 indicate the two different projections considered for the impact analysis on future states (extensively explained in the text and appendixes).

After calibration, the entropy footprints value is shown in Figure 3 (center graph) that also provides results for the dataset of years 1981–2005, varied by 0.2% at maximum.

Comparing *ef** with *ef*, and considering their ratio, a drop in the entropy production can be found, due to the green coverage presence. All this acts by modifying the environmental constraints of temperature (only the humidity changes are considered in Equation (1) for leaf temperature calculation) creating cooler boundary conditions for the building plants, which can operate with a reduced external thermal load during summer.

For all the scenarios, the entropy decreases by almost 1% in respect to the contribution of the Sun radiation hitting the ground. The low value is due to that reference at the denominator, which is relatively high.

Using the percentage between the values of each entropy footprint (i.e., *ef* and *ef**), due to the green coverage, the entropy footprint is 9.2–12.5% depending on the considered scenario, but with lower changes resulting in the projections at different scales (Table 2). These trends are provided in Figure 3 and in the bar graph in Figure 4, with a more direct and easier way to understand and compare.

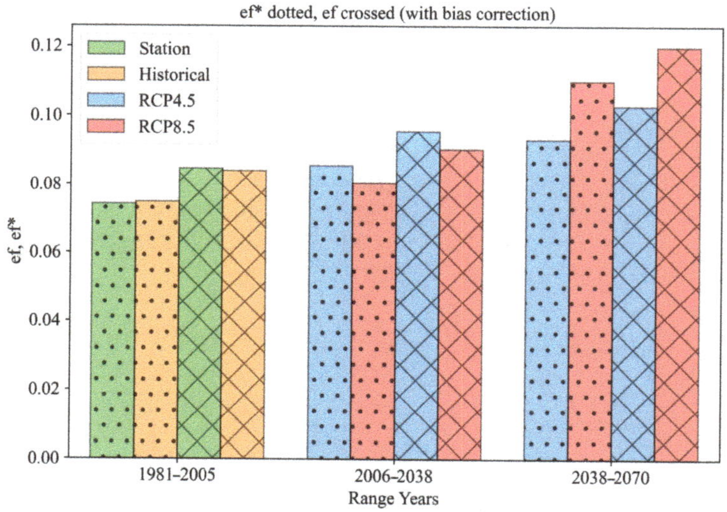

Figure 4. Bar graph of the entropy footprint [-] grouped by data range. *ef** is marked with dots in bars, while *ef* is marked with crosses. Each color indicates a different dataset: green for the Station dataset, orange for the Historical dataset, and blue and red for the two projections considered in the study, RCP4.5 and RCP8.5, respectively. To clarify the interpretation of the different impact of the greenery's contribution, under different Anthropocene forcings for the near and long-term, the two datasets obtained from the projections considered are shown with the same color.

It is important to note that differences between the values of the entropy footprints are small, but statistically significant and important. As a matter of fact, the entropy footprints are close to 10% because they refer to thermal energy coming from the Sun, which, precisely because it exists naturally, can be taken as a basic reference. This is relatively greater than the contribution that buildings and their plant-systems can provide. For this reason, a slight variation (at most 4.3%) in the entropy footprints can be noted.

On the other hand, Table 2 shows the absolute values of the entropy produced, which, for the cases examined alone, reaches the order of magnitude of 10^5 MJ/K. The difference between the scenarios with and without greenery is 3000 MJ/K for a standard TWY. This means that, if the study is extended to the whole urban area, the impact becomes physically and statistically appreciable, confirming the importance of the presence of greenery, especially for those scenarios with more critical conditions.

The effectiveness of the greenery presence is more evident if the absolute value of the difference in entropy released in the environment by the built-up urban area with or without is considered.

In Table 3 the ambient temperature T_a that corresponds to the maximum temperature of each dataset is reported: these values were used to evaluate the entropy difference.

Table 3. Maximum temperature [°C] for each dataset.

Standard	Historical	RCP4.5 2006–2038	RCP4.5 2039–2070	RCP8.5 2006–2038	RCP8.5 2039–2070
36.8	40.0	42.9	41.9	43.3	43.7

ΔS, shown in Figure 5.

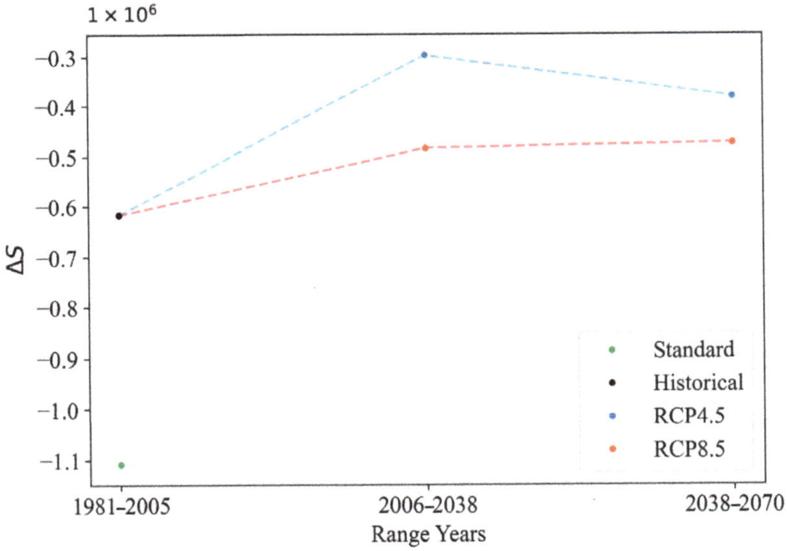

Figure 5. Entropy difference [MJ/kg] calculated for different periods and scenarios and for the hottest day.

As a matter of fact, for RCP4.5 there is a smaller increase in the maximum temperature compared to the RCP8.5 scenario, with a lower overheating effect of the buildings, a greater cooling effect by the plants and therefore a lower energy consumption of the heat pump system (HP) for cooling.

4. Discussion

This study proposes a useful tool, based on the physics of real systems, matching thermodynamics (of First and Second Law) with advanced modeling techniques for evaluating green systems effectiveness to counteract climate change and reduce the entropy of the urban built-up system, ensuring responsive resilience cities. The method can be applied at different urban scales.

The study offers crucial indicators based on thermodynamics, i.e., the entropy footprints, which can be used for quantifying sustainability in urban areas, supporting the decision-making policies of local governments and public administrations for urban and energy planning oriented to sustainability and green energy transition.

Comparing the entropy footprint values of different zones of any urban built-up area, effective interventions can be directly oriented to entropy flows reduction in building–plant systems, in the presence and absence of greenery, referring to dynamic climate change scenarios. The entropy analysis of the urban area allows for energy–environmental sustainability evaluation, in terms of the low environmental impact, energy quality, and durability.

By the method application, quantification of energy–environmental benefits due to entropy flows reduction that effectively counteracts of climate change effects can be assessed: it allows a general but significant evaluation of the lower impact due to air conditioning systems when buildings are immersed in greenery (see Figure 4).

From this perspective, the methodological approach, based on entropy footprint, is innovative because, using the Second Law of Thermodynamics, allows the quantification of greenery presence effects on any urban system by a simplified and easy-to-implement approach without the need for large amounts of measured/detected data or obtained from the literature evidence and/or specific data sources. It is robust–rigorous, as it is validated and confirmed by the important relevant literature evidence that by means of energy analyses application to urban systems (i.e., First Law of Thermodynamics) show the real cooling demand reduction in buildings due to greenery presence (e.g., NBS integration, green roofs and green walls), remote sensing data on surface temperature drops, and measured air temperature differences between green and non-green areas in urban systems [54,81–89].

The innovative contribution of the method also lies in the fact that, through the quantification of the entropy footprints (*ef* and *ef**) and the indicator obtained from their ratio, a homogeneous scale, of physical significance, can be built, identifying the most critical and vulnerable areas on which to intervene as a priority. This allows us to use and apply a physical-entropy-informed criterion for controlling and planning interventions over time. As a matter of fact, results comparison (Table 2) shows that the contribution due to the greenery decreases for different scenarios referred to the different climatological dataset, but at the same time the increase in entropy becomes higher for the scenario with higher emissions (RCP8.5).

However, it must be considered that the proposed method is simplified because all the thermal properties of building materials are unified in the global heat exchange coefficient that is linked to the age class. Moreover, the present study aims at urban-level analyses rather than at the performance/design of building–plant systems and indoor scenarios assessment. In line with the two Laws of Thermodynamics, building energy balances for summer period, consider the temperature levels at which heat exchanges occur. Furthermore, any increase in greenery or energy efficiency interventions and NBS integration on buildings would certainly lead to a lowering of the ambient temperature together with a reduced energy consumption by the cooling systems. The latter, thanks to the energy efficiency of buildings, would entail significant improvements in environmental impacts decrease, i.e., the entropy production reduction (Table 3).

Another limitation/simplification of the method is due to a basic general evaluation of the overall heat exchange of buildings with the external environment as a function of their overall heat exchange coefficient and volume (disregarding their mass and thermal inertia and heat capacity of specific building materials, linked to the accumulation of heat due to solar radiation). This latter choice is motivated by the fact that most available databases and the literature do not provide specific information on the construction and thermo-physical characteristics of buildings.

However, the method proposed can be directly applied by public administrations, all stakeholders, and local governments in the perspective of planning the energy green and sustainability transition for livable and comfortable urban areas and resilient cities.

These aspects are also supported by the recent literature results [90] that show how at least three quarters of the current carbon storage of urban trees in Central and Southern Europe will be seriously threatened by 2070.

With adequate and available data, the entropy footprint evaluation can provide useful quantification of sustainability due to the current conditions of any urban area and effective analyses of different strategies to combat climate change, UHIs, and anthropogenic effects in cities, in the short and long term, considering that conversion of impervious surfaces into green areas has high costs and requires time [91].

In the sustainability, resilience, and transition perspectives, the method implementation could be used as predictive tool. Indeed, it provides a physical-entropy-informed support for practical applications at large scales, focused on selection of suitable species for NBS and integrated green systems, maintenance of urban functional diversity, and permeability [92,93].

More in general, results comparison, obtained at small scale, show that further applications of the method can be for the management, design, and implementation of energy and environmentally sustainable projects, considering the entropy footprints based on the necessary operating times in relation to the developments of climate change effects over the years.

Future research directions will concern the method's application to different urban realities, both in terms of complexity and infrastructures and the presence of different green areas and water surfaces (e.g., fountains, rivers, lakes, etc.) for different climatic contexts. A further interesting development of the research will concern the study of entropy footprint at regional level, in the local context, but also at a wider national level.

Starting from the entropy footprints connected to the different development of green areas, intensive vegetation and strong integration of NBS, and the choice of effective strategies in the medium to short term, the method foresees future developments, including its implementation at a large scale, and of an objective function to be minimized/maximized in relation to the decision-making policies and strategic environmental, energy, and proactive resilience choices of specific managerial groups and stakeholders (taking into account economic aspects).

In this perspective, the research will use AI and implement finalized machine learning techniques, especially oriented towards a detailed study of criticalities and vulnerability of the urban areas.

5. Conclusions

This study proposes an integrated method, simple and easy to implement, to quantify the energy–environmental sustainability of urban areas. The research links applied thermodynamics of Second Law to strategic modeling of biophysical thermo-hygrometric balances of leaf surfaces, high-resolution climate at urban scale, identification and characterization of plants for specific species and vegetation, and spatial georeferencing applied techniques for entropy exchanges evaluation within the urban context.

The focus, i.e., the entropy flow reduction and entropy impact of urban built-up areas assessment, with and without greenery, constitutes the innovative thermodynamic-entropy framework to quantify the benefits of urban greening.

The simple method proposed can thus be easily implemented for energy and entropy analyses on a small and large urban scale.

Given that most databases and the literature lack detailed thermo-physical data on buildings, but provide georeferenced and general attributes (e.g., age-based heat exchange coefficients, volume, surface area, height, system type, and use), the proposed simplified method relies on readily available information. It evaluates the energy and entropy balance of the built environment in relation to green space distribution, without accounting for detailed building-system thermo-physics. Thus, this method can be a crucial tool for energy planning and thermo-hygrometric comfort improvement in urban areas. It allows also programming activities of authorities, local municipal governments, disaster management authorities and the same citizens, to better understand the relationship between the potential for reducing anthropogenic impacts, the UHI mitigation, climate change adaptation, housing factors and urban areas development, urban forestry, and NBS and/or green systems integration. The method can be easily applied for quantifying proactive resilience and sustainability transition, as it provides a homogeneous scale of entropy footprints, i.e., irreversibility, as a measure of sustainability, and a rational criterion for comparing and integrating these evaluations. While not suited for deterministic modeling, this approach might serve as a decision-support tool for evaluating ex-ante and ex-post urban energy strategies, including energy savings, renewable integration, and nature-based solutions, through entropy-based assessment (with entropy footprints) of adaptive sustainability and resilience.

Nevertheless, basic limitations of the method concern its effectiveness and applicability, especially when moving from small scale to large scale (i.e., from single building to group of buildings, block and whole urban area, from city to regional scale). Its reliability depends on the availability, quality, and quantity of information and usable data (e.g., climate and local weather conditions, buildings' features and their air conditioning systems, greenery).

Despite the method, by means of the entropy footprints assessment, which can provide a rational criterion to compare complex innovative and sustainable technology solutions, capable of leading to important irreversibility reduction and energy efficiency increase, its application to different urban areas and then to larger scale of analysis (e.g., regional and national scale) can show possible implementation downsides mainly due to the reliability and typology of necessary data.

Author Contributions: C.B. (Carla Balocco); Writing, review and editing, Supervision, Funding acquisition, Conceptualization, Methodology, Investigation, G.P.; Writing, Methodology, Investigation, Validation, Software, Data curation, M.B.; Writing, Investigation, Validation, Software, Data curation, C.B. (Costanza Borghi); Writing, Investigation, Data curation, S.F.; Writing, Data curation, G.C.; Review and editing, S.M.; Review and editing. All authors have read and agreed to the published version of the manuscript.

Funding: This work was supported by the EU Project, LIFE Energy + LIFE Climate—Project 101157553-LIFE23-CCA-IT-LIFE ESCAPOS "Environmental energy for Strategic CApillary urban POlicieS", funded by the European Union.

Institutional Review Board Statement: Not applicable.

Informed Consent Statement: Not applicable.

Data Availability Statement: The original contributions presented in this study are included in the article. Further inquiries can be directed to the corresponding author.

Conflicts of Interest: The authors declare no conflicts of interest and that they have no competing financial interests or personal relationships that could have appeared to influence the work reported in this paper.

Nomenclature

A_b	dispersing surface of the buildings in the investigated area [m^2]
A_{urban}	urban area considered for the analysis [m^2]
b	correction factor for the entropy weights [-]
SCOP	mean coefficient of performance for heat pumps [-]
D_l	characteristic dimension of the leaves in the wind direction [m]
ef	entropy weight without the presence of the greenery [-]
ef^*	entropy weight with the presence of the greenery [-]
GHI	global horizontal irradiation on a specific site [W/m^2]
h_l	heat transfer coefficient by natural convection between leaves and the air below [W/m^2 °C]
h_s	heat transfer coefficient of natural convection between ground and the air above [W/m^2 °C]
k_1	empirical coefficient [J/m^2·s^{1}/^{2}·°C^1]
k_2	empirical coefficient [s^1/^{2}·m^1]
L_l	latent heat of vaporization at the surface temperature of the leaves [J/kg]
m	air flow rate on the considered surface as a function of the wind velocity [kg/s m^2]
Q_b	cooling power for buildings in the summer period without the presence of greenery [W]
Q_{b-G}	cooling power for buildings in the summer period with the presence of greenery [W]
Q_{b_rel}	thermal power released to the environment without the presence of greenery [W]
$Q_{b\text{-}G\_rel}$	thermal power released to the environment with the presence of greenery [W]
Q_{Sun}	total horizontal Sun radiation hitting the urban area in the summer period [W/m^2]
r_l	evaporation resistance of the leaves [s/m]
RH_a	relative humidity of the air [-]
T_a	ambient temperature [K]
T_{a-g}	air temperature above the ground [K]
T_e	external mean air temperature in the summer period without the presence of greenery [K]
T_{e-G}	external mean air temperature in the summer period with the presence of greenery [K]
T_g	temperature of the ground surface [K]
T_i	indoor temperature for the buildings in the summer period [K]
T_l	surface temperature of the leaves [K]
U	global heat exchange coefficient for the building (W/m^{2}K]
v_w	wind velocity [m/s]
W_l	characteristic dimension of the leaves in the direction transverse to the wind [m]
Geek symbols	
α_a	adsorption coefficient of the air in the visible range [-]
α_g	adsorption coefficient of the ground across the entire spectrum [-]
α_l	absorption coefficient [-] for the leaves across the entire spectrum [-]
α_m	mean absorption coefficient [-] for the urban area across the entire spectrum [-]
ΔS_b	entropy generation due to the buildings without the presence of greenery [J/K]
ΔS_{b-G}	entropy generation due to the buildings with the presence of greenery [J/K]
$\Delta S_{Sun,\ e}$	entropy generation due to Sun radiation hitting the urban area without greenery [J/K]
$\Delta S_{Sun,\ e-G}$	entropy generation due to Sun radiation hitting the urban area with greenery [J/K]
ε_a	emissivity coefficient of the air in the infrared range [-]
$\varepsilon_{effective}$	emissivity between two bodies/surfaces as a function of the emissivity of each [-]
ε_g	emissivity coefficient for the ground in the infrared range [-]
ε_l	emissivity coefficient of leaves in the infrared range [-]
ρ_a	density of the air at ambient temperature [kg/m^3]
ρ_{va}	vapor density at the ambient temperature [kg/m^3]
ρ_{vl}	vapor density at the surface temperature of the leaves [kg/m^3]
σ	Stefan–Boltzmann constant [W/m^{2}K^4]

Appendix A

Appendix A.1

In the application of the present method, the VHR_PRO_IT tool is used, a climatological model with very high spatial and temporal resolution, obtained as a dynamic downscaling starting from an ensemble of climatological projections within the Med-CORDEX project (http://cordex.org/domains/region-12-mediterranean/ (accessed on 15 August 2025)).

In order to quantify the different responses to the different strategies that can be implemented to counteract climate change, two different possible scenarios are considered and defined by the RCP. Those different scenarios, formally adopted by the IPCC, quantify different paths to describe future greenhouse gas concentrations. Over the years, several possible scenarios have been identified, represented by a reduction in greenhouse gas emissions, up to a continuous increase in the same. This translates into a different response to climate change, and the average values of greenhouse gas emissions correspond to a greater increase in the average surface temperature. These pathways are tabulated and based on the expected changes in the radiative forcing values. For this case study, we considered the RCP4.5 and RCP8.5 scenarios, therefore corresponding to an average increase in radiative forcing (4.5 W/m^2) and the most extreme condition (8.5 W/m^2), respectively. The first scenario foresees an increase in emissions that reaches its peak around 2040: that is the most likely a baseline, as it does not take into account strong containment policies at a global level, but only the exhaustible nature of fossil fuels. The RCP8.5 scenario, on the other hand, foresees a continuous increase in emissions throughout the 21st century. Although, since the publication of the IPCC Fifth Assessment Report [83] this hypothesis has been criticized for overestimating coal production, and considered an unreliable and unrealistic long-term scenario, it is still interesting for quantifying climatological trends in the near future (up to 2050) and the impact of the socio-political choices for recent years.

In particular, to evaluate the medium and long-term impact, the two scenarios, available for the years 2006–2070, are divided into two different datasets, namely between the years 2006–2038 and 2039–2070.

On the other hand, when the climate model outputs (e.g., temperature, precipitation) are compared in historical mode and real-word measurements from fixed weather stations, systematic discrepancies often arise. These differences, known as bias, occur due to imperfect parametrization, missing physical process, or limitations in model resolution. As a matter of fact, model data, which describe an average behavior over the area defined by the resolution of the model, generally fails to describe the hyperlocal nature of the atmospheric dynamic characteristics of the zone where the fixed station is located.

To quantify these differences, a data processing approach is applied in historical mode, aligning model outputs temporally with the records from weather stations. Although the model data used has a spatial resolution of 2.2 km, this method enables a direct comparison with the time-synchronized data aggregated in the standard TWY file.

This model data allows the quantification of different responses to the balance equations, in particular the estimated temperature above the ground, T_{a-g}. This step is crucial, because a bias in these temperatures is directly transmitted into the evaluation of the entropy footprints *ef* and *ef** which strictly depend on T_{a-g}. To address this bias and having projection results comparable to what is measured by the weather stations, a bias correction process to the values of *ef* and *ef** obtained in the TWY file from the model data, is implemented. This procedure is based on correcting the estimated entropy footprints by a value b that considers the differences in entropy footprints between the historical and weather station datasets. The correction is performed by shifting the entropy footprints estimated from the model data by a factor b:

$$ef_{model} = ef_{model} + b \quad \text{(A1a)}$$

$$ef^*_{model} = ef^*_{model} + b \quad \text{(A1b)}$$

where the subscript model in the two equations indicates the different modes: historical, RCP4.5 and RCP8.5. The bias correction b was obtained as a weighted average between the differences in the entropy footprint in the presence and absence of green cover between the two temporal overlapping datasets, i.e.,

$$b = \frac{ef_{station} - ef_{historical}}{2} + \frac{ef^*_{station} - ef^*_{historical}}{2} \quad \text{(A2)}$$

Before applying the balances and quantifying the average entropic contributions in the different configurations (with green and without green), for each dataset, it was necessary to generate a specific TWY, so as to transform a dataset with an average duration of 30 years, into an annual dataset, but with the typical characteristics of the different years. A detailed description of the procedure used to generate the TWY is provided in Appendix A.2.

Appendix A.2

The typical weather year, i.e., the TWY weather data, is composed of 12 separate months of data, each chosen to be the most "typical" month from the total years of data. TMY data selection is carried out using the Sandia method [84] which is an empirical approach that selects individual months from different years of the period of record. Given a range of N_y data years, we analyze all the m-th months in the N_y years. For instance, suppose m = 1, January, all the N_y Januarys are examined and the one judged "most typical" is selected to be included in the TMY. The other months of the year are treated in a similar manner, and then the 12 selected typical months are concatenated to create a complete year. The typical month is selected based on nine daily indices consisting of the maximum, minimum, and mean dry bulb and dew point temperatures; the maximum and mean wind velocity; and the total global horizontal solar radiation. For each month of the calendar year, we select the year with cumulative distribution functions (CDFs) for the daily indices that are the closest to the long-term CDFs, computed considering all the m-th of all years in the dataset. The CDF gives the proportion of values that are less than or equal to a specified value of an index. Candidate monthly CDFs are compared to the long-term CDFs by using the following Finkelstein-Schafer (Δ) statistics for each index:

$$\Delta^{(j)} = \frac{1}{n}\sum\nolimits_{i=1}^{N} \delta_i{}^{(j)} \quad \text{(A3)}$$

where $\delta_i{}^{(j)}$ is the absolute difference between the long-term CDF and the candidate month, n is the number of daily readings in a month and the index j runs over the different variables. Because some indices are judged more important than others, a weighted sum of the $\Delta^{(j)}$ statistics, obtained considering all the different variables, is used to select the candidate month that has the lowest weighted sums:

$$W = \sum\nolimits_j w_j \cdot \Delta^{(j)} \quad \text{(A4)}$$

where w_j is the weighting factor, listed in Table A1.

A typical month is then selected by choosing the months of the year with the lowest w values, e.g., the one with the smallest deviation from the long-term CDF is applied.

Table A1. Weighting factors used to compute the weighted sum W.

Weather Index	w_j	Weather Index	w_j
Maximum dry bulb temperature	1/24	Mean dew bulb temperature	2/24
Minimum dry bulb temperature	1/24	Maximum wind velocity	2/24
Mean dry bulb temperature	2/24	Mean wind velocity	2/24
Maximum dew bulb temperature	1/24	Total horizontal solar radiation	11/24

In particular, the external air vapor content per unit kg of dry air (also known as absolute humidity) was calculated because it provides information on the air enthalpy variation in presence or in absence of greenery and the connected work of the air conditioning systems.

In Figure A1 the probability distribution of the absolute humidity of the air for the hot hours (HH) values in the different datasets, is shown.

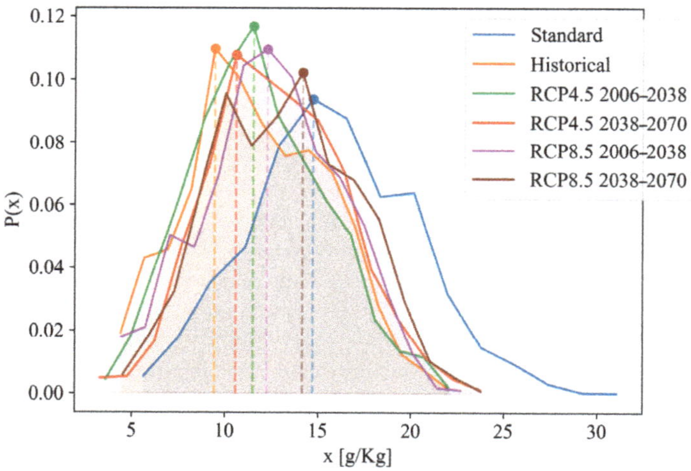

Figure A1. Probability distribution of the air absolute humidity for HH values in different datasets.

If the modes of the probability distributions (marked with a vertical dashed line in Figure A1) are compared, it is possible to notice the important trend of the absolute humidity distribution which has increasingly increasing values for the different models and over time.

References

1. Tamang, B. Global Climate Change: Challenges, Opportunities, and Multilateral Strategies for Sustainable Development. *NPRC J. Multidiscip. Res.* **2024**, *1*, 65–76. [CrossRef]
2. Afifa; Arshad, K.; Hussain, N.; Ashraf, M.H.; Saleem, M.Z. Air Pollution and Climate Change as Grand Challenges to Sustainability. *Sci. Total Environ.* **2024**, *928*, 172370. [CrossRef] [PubMed]
3. Malakar, K.D.; Kumar, M.; Anand, S.; Kuzur, G. Climate Crisis and Socio-Ecological Sustainability. In *Climate Change and Socio-Ecological Transformation*; Springer: Singapore, 2023; pp. 439–494.
4. Cornell, S.E.; Gupta, A. Is Climate Change the Most Important Challenge of Our Times? In *Contemporary Climate Change Debates*; Routledge: Oxfordshire, UK, 2019; pp. 6–20.

5. Meerab, K. Addressing the Global Environmental Crisis: A Comprehensive Review of Current Strategies and Challenges. *Int. Res. J. Arts Soc. Sci.* **2023**, *11*, 1–4. [CrossRef]
6. Francini, S.; Chirici, G.; Chiesi, L.; Costa, P.; Caldarelli, G.; Mancuso, S. Global Spatial Assessment of Potential for New Peri-Urban Forests to Combat Climate Change. *Nat. Cities* **2024**, *1*, 286–294. [CrossRef]
7. Gao, J.; O'Neill, B.C. Mapping Global Urban Land for the 21st Century with Data-Driven Simulations and Shared Socioeconomic Pathways. *Nat. Commun.* **2020**, *11*, 2302. [CrossRef] [PubMed]
8. IISD. International Day of Forests. Available online: https://sdg.iisd.org/events/international-day-of-forests-2018/ (accessed on 12 August 2025).
9. Netto, V.M.; Meirelles, J.; Ribeiro, F.L. Cities and Entropy: Assessing Urban Sustainability. In *Sustainability Assessment of Urban Systems*; Cambridge University Press: Cambridge, UK, 2020; p. 438. Available online: https://www.cambridge.org/core/books/abs/sustainability-assessment-of-urban-systems/cities-and-entropy-assessing-urban-sustainability-as-a-problem-of-coordination/83D93195E53A8B8579E974CEF9AEEB81#chapter (accessed on 12 August 2025).
10. Purvis, B.; Mao, Y.; Robinson, D. Entropy and Its Application to Urban Systems. *Entropy* **2019**, *21*, 56. [CrossRef]
11. Schneider, E.D.; Kay, J.J. Life as a Manifestation of the Second Law of Thermodynamics. *Math. Comput. Model.* **1994**, *19*, 25–48. [CrossRef]
12. Roach, T.N.F. Use and Abuse of Entropy in Biology: A Case for Caliber. *Entropy* **2020**, *22*, 1335. [CrossRef]
13. Zhao, S.; Ren, Y.; Mao, C.; Yue, A. Are Cities Healthy? A City Health Diagnose Framework from the Perspective of Living Organism. *Ecol. Indic.* **2024**, *160*, 111834. [CrossRef]
14. Modonesi, C.; Serafini, S.; Giuliani, A. Urban Systems Between the Environment, Human Health and Society: An Overview. *Systems* **2025**, *13*, 487. [CrossRef]
15. Zhang, Y.; Yang, Z.; Li, W. Analyses of Urban Ecosystem Based on Information Entropy. *Ecol. Modell.* **2006**, *197*, 1–12. [CrossRef]
16. Rondinel-Oviedo, D.R.; Keena, N. Entropy and Cities: A Bibliographic Analysis towards More Circular and Sustainable Urban Environments. *Entropy* **2023**, *25*, 532. [CrossRef]
17. Al Mashagbah, A.F. The Use of GIS, Remote Sensing and Shannon's Entropy Statistical Techniques to Analyze and Monitor the Spatial and Temporal Patterns of Urbanization and Sprawl in Zarqa City, Jordan. *J. Geogr. Inf. Syst.* **2016**, *8*, 293–300. [CrossRef]
18. Gaur, A.; Eichenbaum, M.K.; Simonovic, S.P. Analysis and Modelling of Surface Urban Heat Island in 20 Canadian Cities under Climate and Land-Cover Change. *J. Environ. Manag.* **2018**, *206*, 145–157. [CrossRef]
19. Mohajerani, A.; Bakaric, J.; Jeffrey-Bailey, T. The Urban Heat Island Effect, Its Causes, and Mitigation, with Reference to the Thermal Properties of Asphalt Concrete. *J. Environ. Manag.* **2017**, *197*, 522–538. [CrossRef] [PubMed]
20. Wu, Q.; Huang, Y.; Irga, P.; Kumar, P.; Li, W.; Wei, W.; Shon, H.K.; Lei, C.; Zhou, J.L. Synergistic Control of Urban Heat Island and Urban Pollution Island Effects Using Green Infrastructure. *J. Environ. Manag.* **2024**, *370*, 122985. [CrossRef] [PubMed]
21. Kleidon, A.; Malhi, Y.; Cox, P.M. Maximum Entropy Production in Environmental and Ecological Systems. *Philos. Trans. R. Soc. B Biol. Sci.* **2010**, *365*, 1297–1302. [CrossRef] [PubMed]
22. Aoki, I. Entropy Budgets of Deciduous Plant Leaves and a Theorem of Oscillating Entropy Production. *Bull. Math. Biol.* **1987**, *49*, 449–460. [CrossRef]
23. Chen, Y.; Wang, J.; Feng, J. Understanding the Fractal Dimensions of Urban Forms through Spatial Entropy. *Entropy* **2017**, *19*, 600. [CrossRef]
24. Free, G.; Poikane, S.; Solheim, A.L.; Bussettini, M.; Bradley, C.; Smith, J.; Caroni, R.; Bresciani, M.; Pinardi, M.; Giardino, C.; et al. Climate Change and Ecological Assessment in Europe under the WFD—Hitting Moving Targets with Shifting Baselines? *J. Environ. Manag.* **2024**, *370*, 122884. [CrossRef]
25. Repo, A.; Albrich, K.; Jantunen, A.; Aalto, J.; Lehtonen, I.; Honkaniemi, J. Contrasting Forest Management Strategies: Impacts on Biodiversity and Ecosystem Services under Changing Climate and Disturbance Regimes. *J. Environ. Manag.* **2024**, *371*, 123124. [CrossRef]
26. Pelorosso, R.; Gobattoni, F.; Leone, A. The Low-Entropy City: A Thermodynamic Approach to Reconnect Urban Systems with Nature. *Landsc. Urban Plan.* **2017**, *168*, 22–30. [CrossRef]
27. Fistola, R.; Gargiulo, C.; La Rocca, R.A. Rethinking Vulnerability in City-Systems: A Methodological Proposal to Assess "Urban Entropy". *Environ. Impact Assess. Rev.* **2020**, *85*, 106464. [CrossRef]
28. Fistola, R. Urban Entropy vs Sustainability: A New Town Planning Perspective. In Proceedings of the 7th International Conference on Urban Regeneration and Sustainability, Ancona, Italy, 7 May 2012; pp. 195–204.
29. Sun, X. Green City and Regional Environmental Economic Evaluation Based on Entropy Method and GIS. *Environ. Technol. Innov.* **2021**, *23*, 101667. [CrossRef]
30. Prigogine, I.; Stengers, I. La Nouvelle Alliance: Métamorphose de La Science. *Contemp. Fr. Civiliz.* **1979**, *7*, 305.

31. Pulselli, M.R.; Tiezzi, E. *Città Fuori Dal Caos*; Donzelli Editor, Series: Essays; Nature and Artifact: Rome, Italy, 2008; ISBN 9788860362582.
32. Zuo, S.; Dai, S.; Ju, J.; Meng, F.; Ren, Y.; Tian, Y.; Wang, K. The Importance of the Functional Mixed Entropy for the Explanation of Residential and Transport CO_2 Emissions in the Urban Center of China. *J. Clean. Prod.* **2022**, *380*, 134947. [CrossRef]
33. JRC. Urban Heat Islands: Managing Extreme Heat to Keep Cities Cool. Available online: https://joint-research-centre.ec.europa.eu/jrc-news-and-updates/urban-heat-islands-managing-extreme-heat-keep-cities-cool-2024-07-22_en (accessed on 22 July 2024).
34. Ghanghermeh, A.; Roshan, G.; Orosa, J.A.; Costa, Á.M. Analysis and Comparison of Spatial–Temporal Entropy Variability of Tehran City Microclimate Based on Climate Change Scenarios. *Entropy* **2018**, *21*, 13. [CrossRef]
35. Cabral, P.; Augusto, G.; Tewolde, M.; Araya, Y. Entropy in Urban Systems. *Entropy* **2013**, *15*, 5223–5236. [CrossRef]
36. Rapf, M.; Kranert, M. Irreversible Entropy to Account for Environmental Impacts and Sustainability. *Procedia CIRP* **2021**, *98*, 601–606. [CrossRef]
37. Giorgi, F.; Jones, C.; Asrar, G. Addressing Climate Information Needs at the Regional Level: The CORDEX Framework. *Organ. Bull.* **2009**, *58*, 175–183.
38. Jacob, D.; Petersen, J.; Eggert, B.; Alias, A.; Christensen, O.B.; Bouwer, L.M.; Braun, A.; Colette, A.; Déqué, M.; Georgievski, G.; et al. EURO-CORDEX: New High-Resolution Climate Change Projections for European Impact Research. *Reg. Environ. Change* **2014**, *14*, 563–578. [CrossRef]
39. Rizwan, A.M.; Dennis, L.Y.C.; Liu, C. A Review on the Generation, Determination and Mitigation of Urban Heat Island. *J. Environ. Sci.* **2008**, *20*, 120–128. [CrossRef]
40. Dee, D.P.; Uppala, S.M.; Simmons, A.J.; Berrisford, P.; Poli, P.; Kobayashi, S.; Andrae, U.; Balmaseda, M.A.; Balsamo, G.; Bauer, P.; et al. The ERA-Interim Reanalysis: Configuration and Performance of the Data Assimilation System. *Q. J. R. Meteorol. Soc.* **2011**, *137*, 553–597. [CrossRef]
41. Kotlarski, S.; Keuler, K.; Christensen, O.B.; Colette, A.; Déqué, M.; Gobiet, A.; Goergen, K.; Jacob, D.; Lüthi, D.; van Meijgaard, E.; et al. Regional Climate Modeling on European Scales: A Joint Standard Evaluation of the EURO-CORDEX RCM Ensemble. *Geosci. Model Dev.* **2014**, *7*, 1297–1333. [CrossRef]
42. Boberg, F.; Christensen, J.H. Overestimation of Mediterranean Summer Temperature Projections Due to Model Deficiencies. *Nat. Clim. Change* **2012**, *2*, 433–436. [CrossRef]
43. Colmet-Daage, A.; Sanchez-Gomez, E.; Ricci, S.; Llovel, C.; Borrell Estupina, V.; Quintana-Seguí, P.; Llasat, M.C.; Servat, E. Evaluation of Uncertainties in Mean and Extreme Precipitation under Climate Change for Northwestern Mediterranean Watersheds from High-Resolution Med and Euro-CORDEX Ensembles. *Hydrol. Earth Syst. Sci.* **2018**, *22*, 673–687. [CrossRef]
44. Derdour, S.; Ghenim, A.N.; Megnounif, A.; Tangang, F.; Chung, J.X.; Ayoub, A.B. Bias Correction and Evaluation of Precipitation Data from the CORDEX Regional Climate Model for Monitoring Climate Change in the Wadi Chemora Basin (Northeastern Algeria). *Atmosphere* **2022**, *13*, 1876. [CrossRef]
45. Prein, A.F.; Langhans, W.; Fosser, G.; Ferrone, A.; Ban, N.; Goergen, K.; Keller, M.; Tölle, M.; Gutjahr, O.; Feser, F.; et al. A Review on Regional Convection-permitting Climate Modeling: Demonstrations, Prospects, and Challenges. *Rev. Geophys.* **2015**, *53*, 323–361. [CrossRef] [PubMed]
46. IPCC. Emission Scenarios. Available online: https://www.ipcc.ch/report/emissions-scenarios (accessed on 7 March 2025).
47. Balocco, C.; Grazzini, G. Thermodynamic Parameters for Energy Sustainability of Urban Areas. *Sol. Energy* **2000**, *69*, 351–356. [CrossRef]
48. Gates, D. *Physical Processes in Ecosystems*; University of Washington Center for Quantitative Science: Seattle, WA, USA, 2020.
49. Holman, J.P. *Heat Transfer*; McGraw-Hill: New York, NY, USA, 2010; ISBN 0–07–352936–2.
50. Gazzetta Ufficiale. Norme per L'attuazione Del Piano Energetico Nazionale in Materia Di Uso Razionale Dell'energia, Di Risparmio Energetico e Di Sviluppo Delle Fonti Rinnovabili Di Energia. 1991. Available online: https://www.gazzettaufficiale.it/gunewsletter/dettaglio.jsp?service=1&datagu=2010-04-21&task=dettaglio&numgu=92&redaz=010G0078&tmstp=1273049012988 (accessed on 12 August 2025).
51. Gazzetta Ufficiale. Regolamento Recante Norme per la Progettazione, L'installazione, l'esercizio e La Manutenzione Degli Impianti Termici Degli Edifici Ai Fini Del Contenimento Dei Consumi Di Energia 1993. Available online: https://www.gazzettaufficiale.it/eli/id/1993/10/14/093G0451/sg (accessed on 12 August 2025).
52. *UNI UNI/TS 11300-1:2014*; Energy Performance of Buildings-Part 1: Evaluation of Energy Need for Space Heating and Cooling. British Standards Institution (BSI): London, UK, 2014.
53. WMO. *Guidance to Measuring, Modelling and Monitoring the Canopy Layer Urban Heat Island*; WMO: Geneva, Switzerland, 2023; ISBN 9789263112.

54. Pena Acosta, M.; Vahdatikhaki, F.; Santos, J.; Dorée, A.G. A Comparative Analysis of Surface and Canopy Layer Urban Heat Island at the Micro Level Using a Data-Driven Approach. *Sustain. Cities Soc.* **2023**, *99*, 104944. [CrossRef]
55. Raffa, M.; Adinolfi, M.; Reder, A.; Marras, G.F.; Mancini, M.; Scipione, G.; Santini, M.; Mercogliano, P. Very High Resolution Projections over Italy under Different CMIP5 IPCC Scenarios. *Sci. Data* **2023**, *10*, 238. [CrossRef] [PubMed]
56. CMCC DDS. Climate Projections RCP4.5 and RCP8.5 Downscaled @2.2 Km over Italy. Available online: https://doi.org/10.25424/CMCC-J90A-5P12 (accessed on 7 August 2025).
57. Douss, R.; Farah, I.R. Extraction of Individual Trees Based on Canopy Height Model to Monitor the State of the Forest. *Trees For. People* **2022**, *8*, 100257. [CrossRef]
58. Mohan, M.; Leite, R.V.; Broadbent, E.N.; Wan Mohd Jaafar, W.S.; Srinivasan, S.; Bajaj, S.; Dalla Corte, A.P.; do Amaral, C.H.; Gopan, G.; Saad, S.N.M.; et al. Individual Tree Detection Using UAV-Lidar and UAV-SfM Data: A Tutorial for Beginners. *Open Geosci.* **2021**, *13*, 1028–1039. [CrossRef]
59. Popescu, S.C.; Wynne, R.H. Seeing the Trees in the Forest. *Photogramm. Eng. Remote Sens.* **2004**, *70*, 589–604. [CrossRef]
60. Mohan, M.; Mendonça, B.A.F.; de Silva, C.A.; Klauberg, C.; de Saboya Ribeiro, A.S.; Araújo, E.J.G.; de Monte, M.A.; Cardil, A. Optimizing Individual Tree Detection Accuracy and Measuring Forest Uniformity in Coconut (*Cocos nucifera* L.) Plantations Using Airborne Laser Scanning. *Ecol. Modell.* **2019**, *409*, 108736. [CrossRef]
61. Yu, K.; Hao, Z.; Post, C.J.; Mikhailova, E.A.; Lin, L.; Zhao, G.; Tian, S.; Liu, J. Comparison of Classical Methods and Mask R-CNN for Automatic Tree Detection and Mapping Using UAV Imagery. *Remote Sens.* **2022**, *14*, 295. [CrossRef]
62. Breiman, L. Random Forests. *Mach. Learn.* **2001**, *45*, 5–32. [CrossRef]
63. Francini, S.; Schelhaas, M.-J.; Vangi, E.; Lerink, B.; Nabuurs, G.-J.; McRoberts, R.E.; Chirici, G. Forest Species Mapping and Area Proportion Estimation Combining Sentinel-2 Harmonic Predictors and National Forest Inventory Data. *Int. J. Appl. Earth Obs. Geoinf.* **2024**, *131*, 103935. [CrossRef]
64. Francini, S.; Cocozza, C.; Hölttä, T.; Lintunen, A.; Paljakka, T.; Chirici, G.; Traversi, M.L.; Giovannelli, A. A Temporal Segmentation Approach for Dendrometers Signal-to-Noise Discrimination. *Comput. Electron. Agric.* **2023**, *210*, 107925. [CrossRef]
65. Hermosilla, T.; Bastyr, A.; Coops, N.C.; White, J.C.; Wulder, M.A. Mapping the Presence and Distribution of Tree Species in Canada's Forested Ecosystems. *Remote Sens. Environ.* **2022**, *282*, 113276. [CrossRef]
66. Lindquist, M.A. The Statistical Analysis of FMRI Data. *Stat. Sci.* **2008**, *23*, 439–464. [CrossRef]
67. Fuchs, M.; Tanner, C.B. Infrared Thermometry of Vegetation 1. *Agron. J.* **1966**, *58*, 597–601. [CrossRef]
68. Mashabatu, M.; Ntshidi, Z.; Dzikiti, S.; Jovanovic, N.; Dube, T.; Taylor, N.J. Deriving Crop Coefficients for Evergreen and Deciduous Fruit Orchards in South Africa Using the Fraction of Vegetation Cover and Tree Height Data. *Agric. Water Manag.* **2023**, *286*, 108389. [CrossRef]
69. Tiralla, N.; Panferov, O.; Kreilein, H.; Olchev, A.; Ali, A.A.; Knohl, A. Quantification of Leaf Emissivities of Forest Species: Effects on Modelled Energy and Matter Fluxes in Forest Ecosystems. *Geogr. Environ. Sustain.* **2019**, *12*, 245–258. [CrossRef]
70. Comune di Firenze Piano Strutturale e Piano Operativo. Available online: https://www.comune.fi.it/dalle-redazioni/piano-operativo-e-piano-strutturale-2023 (accessed on 7 March 2025).
71. Istat. Basi Territoriali: Anni 1991, 2001, 2011 E 2021. Note Legali: Responsabilità E Licenza. 2024. Available online: https://www.istat.it/wp-content/uploads/2024/07/Descrizione-dei-dati-Basi-territoriali.pdf (accessed on 7 March 2025).
72. Grazzini, G.; Gonnelli, A. Il Piano Energetico Ambientale Comunale di Firenze. In Proceedings of the II Congresso Nazionale AIGE, Pisa, Italy, 4–5 September 2008.
73. Regione Toscana Geoscopio—Cartoteca. Available online: https://www502.regione.toscana.it/geoscopio/cartoteca.html (accessed on 7 March 2025).
74. Martinez, A.; de la, I.; Labib, S.M. Demystifying Normalized Difference Vegetation Index (NDVI) for Greenness Exposure Assessments and Policy Interventions in Urban Greening. *Environ. Res.* **2023**, *220*, 115155. [CrossRef]
75. Hawryło, P.; Francini, S.; Chirici, G.; Giannetti, F.; Parkitna, K.; Krok, G.; Mitelsztedt, K.; Lisańczuk, M.; Stereńczak, K.; Ciesielski, M.; et al. The Use of Remotely Sensed Data and Polish NFI Plots for Prediction of Growing Stock Volume Using Different Predictive Methods. *Remote Sens.* **2020**, *12*, 3331. [CrossRef]
76. Chicco, D.; Jurman, G. The Advantages of the Matthews Correlation Coefficient (MCC) over F1 Score and Accuracy in Binary Classification Evaluation. *BMC Genomics* **2020**, *21*, 6. [CrossRef]
77. Matthews, B.W. Comparison of the Predicted and Observed Secondary Structure of T4 Phage Lysozyme. *BBA—Protein Struct.* **1975**, *405*, 442–451. [CrossRef]
78. Rubio, E. Emissivity Measurements of Several Soils and Vegetation Types in the 8–14, Mm Wave Band: Analysis of Two Field Methods. *Remote Sens. Environ.* **1997**, *59*, 490–521. [CrossRef]
79. Turner, N.C.; Heichel, G.H. Stomatal Development and Seasonal Changes in Diffusive Resistance of Primary and Regrowth Foliage of Red Oak (*Quercus rubra* L.) and Red Maple (*Acer rubrum* L.). *New Phytol.* **1977**, *78*, 71–81. [CrossRef]

80. WMO. *Comité Météorologique International, Procès-Verbaux de La Session de Paris*; Imprimerie La Concorde: El Baten, Tunisia, 1–12 July 1946.
81. Du, H.; Cai, W.; Xu, Y.; Wang, Z.; Wang, Y.; Cai, Y. Quantifying the Cool Island Effects of Urban Green Spaces Using Remote Sensing Data. *Urban For. Urban Green.* **2017**, *27*, 24–31. [CrossRef]
82. Arellano, B.; Roca, J. Assessing Urban Greenery Using Remote Sensing. In Proceedings of the SPIE—The International Society for Optical Engineering, Conference: SPIE Optical Engineering + Applications, San Diego, CA, USA, 21–26 August 2022. Proceedings Volume 12232, Earth Observing Systems XXVII.
83. Sharmin, M.; Tjoelker, M.G.; Pfautsch, S.; Esperon-Rodriguez, M.; Rymer, P.D.; Power, S.A. Tree Crown Traits and Planting Context Contribute to Reducing Urban Heat. *Urban For. Urban Green.* **2023**, *83*, 127913. [CrossRef]
84. Speak, A.; Montagnani, L.; Wellstein, C.; Zerbe, S. The Influence of Tree Traits on Urban Ground Surface Shade Cooling. *Landsc. Urban Plan.* **2020**, *197*, 103748. [CrossRef]
85. He, T.; Hu, Y.; Guo, A.; Chen, Y.; Yang, J.; Li, M.; Zhang, M. Quantifying the Impact of Urban Trees on Land Surface Temperature in Global Cities. *ISPRS J. Photogramm. Remote Sens.* **2024**, *210*, 69–79. [CrossRef]
86. Jin, J.; Chen, W.Y.; Jia, B.; Wang, C. Cooling Effect of Urban Greenery: A Bibliometric Analysis. *Urban For. Urban Green.* **2024**, *99*, 128453. [CrossRef]
87. Morakinyo, T.E.; Dahanayake, K.W.D.K.C.; Ng, E.; Chow, C.L. Temperature and Cooling Demand Reduction by Green-Roof Types in Different Climates and Urban Densities: A Co-Simulation Parametric Study. *Energy Build.* **2017**, *145*, 226–237. [CrossRef]
88. Meili, N.; Zheng, X.; Takane, Y.; Nakajima, K.; Yamaguchi, K.; Chi, D.; Zhu, Y.; Wang, J.; Qiu, Y.; Paschalis, A.; et al. Modeling the Effect of Trees on Energy Demand for Indoor Cooling and Dehumidification Across Cities and Climates. *J. Adv. Model. Earth Syst.* **2025**, *17*, e2024MS004590. [CrossRef]
89. Mohammed, A.; Khan, A.; Khan, H.S.; Santamouris, M. Cooling Energy Benefits of Increased Green Infrastructure in Subtropical Urban Building Environments. *Buildings* **2023**, *13*, 2257. [CrossRef]
90. Lloyd, G.; Ossola, A.; Burley, H.; Evans, K. Climate Change Threatens Carbon Storage in Europe's Urban Trees. *Urban For. Urban Green.* **2024**, *101*, 128532. [CrossRef]
91. Croeser, T.; Garrard, G.E.; Visintin, C.; Kirk, H.; Ossola, A.; Furlong, C.; Clements, R.; Butt, A.; Taylor, E.; Bekessy, S.A. Finding Space for Nature in Cities: The Considerable Potential of Redundant Car Parking. *NPJ Urban Sustain.* **2022**, *2*, 27. [CrossRef]
92. Pattnaik, N.; Honold, M.; Franceschi, E.; Moser-Reischl, A.; Rötzer, T.; Pretzsch, H.; Pauleit, S.; Rahman, M.A. Growth and Cooling Potential of Urban Trees across Different Levels of Imperviousness. *J. Environ. Manag.* **2024**, *361*, 121242. [CrossRef] [PubMed]
93. Sjöman, H.; Watkins, H.; Kelly, L.J.; Hirons, A.; Kainulainen, K.; Martin, K.W.E.; Antonelli, A. Resilient Trees for Urban Environments: The Importance of Intraspecific Variation. *Plants People Planet* **2024**, *6*, 1180–1189. [CrossRef]

atmosphere

MDPI

Article

Estimating Urban Linear Heat (UHI_{ULI}) Effect Along Road Typologies Using Spatial Analysis and GAM Approach

Elahe Mirabi [1,*], Michael Chang [1], Georgy Sofronov [2] and Peter Davies [1]

[1] School of Natural Sciences, Macquarie University, Sydney, NSW 2109, Australia; michael.chang@mq.edu.au (M.C.); peter.davies@mq.edu.au (P.D.)
[2] School of Mathematical and Physical Sciences, Macquarie University, Sydney, NSW 2109, Australia; georgy.sofronov@mq.edu.au
* Correspondence: elahe.mirabi@hdr.mq.edu.au

Abstract: The urban heat island (UHI) effect significantly impacts urban environments, particularly along roads, a phenomenon known as urban linear heat (UHI_{ULI}). Numerous factors contribute to roads influencing the UHI_{ULI}; however, effective mitigation strategies remain a challenge. This study examines the relationship between canopy cover percentage, normalized difference vegetation index, land use types, and three road typologies (local, regional, and state) with land surface temperature. This study is based on data from the city of Adelaide, Australia, using spatial analysis, and statistical modelling. The results reveal strong negative correlations between land surface temperature and both canopy cover percentage and normalized difference vegetation index. Additionally, land surface temperature tends to increase with road width. Among land use types, land surface temperature varies from highest to lowest in the order of parkland, industrial, residential, educational, medical, and commercial areas. Notably, the combined influence of the road typology and land use produces varying effects on land surface temperature. Canopy cover percentage and normalized difference vegetation index consistently serve as dominant cooling factors. The results highlight a complex interplay between built and natural environments, emphasizing the need for multi-factor analyses and a framework based on the local climate and the type of roads (local, regional, and state) to effectively evaluate UHI_{ULI} mitigation approaches.

Keywords: urban linear heat (UHI_{ULI}); road typology; spatial analysis; generalized additive model (GAM)

1. Introduction

The urban heat island (UHI) effect is a well-documented phenomenon impacting cities [1] and its effects can be exacerbated along urban linear infrastructure (ULI). Urban linear heat (UHI_{ULI}) [2] is the term used to describe the UHI effect along transportation corridors and pipeline networks. Collectively, the UHI impacts across boundary, canyon, surface, and subsurface layers [3] and multiple studies have reported increasing tree canopy can provide a cooling benefit via shading and evapotranspiration [4–6]. ULI comprises a significant proportion of the land use of cities, with roads alone occupying up to one-third of the urban area [7]. The area that roads occupy is increasing as cities expand and densify which is also influenced by rising car ownership and use [8,9]. The corollary of the increase in road area is a decrease in urban green spaces, where roads and driveways expand into the vegetated spaces, such as nature strips and gardens [10–12]. These dark and impervious materials used for road construction prevent water infiltration, increase the absorption of solar radiation and trap heat [13–15]. This creates a positive heat feedback loop that magnifies the energy budget of cities [16], driving the UHI_{ULI} [17].

A variety of factors determine how UHI_{ULI} forms along roads, particularly the width of road (road typology), materials and colours [10–12]. Wider roads receive more solar radiation and thus gaining more heat than narrower (often local) roads. Wide roads (such as regional and state highways) typically accommodate more cars and thus are associated with higher anthropogenic heat emissions from vehicles [18–24]. Mitigating heat along roadways has proven to be a challenging task [25], with various approaches turning to applying high-albedo and permeable materials, greening adjacent land uses and seeking different designs (morphology-related strategies) to lessen heat and radiation gains [26–29].

The most recommended approach to mitigate UHI_{ULI} is to integrate tree canopy cover along roads [30–32]. Canopy trees have been reported to provide a cooling benefit of between 1 °C and 30 °C [5,6]. The effectiveness of canopy tree cooling is influenced by a number of factors including the geometry of crown and leaves, type of tree, density of leaves, connectivity or clustering of tree, and watering schedule [4,31–43]. Effectiveness of cooling is also influenced by the width of the road. This creates a physical limitation of the canopy trees in offering shade across the width of the road surface through reducing the sky view factor (SVF). A wider road has an increased SVF, which means more of the sky is visible from a specific point at ground level and potentially allowing for greater solar radiation to reach the road surface [4,33–35]. In addition, the density of leaves and the percentage of canopy cover of a given area can also reduce wind flow, ventilation, and humidity levels [32,36–38,42,43]. The literature identifies various factors that may influence the UHI as a result of road networks, as listed in Table 1 [44].

The canopy cover percentage (CCP) is a significant variable influencing UHI_{ULI} mitigation along roads [45–48]. CCP represents the proportion of ground area covered by the vertical projection of tree crowns onto the ground, expressed as a percentage. This has led to some agencies introducing policies to achieve a minimum CCP along roads (trees on the sides with their crowns extending over the lanes) [49]. However, the relationship between the cooling effect of CCP and road width is complicated and often overlooked. Road width can be conceptualized by the number of trafficable lanes [50]. From a UHI_{ULI} mitigation perspective, roads can be divided into three categories: local, regional, and state roads [44]. This roughly equates to two lanes for local roads, four lanes for regional roads, and six lanes or more for state roads [44]. The cooling effects of canopy trees linked to the CCP across this three road categorization forms the basis for the inquiry of this paper. Within this framework, three research questions are examined:

1. How does the UHI_{ULI} change with respect to the road hierarchy in relation to variations in CCP?
2. Is there any optimum value for CCP to obtain the most efficient cooling effect with respect to the road hierarchy?
3. To what extent do other factors beyond canopy cover (Table 1: NDVI, land use type, geographic coordinates and elevation) contribute to UHI_{ULI} mitigation (LST) relative to the road hierarchy?

Table 1. Contributing factors to UHI_{ULI} in this study.

Variable	Definition	Impact	References
LST	Land surface temperature	Surface energy and radiation balance	[51]
CCP	Canopy cover percentage	Terrain roughness, shading, and evapotranspiration	[44,51,52]
NDVI	Normalized difference vegetation index	Land cover properties	[53,54]
Road Typology	Local, regional, and state roads based on width	SVF, AR, impervious surface, albedo, anthropogenic heat (engine combustion)	[44]
Land Type	Land use and land cover	Anthropogenic heat (cooling/heating, industry process), terrain roughness	[51,55]
Geographic information	Latitude, longitude and elevation	Mean annual heat flux density	[56]

The city of Adelaide in Australia was chosen as the study site due to its comprehensive data availability relevant to urban heat studies. Adelaide's diverse urban landscape, coupled with its susceptibility to heat waves, makes it a valuable location to analyze urban heat island (UHI) effects and spatial trends of urban hot spots, as seen in the project of "Urban Heat and Tree Mapping of Adelaide Metropolitan Area" [57]. By focusing on Adelaide, this study provides critical insights into urban heat dynamics and contributes to the broader discourse on climate resilience in metropolitan areas. More background information about Adelaide is given in Section 2 below.

2. Materials and Methods

2.1. Study Site

The site for this study is Adelaide, the capital city of South Australia, situated at an average elevation of 50 m, latitude of 34.95° S, and longitude of 138.52° E on the southern coast of Australia [58] (Figure 1). This site was chosen as it contained a high-resolution dataset 'Urban Heat and Tree Mapping of Adelaide' [59] to examine the relationships between urban heat and the other variables (see Table 1). This marks the first time that tree canopy and urban heat data have been captured for the entire metropolitan Adelaide area in a single, regionally consistent dataset, offering a comprehensive view of the urban environment. The Köppen–Geiger climate classification system categorizes Adelaide as having a temperate climate with dry and hot summers (Csa) [60]. This is one of the most widely used climate classification systems, based on the principle that climate is a primary driver of vegetation coverage. It helps contextualize the region's climatic conditions by considering temperature, aridity, and precipitation patterns, making it a valuable tool for understanding climate–vegetation relationships and assessing climate change impacts. Csa classification is defined as follows:

- C: Air temperature of the hottest month is 10 °C and above, and air temperature of the coldest month is between 0 and 18 °C;
- s: The precipitation of the driest month in summer is less than 40 mm and also less than one-third of the precipitation of the wettest month in winter; and
- a: Air temperature of the hottest month in summer is higher than 22°.

Extreme heat days (above 35 °C) are increasing over and above the long-term average. Notably, between 1980 and 2010, the average temperature was 4.3 °C higher when compared to the period between 1950 and 1980. This warning trend has continued with the period from 2011 to 2024 being higher again by 1.85 °C compared to the 1980–2010 period. Adelaide reached its highest temperature record of 46.6 °C, which occurred in January 2019 [61], and since then, as shown in Figure 2, it has experienced the dry summer, with higher temperatures during the days and nights, very close to the average [62]. Extreme heat (air temperature (Ta) higher than 35 °C) occurring on 17.5 days between 1961 and 1990 increased to 25.1 during a 9-year period from 2000 to 2009 and is also predicted to keep rising to 51 days by 2090 [63,64]. Figure 2 summarizes the annual temperature data for Adelaide (based on the weather observations drawn from Adelaide Airport station ID: 023034 [63]) from 1955 to 2024. Extreme heat occurred in Adelaide during the summers of 2022 and 2023, which indicates the area needs to mitigate extreme temperatures. For this study, a 20 km^2 area of Adelaide city was chosen as a representative study site that contains three road typologies and diverse land use types (see Table 2).

Figure 1. location of the study area in Adelaide, city, Australia.

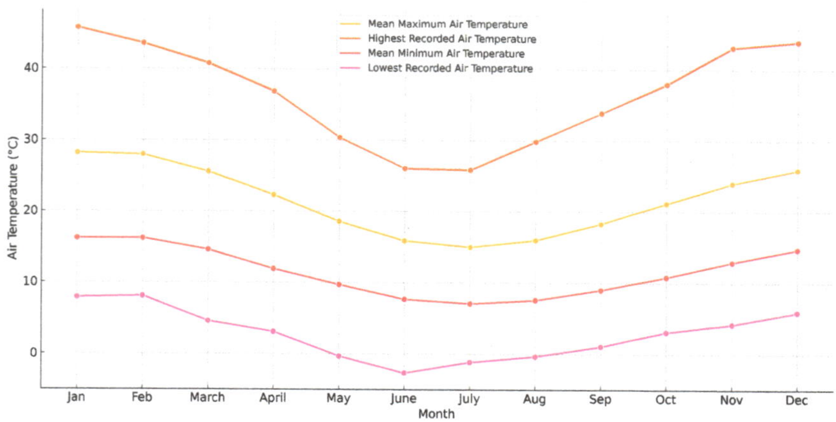

Figure 2. Adelaide average annual thermal statistics between 1955 and 2024 (based on [63]).

While Adelaide's climate is classified as temperate with hot, dry summers (Csa) under the Köppen–Geiger system [60], this broad classification does not fully capture the city's complex intra-urban thermal behaviour. Due to its geographic position in the lower mid-latitudes, Adelaide experiences extreme heatwaves driven by synoptic patterns originating in the mid-latitudes and influenced by tropical interactions [65–68]. These heatwaves are frequently associated with synoptic blocking systems, which prolong high-temperature episodes [66]. Previous studies have detected intra-urban heat island (IUHI) effects within Adelaide [69,70], although the city's topography and the lack of a suitable rural station have made traditional UHI assessments difficult. Rogers et al. [71] showed that Adelaide's UHI intensifies under heatwave conditions—particularly during the night—indicating that urban morphology plays a critical role in local thermal amplification. Their findings suggest that during heatwaves, nocturnal UHIs are stronger than usual, while daytime UHIs may moderate slightly due to altered energy balance. These observations highlight the inadequacy of relying solely on general climate classifications and underscore the importance of mapping and analyzing the spatial distribution of heat within urban areas. The thermal heterogeneity observed across Adelaide reinforces the relevance of local-to-urban-scale studies that integrate land surface temperature, land use, and greening variables [72].

Table 2. Summary of land use–land cover in the study area.

Land Use–Land Cover Type	Area (km^2)	Percentage (%)
Constructed buildings	7.39	35.74
Canopy cover	5.61	27.13
Local roads (Type 1)	3.89	18.81
Regional roads (Type 2)	0.27	1.31
State roads (Type 3)	0.44	2.13
Residential (Type 1)	17.34	83.85
Commercial (Type 2)	1.24	5.99
Educational (Type 3)	0.50	2.42
Medical (Type 4)	0.39	1.89
Industrial (Type 5)	0.01	0.05
Parkland (Type 6)	0.98	4.74
Others	0.14	0.68

2.2. *Methodology*

2.2.1. Data Acquisition

Various datasets were used to classify road typologies, land types, and to quantify LST, CCP, and NDVI.

In South Australian Government Data Directory (Data.SA), a Government web portal, the spatial layers relevant to CCP, LST, and NDVI were obtained [73]. The data produced in this portal include thermal imagery, tree canopy boundaries (height > 3 m), green space (growing vegetation < 2 m, emerging trees and trees between 2 and 3 m in height, shrubs, and other mid-story), vegetation greenness using NDVI, permeable and impermeable surfaces, building footprints, and green space extents. The height data was determined via Light Detection and Ranging (LiDAR) with an average point density of 8 points per m^2, a remote sensing approach scanning the surface of the Earth. The Heat Map was generated from data collected at 3000 m altitude under a clear sky from a purpose-built aircraft fitted with an aerial thermal sensor. Thermal imagery has been captured with a spatial resolution of 2 m $\times$ 2 m [74]. Each pixel shows the average LST as measured. The spatial resolution for the Canopy Height Model (CHM) is 0.5 m $\times$ 0.5 m. In pursuing our objectives, we used the Heat Map captured during the day on 10th March 2018 between 11 a.m. and 3 p.m., and the layer of tree canopy > 3 m height. This is a two-dimensional spatial layer derived from CHM representing the tree canopy above 3 m from point cloud captured by LiDAR.

Australian Bureau of Statistics (ABS) [75] includes mesh blocks (the smallest geographic areas containing between 30 to 60 dwellings) that were used to obtain land uses. These were categorized as residential, commercial, industrial, parkland, education, hospital/medical, transport, water, and other categories where they could not be easily placed in one of the above classifications [76]. The ABS digital boundaries are produced in two formats: OGC GeoPackage or ESRI shapefile, and in Geocentric Datum of Australia 2020 (GDA2020) and Geocentric Datum of Australia 1994 (GDA94). The shapefile of mesh blocks, which were created in the year 2021 and GDA2020 (MB_2021_AUST_GDA2020), was used for the land use–land cover (LULC) data required for this research [76].

Open Street Maps (OSM) [77] is a collaborative, open-source, up-to-date, and accurate dataset creating and providing free maps of the world including roads, rivers, railways, and points of interest. The road data (polylines) for Adelaide from OSM (AdelaideOSM_2023) was extracted comprising road types, length, name, and ID. This data was used to classify roads into simplified road typology for analysis to fulfill the aims of this research.

2.2.2. Data Preparation

The input data inclusive of Mesh Blocks, OSM Adelaide Roads, and Tree Canopy are in vector format, while NDVI, Elevation, and Thermal Day Imagery are in rasters. These layers were used as essential inputs for processing through Geographic Information Systems (GIS) in the map projection of GDA2020_MGA_ZONE54 (Geocentric Datum of Australia 2020 for the research zone). Different sources and layers were used to prepare the data (Figure 3):

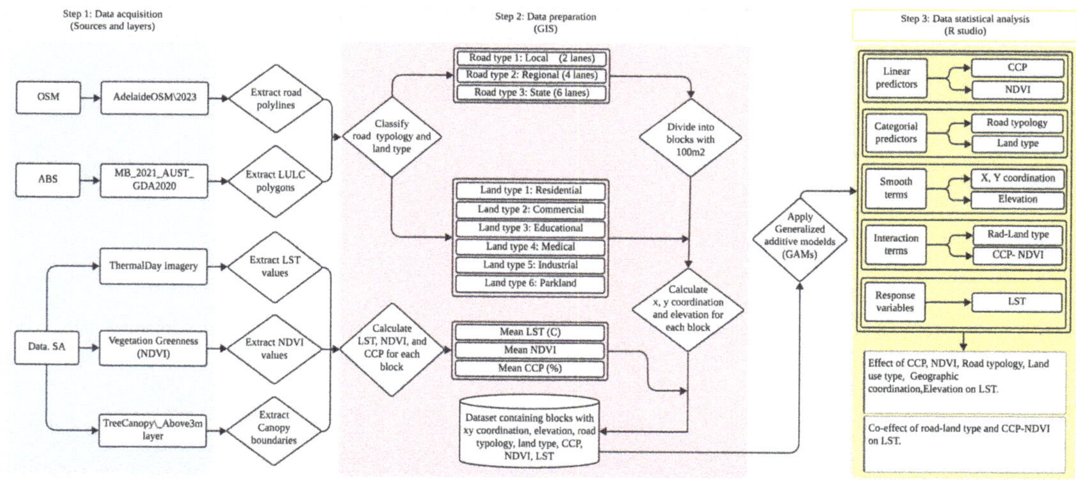

Figure 3. Data flow diagram of steps applied in this study.

1. OSM Adelaide Roads were categorized into three classifications based on the three-tier hierarchy introduced in [44] as Figure 4a shows. The term "typology" means the number of traffic lanes (width of roads). Type 1, 2, and 3 indicate local (2 lanes), regional (4 lanes), and state roads (6 lanes and over).
2. To analyze the impact of various factors on UHI_{ULI} mitigation, each road typology was segmented into uniform blocks at 100 m^2 (in terms of area not dimensions (Figure 5)).

This segmentation ensured that contributing factors could be quantified with minimal complexity and higher accuracy, leading to consistent results. The choice of 100 m^2 blocks was driven by the need to isolate the cooling effects of identified factors while minimizing other influences such as orientation. To validate this approach, block sizes of 100, 500, and 1000 m^2 were examined. However, for larger blocks, multiple land uses were often encompassed within a single unit, which complicated the assessment of land type effects. As a result, the 100 m^2 block size was adopted as the optimal unit. In total, 4433 blocks were created and each block was assigned an ID, X and Y coordinates, elevation, and road typology.

3. From the data in the Mesh Block layer, 7 types of LULC were identified, including land type 1: residential, type 2: commercial, type 3: education, type 4: medical, type 5: industrial, and type 6: parkland. Each block of road typology was allocated a specific land type (Figure 4b). Our comparison is based on land use types adjacent to roads not as the built-up areas. Building areas (rooftops) were removed from the road areas considered in this analysis (explanation for building surface temperature was added).
4. TreeCanopy_Above3m is a vector layer (Figure 4c) that was used for measuring CCP. With the area of each block being 100 m^2, CCP was calculated for each block based on road typology and integrated with land type.
5. NDVI values (mean NDVI) were calculated from the relevant layer (spatial resolution of 3 m × 3 m) in the Adelaide Heat Map dataset for each block with the certain CCP and land type (see Figure 4d). NDVI is included alongside CCP to assess not only the extent and shading of tree canopy cover but also its density, which influence evapotranspiration. While CCP measures the quantity of canopy cover, NDVI reflects canopy quality, distinguishing healthy or stressed, dense or sparse ones. Since areas with similar CCP can have varying cooling effects depending on tree health and greenness, incorporating NDVI provides a more accurate understanding of how canopy cover impacts LST [53,78,79].
6. Mean LST (Figure 4e) for each block was extracted from ThermalDay imagery (spatial resolution of 2 m × 2 m) in GIS. This resulted in the average surface temperature of each block. The data was compiled into a database for further analysis.

2.2.3. Data Analysis

To determine the correlations between the contributing factors in this research, generalized additive models (GAMs) were utilized. With this statistical approach, we could analyze the key variables of LST with respect to various road typologies and adjacent land use types. GAMs extend generalized linear models (GLMs) by replacing linear relationships between predictors and the response variable with flexible, smooth functions of the covariates [80]. This flexibility enables GAMs to model nonlinear relationships between the predictors and the response variable while preserving the straightforward and understandable nature of additive models [81]. In GAMs, each predictor can be modeled using a smooth function, often represented using splines, like cubic splines, or other smoothing techniques. The overall model remains additive, meaning that the contribution of each predictor to the response variable is added together, which simplifies interpretation compared to more complex models like random forests or neural networks [82]. GAMs are particularly useful when nonlinear patterns are expected in the data, but overfitting needs to be controlled by carefully managing the smoothness of the functions [83].

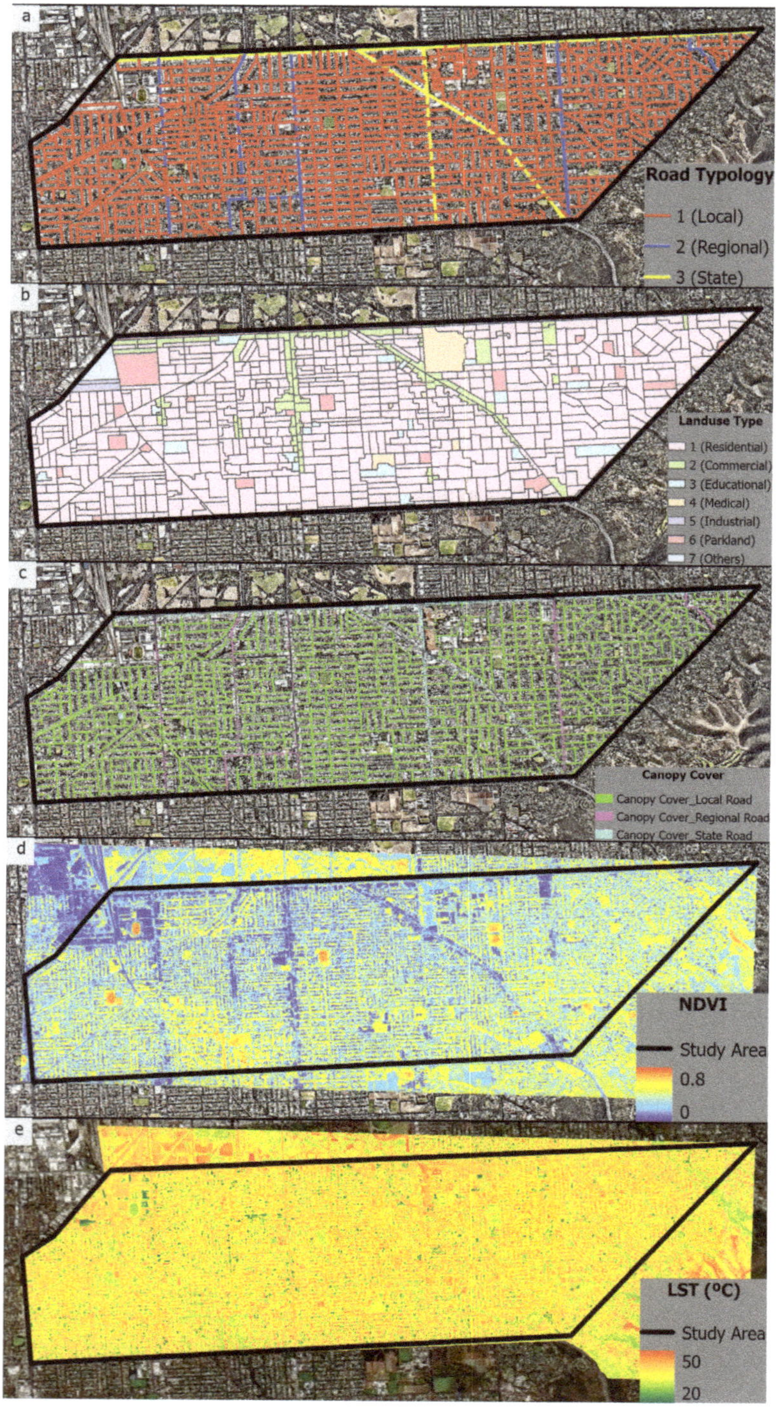

Figure 4. Details of the research zone in Adelaide, Australia; (**a**) road typology, (**b**) land use type, (**c**) canopy cover, (**d**) NDVI, and (**e**) LST.

Figure 5. Graphical illustration of a 100 m^2 blocks for three road typologies.

As different factors contribute to UHI_{ULI}, the GAM model was applied in this study to predict the LST based on several predictor variables (so-called independent variables in some regression models). In the prediction model, both linear predictors (CCP and NDVI) and categorical predictors (road typology and land types) were analyzed under two conditions. The first was without interaction, where each predictor was examined independently to assess its individual effect on LST. This included evaluating how CCP, NDVI, road typology, and land type contribute separately to variations in LST. The second was with interaction, where combined effects of the predictor pairs were explored to understand potential interactions. This involved analyzing the joint influence of CCP and NDVI, and joint influence of road typology and land use type to determine whether their co-effect amplifies or mitigates LST differently or not. However, to determine if there was an association between road typology and land type, which are categorical factors, Pearson's Chi-squared test was applied and proved that the *p*-value is less than 2.2×10^{-16}. This indicates a highly significant association between road typology and land type, meaning certain land type categories are more likely to occur with specific road typology categories. To reveal any dependency between NDVI and CCP, which are linear predictors, linear regression was applied, and as Figure 6 shows, there is a positive correlation between the CCP and NDVI values, meaning as CCP increases, NDVI values tend to rise and areas with more tree cover have healthier or denser vegetation. These associations highlight the need to consider the dependency between these variables to ensure accurate and reliable model estimates.

Given the explanation above, GAM was assigned to a model with spatial coordinates (X, Y) and elevation as smooth functions, CCP and NDVI as linear predictors, road typology (local, regional, and state) and land use types (residential, commercial, educational, medical, industrial, and parkland) as categorical predictor variables, and LST as a dependent or response variable. As we should examine the co-effects of linear and categorical predictors, the model also considered the interaction between these variables (Figure 3).

The model uses a t-distribution rather than a normal distribution to be more suitable for modeling the data as it has heavier trails compared to a normal distribution. The identity link function is also used to predict the response variable (LST) directly without transformation, and it is used to understand and the results.

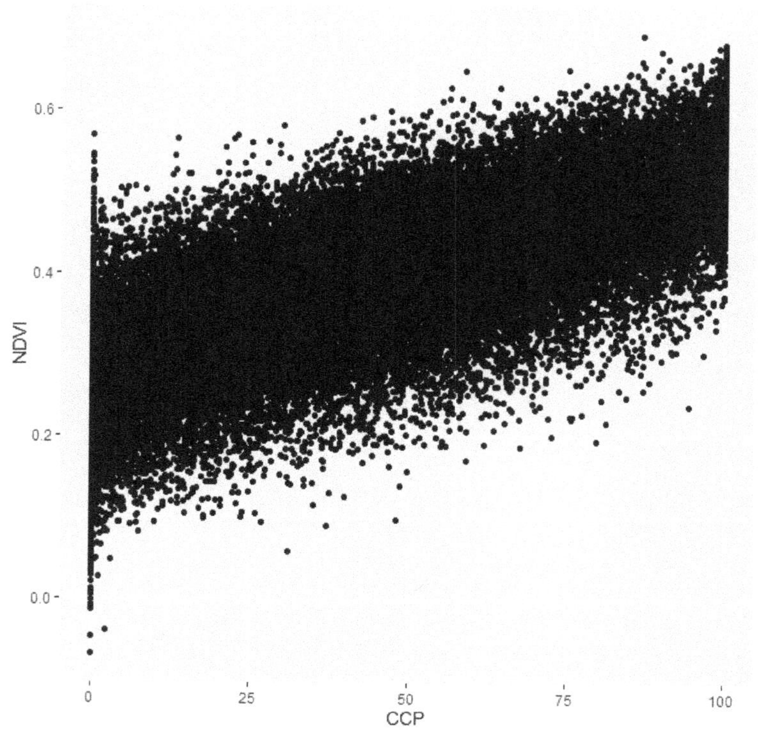

Figure 6. The scatter plot of NDVI and CCP.

```
Model:
GAM(LST ~ s(X,Y) + CCP*NDVI + factor(RoadType)*factor(LandType)
    + s(Elevation), family = scat(link = "identity"))
```

The deviance for the model explains 63.9% of deviance in the response variable (LST), which is quite high. In the model, ~ means LST is a response variable depending on other factors; CCP(*NDVI) and factor (RoadType) * factor (LandType) indicate interactions between two linear predictors and categorical factors, respectively. The s(X,Y) and s(Elevation) indicate smooth functions of spatial coordination and elevation allowing the model to capture complex and nonlinear relationships with LST. The inclusion of s(X, Y) helps account for localized spatial variations in LST that may arise from the categorical and linear factors, while s(Elevation) captures potential nonlinear temperature changes associated with altitude. By using smooth functions, the model gains flexibility, improving its ability to accurately predict LST compared to a purely linear approach. For road typology and land type, as they are categorical variables, local road (road type 1) and residential land type (land type 1) were considered as the baselines whose effect is compared with that of the other types. The temperature difference for the statistical model throughout this paper is per 100 m^2. The "family= scat" stands the scaled t-distribution which is useful for this paper's model as the response variable (LST) has outliers. By applying this, t-distribution is less affected by extreme LST values and gives more stable estimates than a normal model. The "link= identity" is a standard identity link function, assuming a direct, untransformed relationship between the continuous response variable (LST) and the linear predictor and keeping. So, using family = scat (link = "identity") helps the model provide robustness while ensuring straightforward interpretability.

Figure 7 and Table 3 present the distribution of the three continuous primary variables used in this study—LST, CCP, and NDVI—and provide a comprehensive overview of the landscape structure and thermal environment of the study area. The LST has a nearly normal distribution centred around a mean of 37.29 °C (SD = 3.21 °C), with values ranging from 21.76 °C to 49.09 °C. The median (37.36 °C), first quartile (Q1 = 34.91 °C), and third quartile (Q3 = 39.74 °C) are close to the mean, confirming the symmetric nature of the distribution. This suggests relatively consistent surface temperatures across the study area, potentially influenced by the uniformity of land use or surface materials. The CCP distribution is markedly right-skewed, with a significant proportion of observations clustered at or near 0%. This is reflected in the descriptive statistics, where the mean canopy cover is 41.18%, the median is slightly lower at 39.98%, and the first quartile is only 12.85%. Despite the maximum reaching 100%, the high standard deviation (30.54) and wide interquartile range indicate substantial spatial variability in canopy cover. The NDVI variable exhibits a moderately symmetric distribution with a peak around 0.38, closely aligned with its mean (0.37) and median (0.38). The NDVI values span from −0.07 to 0.69, with the interquartile range (Q1 = 0.29, Q3 = 0.46) capturing moderately vegetated areas. The relatively small standard deviation (0.12) suggests lower variability compared to CCP, indicating that while vegetation density varies, it is more consistent than canopy extent.

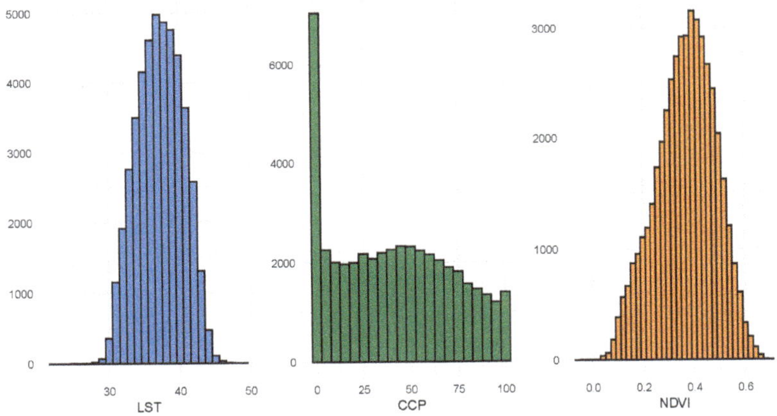

Figure 7. Histograms of LST, CCP, and NDVI (Q8).

Table 3. Main statistics of the variables.

Variable	Mean	Max	Min	Median	SD	Q1	Q3
CCP	41.18	100	0	39.98	30.54	12.85	65.54
NDVI	0.37	0.69	−0.07	0.38	0.12	0.29	0.46
LST	37.29	49.09	21.76	37.36	3.21	34.91	39.74

3. Results

In the study area within the Adelaide region, the findings indicate that average LSTs are influenced by three types of variables—linear (CCP and NDVI), categorical (road typology, and land type), and smooth terms (spatial coordinates and elevation)—though the level of significance varies among them. Overall, the interaction (co-effect) between linear variables and between categorical variables contributes to changes in LST, and consequently (UHI_{ULI}), highlighting the importance of the individual effects of the predictors and their co-effects. Table 4 provides a summary of the variables analyzed, including their descriptions and statistical significance. The baseline is road type 1 (local roads), land type 1 (residential land use), CCP = 0 (no tree canopy), NDVI = 0, and LST = 42.78 °C. The baseline serves as a reference point for interpreting the effects of the variables in the model. In other words, all other predictors or effects in the model are deviations from this baseline. As an example, the cooling effect of road type 2 (regional road) and land type 2 (commercial) combination is 0.15 °C less than the baseline per 100 m^2. In Table 4, the estimate represents the effect size of each predictor variable on LST, indicating both the direction (negative or positive) and magnitude of the relationship. The *p*-value assesses the statistical significance of the predictor's contribution to LST changes, and is categorized into three levels: highly significant ($p < 0.001$), moderately significant ($p < 0.01$), and weakly significant ($p < 0.05$). Small *p*-values suggest strong evidence of influence on LST. The standard error (Std. Error) quantifies the uncertainty in the estimate, where smaller values indicate greater reliability and precision, while larger values suggest higher uncertainty.

To assess the individual contribution of each predictor variable to LST, a set of partial models was also developed, each containing only one main predictor or interaction term. We then evaluated their performance using two statistical indicators: Adjusted R^2 and Deviance Explained (%). These metrics reflect the extent to which each variable, in isolation, accounts for the variability in LST, with Deviance Explained also indicating the contribution level of each predictor variable. Table 5 summarizes the outcomes of these assessments. The level of contribution is visually highlighted using color shading: darker green tones represent stronger contributions, while darker red tones indicate weaker contributions. Among all predictors, the interaction term CCP * NDVI showed the strongest explanatory contribution, with an Adjusted R^2 of 0.68 and 68.3% deviance explained. This highlights the substantial role of both the quantity and quality of vegetation in regulating LST. Other strong contributors include CCP alone (Adjusted R^2 = 0.65 and 65.3% deviance explained) and NDVI (Adjusted R^2 = 0.54 and 54.1% deviance explained). These results confirm that vegetation cover, especially when both CCP and NDVI are considered, is a critical factor in mitigating urban heat. In contrast, spatial coordinates s(X, Y), land type, and road type, as well as their interaction, exhibited notably weaker contributions, with deviance explained values below 10%. Elevation had a negligible effect (Adjusted R^2 = 0.007; 0.79% deviance explained), suggesting its influence on LST is minimal in this urban context.

Table 4. Summary of predictor variables and their parametric details.

Predictor Variables	Description	Estimate	Std.Error	*p*-Value
Intercept	Baseline	42.78	0.05	$<2 \times 10^{-16}$
CCP	Canopy cover percentage	−0.05	0.001	$<2 \times 10^{-16}$
NDVI	Normalized difference vegetation index	−7.22	0.15	$<2 \times 10^{-16}$
CCP-NDVI	Interaction effects	−0.03	0.002	$<2 \times 10^{-16}$
RoadType 1	Local Road	Baseline	Baseline	Baseline
RoadType 2	Regional Road	−0.62	0.04	$<2 \times 10^{-16}$
RoadType 3	State Road	−0.77	0.05	$<2 \times 10^{-16}$
LandType 1	Residential	Baseline	Baseline	Baseline
LandType 2	Commercial	−0.37	0.05	$<2 \times 10^{-16}$
LandType 3	Educational	−0.13	0.08	0.08
LandType 4	Medical	−0.15	0.07	0.04
LandType 5	Industrial	0.05	0.15	0.66
LandType 6	Parkland	0.11	0.06	0.08

Table 4. *Cont.*

Predictor Variables	Description	Estimate	Std.Error	*p*-Value
RoadType2-LandType 2	Interaction effects	0.15	0.09	0.09
RoadType3-LandType 2	Interaction effects	0.6	0.08	$<2 \times 10^{-16}$
RoadType2-LandType 3	Interaction effects	0.8	0.24	0.00074
RoadType3-LandType 3	Interaction effects	−1.03	0.27	0.00014
RoadType3-LandType 4	Interaction effects	0.51	0.15	0.00092
RoadType2-LandType 6	Interaction effects	0.34	0.17	0.04
RoadType3-LandType 6	Interaction effects	1.28	0.1	$<2 \times 10^{-16}$
X,Y	Geographic coordinates	-	-	$<2 \times 10^{-16}$
Elevation	Height above sea level	-	-	$<2 \times 10^{-16}$

Table 5. Summary of contribution levels of predictor variables.

Model Numbers	Predictor Variables	Adjusted R^2	Deviance Explained	Contribution Level
1	CCP * NDVI	0.68	68.3%	Strong
2	CCP	0.65	65.3%	Strong
3	NDVI	0.54	54.1%	Strong
4	LandType * RoadType	0.09	9.9%	week
5	s(X, Y)	0.09	9.37%	Weak
6	RoadType	0.06	6.03%	Weak
7	LandType	0.06	6.03%	Weak
8	s(Elevation)	0.007	0.79%	Weak

Note: Row colors indicate contribution strength—dark green to light green show high to lower strong contribution, and dark red to light red show high to lower weak contribution. Variables combined using an asterisk (*) represent interaction terms.

3.1. Effect of Spatial Variation and Elevation on LST

The analysis shows that the location coordinate is significant to LST (Figure 8). The isolines in Figure 8 represent three-dimensional data in two dimensions and provide evidence for spatial variation of LST across the x and y axes. The pattern suggests that the study area has fluctuating LST values in relation to the solid black line, which represents the reference value of 27.6 °C. The dashed green lines indicate higher LST values, while the dashed red lines indicate lower LST compared to the reference line. These dashed lines correspond to one standard error above (green) and below (red) the reference value, illustrating where the LST is higher or lower, respectively. The clustering of lines suggests rapid changes or steep gradients in LST, while more widely spaced lines indicate more gradual changes in LST. The enclosed loops in certain regions represent the local maximum or minimum in LST. These changes in LST might reflect the influence of localized factors, such as the presence of vegetation (for example, parks or gardens).

The small *p*-value for elevation suggests that elevation is an important factor in predicting LST; however, the relationship is not linear and may be more complex, with LST potentially increasing or decreasing at different rates in different elevations. Higher elevations are generally expected to be cooler due to decreased atmospheric pressure, and vice versa [56]. Elevation in the area ranges from 23 to 178 m and the graph shows an almost uniform spread of LST along this area, with clustering of LST between 30 and 45 °C. This suggests that there is a moderate correlation between elevation and LST, as mentioned in other studies [56,84,85].

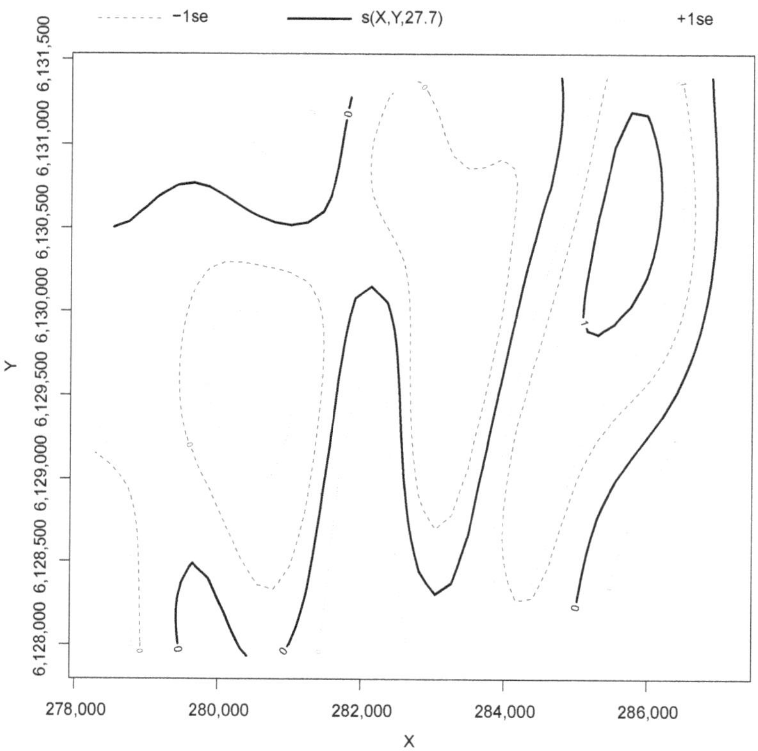

Figure 8. Spatial variability by coordinates on LST.

3.2. *Effect of Road Typology and Land Use and Land Cover on LST*

Figure 9 shows the median LST across the three road typologies. The data suggests that a difference in surface temperature exists relative to the road type. Local roads (type 1) reported a median LST of 36.98 (range of 25.21 to 49.09 °C with a mean LST of 36.95 °C), showing the lowest variability compared to the other road types. Regional roads (type 2) reported a slightly higher median LST of 39.51 °C (range of 28.41 to 45.6 °C, mean LST = 38.70 °C). For state roads, the median LST was 39.98 °C (range of 21.76 to 46.79 °C, mean 39.35 °C). Regional and state roads reported higher LST values, with a greater concentration on LST in the middle to upper range compared to local roads. The statistical model and Table 4 suggest that regional and state roads generally have lower temperatures of 0.62 and 0.77 °C (Estimate column in Table 4) than local roads. The effect of road typology (width of roads) on LST is statistically significant (small *p*-value in Table 4) with precise estimates (small Std.error in Table 4).

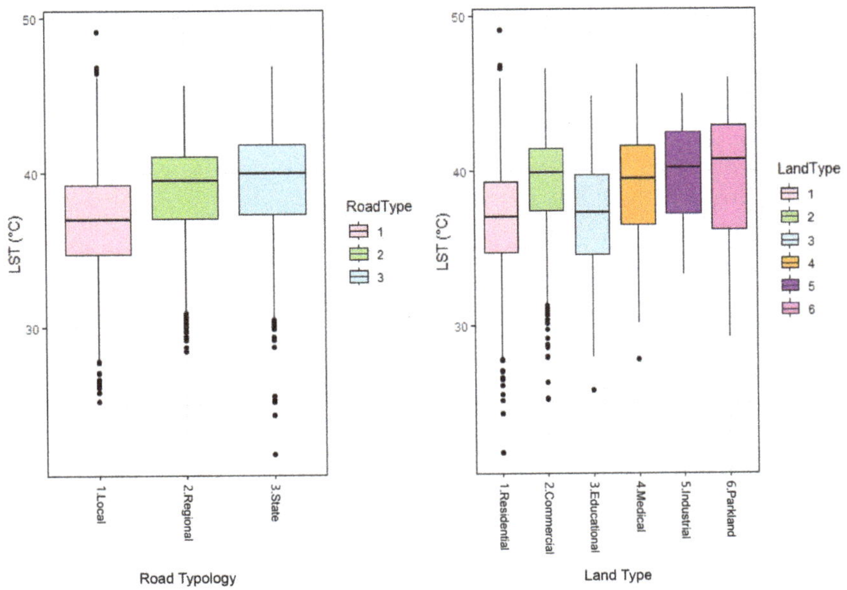

Figure 9. Distribution of LST over road typology (**left**) and land type (**right**).

Figure 9 represents LST variation across land types. For residential land use (Type 1), the median LST is 37.04 °C, and it has a wide range from 21.76 to 49.09 °C. Commercial land use (Type 2) has a higher median of 39.91 °C, with a minimum of 25.21 and maximum of 46.54 °C. Educational land use (Type 3) has a minimum, median, and maximum LST of 25.76, 37.32, and 44.79 °C. The median for medical land use (Type 4) is 39.5 °C, showing slightly higher temperatures compared to educational land use and also with a higher range from 27.78 to 46.84 °C. Industrial land use (Type 5) and parkland (Type 6) show the highest LST with a median of 40.18 and 40.73 °C and also maximum of 40.18 and 40.73 °C, respectively. The LST variation for parkland, however, is larger, starting from 29.9 °C compared to that of industrial land use with 33.19 °C. Parkland includes a greater proportion of lower LST. The statistical model and Table 4 also show that for commercial land use, LST is about 0.37 °C lower than areas classified as residential and this effect is statistically significant and reliable, assuming all other variables remain constant. Educational land use is associated with a 0.13 °C lower LST compared to residential. This effect is not as large as commercial land use, showing only marginal significance and less reliability. Medical land use is, similarly, associated with a 0.15 °C decrease in LST compared to residential land use, and the results showed a moderate significance. For industrial land use, the coefficient is 0.06 °C, meaning that LST is slightly higher than residential land use. However, this difference is neither statistically significant nor highly reliable, suggesting that any observed increase in LST could be due to random variation rather than a consistent pattern. In contrast, parkland is associated with a 0.11 °C increase in LST compared to residential land use. Although this result is marginally significant, its high precision indicates that the observed difference is likely consistent and may reflect a real-world phenomenon. This could mean that parkland, despite being a green space, might absorb or retain heat more than what is expected due to factors like vegetation species, soil composition, the presence or patterns of irrigation, or lack of canopy shading impacting on the health of the underlying ground cover.

Figure 10 illustrates the effect of interaction between road typology and land type on mean, minimum, and maximum LST in the research area. Local–residential type (1-1) has the highest maximum LST (49.1 °C) but the second lowest mean LST (36.8 °C). However, the lowest minimum LST of residential belongs to the state–residential type (3-1) with 21.8 °C. Although local–parkland type (1-6) has the lowest mean LST (36.4 °C), this value is the highest for state–parkland (3-6) (41.2 °C). State–education type (3-3) has both the lowest mean and maximum LST with 36.4 and 41.7 °C, respectively. Local–industry type (1-5) has the highest minimum LST of 33.2 °C and a high mean and maximum LST value (39.8 and 45 °C). Interestingly, the maximum value for this combination is similar to that of local–parkland (1-6) and is even lower than that of state and regional–parkland types (3-6 and 2-6). The combination of parkland (6) with three road typologies does not have the lowest minimum LST (22.9 °C) when compared to the range of minimum LST values from 22.5 to 32.5 °C.

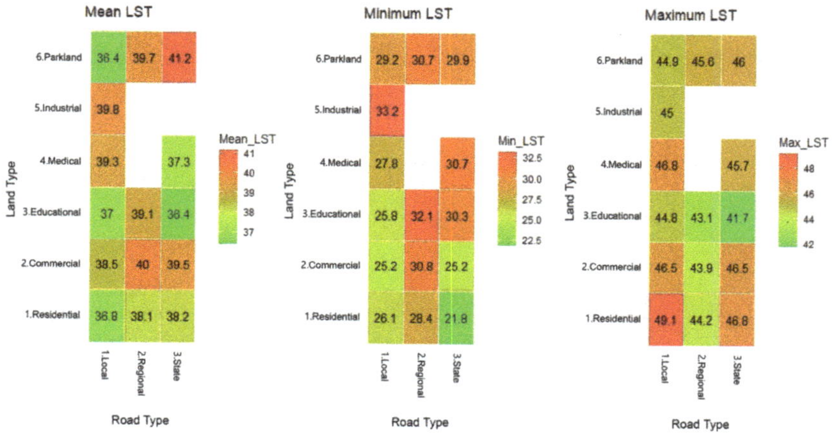

Figure 10. The co-effect of road typology and land type on LST (note: 1. white spaces mean no data; 2. unit for LST is °C; 3. scale is different for each heat map).

As discussed in Section 2.2.3 and presented in Table 4, local road and residential land type are considered as a baseline (estimate = 0) to enable a comparison of other combinations impacting LST. The co-effects (Table 4) on LST are listed as follows:

Regional roads with adjoining commercial land use have a coefficient of 0.15, while the individual coefficients of regional road and commercial land type are −0.62 and −0.37, respectively. That means for this combination, LST is calculated by summing their individual and the interaction coefficients. In doing so, the effect will be adding these values together (−0.62 −0.37 +0.15), resulting in a total effect of −0.84. This indicates that on average, the combination of a regional road with commercial land use reduces LST by 0.84 °C when compared to local road with residential land use. The effect of this combination is marginally significant (typically not highly significant (at the 5% level) but marginally significant (at the 10% level)) and the estimate is relatively precise. This suggests a weak effect that lacks precision, meaning that the effect in reality might be different.

Regional roads with educational land use have a coefficient of 0.8, showing a 0.05 °C increase in LST (−0.62 −0.13 +0.8) for the interaction compared to that for local roads and residential land use. This result is highly significant but has relatively weak precision. This indicates variability in this effect, meaning that while road type 2 in land type 3 generally tends to increase LST, the actual increase may fluctuate around this estimate.

Regional roads with parkland land use show a coefficient of 0.34 that indicates a 0.17 °C decrease (−0.62 +0.11 +0.34) in LST due to the interaction compared to the baseline with moderate significance and reliability. However, the relatively high standard error implies some variability in this effect, meaning that the precise decrease in LST could range slightly.

State roads with a commercial land use are associated with a coefficient of 0.6, causing a 0.54 °C decrease in LST (−0.77 −0.37 +0.6) compared to local roads with residential land use. This combination has very strong significance and relatively strong estimated precision. This low error suggests a higher confidence in the impact on LST, meaning road type 3 (state) in land type 2 (commercial) is consistently associated with lower temperatures.

State roads with an educational land use reported a coefficient of −1.03 and are associated with a 1.93 °C decrease in LST (−0.77 −0.13 −1.03) compared to the combination of local roads and residential land use with strong significance and weak precision. The relatively large error means that while road type 3 (state) with land type 3 (educational) is associated with cooling, the precise amount of cooling varies more in this combination than in others.

State roads with medical land use reports a coefficient of 0.51, indicating a 0.41 °C decrease (−0.77 −0.15 +0.51) in LST for this interaction compared to that for local roads and residential land use, and the result is highly significant but not with a high precision. The observed value for the interaction term might be due to increased heat retention in areas with road type 3 (state) and land type 4 (medical). However, the effect size varies moderately, meaning the temperature reduction is significant but not consistent across all locations.

State roads with parkland land use report a coefficient of 1.28, showing a 0.62 °C increase in LST (−0.77 +0.11 +1.28) compared to the interaction between local roads and residential land use. This is highly significant with moderate reliability. This strong and relatively stable effect is likely due to the characteristics of road type 3 (state) and land type 6 (parkland) areas that are more exposed to sun and retain more heat.

3.3. *Effect of CCP and NDVI on LST*

Figure 11 illustrates that both CCP and NDVI appear to be negatively correlated with LST. This suggests that areas with higher values of CCP and NDVI are associated with lower LST, and increasing CCP and NDVI along all road typologies is an effective strategy for reducing UHI_{ULI}. However, the relationship between NDVI and LST appears to exhibit a stronger rate of change (steeper gradient) compared to the relationship between CCP and LST. This could indicate that NDVI, at lower values, is more sensitive to changes in LST. NDVI measures the difference between near-infrared light, which vegetation strongly reflects, and red light, which vegetation absorbs for photosynthesis. Lower NDVI values (typically up to 0.5) suggest stressed, sparse, or unhealthy vegetation [86]. NDVI is a well-known indicator of vegetation health and greenness [87,88]. This result suggests that if linear green infrastructure is to be applied as a UHI cooling strategy, the greatest gains will be made if this vegetation is supported by irrigation to improve its health, particularly where hot and dry summer climates prevail. The results also suggest that CCP and NDVI may have a more significant role in attenuating UHI along linear green infrastructure. Table 4 reports, for every 1% increase in CCP, LST decreases by about 0.05 °C, holding other variables constant. For every unit increase of 0.1 in NDVI, LST decreases by about 0.72 °C. Regarding Table 4, the estimated coefficient for CCP is more reliable than that of NDVI, as indicated by its smaller standard error. However, both variables exhibit a highly significant cooling effect on LST ($p < 0.001$). Additionally, CCP and NDVI are strongly interrelated, meaning that an increase in canopy cover generally leads to a higher NDVI and vice versa.

Figure 12 illustrates the 3D scatter plot of co-effect of CCP and NDVI on LST within road typology. The cluster of points suggests that higher LST values (red points) tend to occur when CCP is lower and potentially when NDVI is also lower. This reflects the absence of vegetation, which often leads to higher temperatures. Conversely, lower LST values (green points) are associated with higher CCP and higher NDVI values, demonstrating that areas with more canopy cover and greener or denser vegetation tend to be cooler. The cooling effect from the trees may still help keep LST lower than areas with no canopy at all, but LST could be higher than in areas where both CCP and NDVI are low. The LST could be relatively low even with lower CCP. Table 4 reports the output from the statistical model where the interaction term between CCP and NDVI provides a negative coefficient of −0.03. This means that increasing both CCP and NDVI simultaneously by one unit declines LST by 7.75 °C (−0.5 −7.22 −0.03 = −7.75), which is more effective than CCP (−0.5) and NDVI (−7.22) separately, and the estimated coefficient is highly precise.

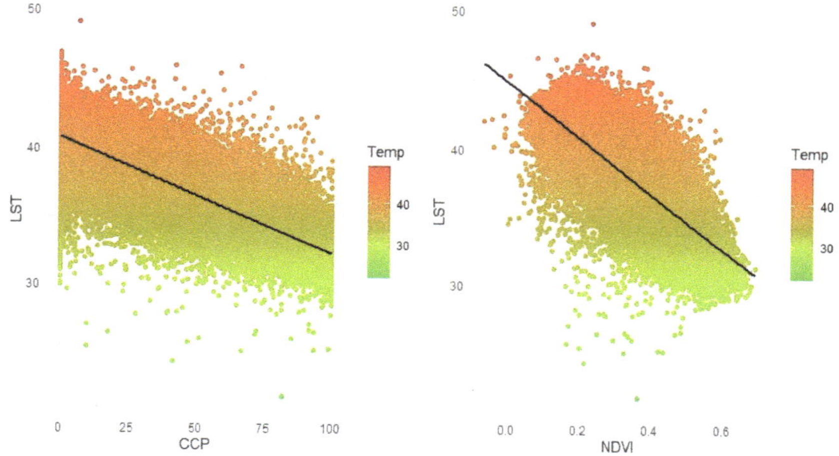

Figure 11. The relationship of CCP and NDVI with LST (note: unit for LST is °C).

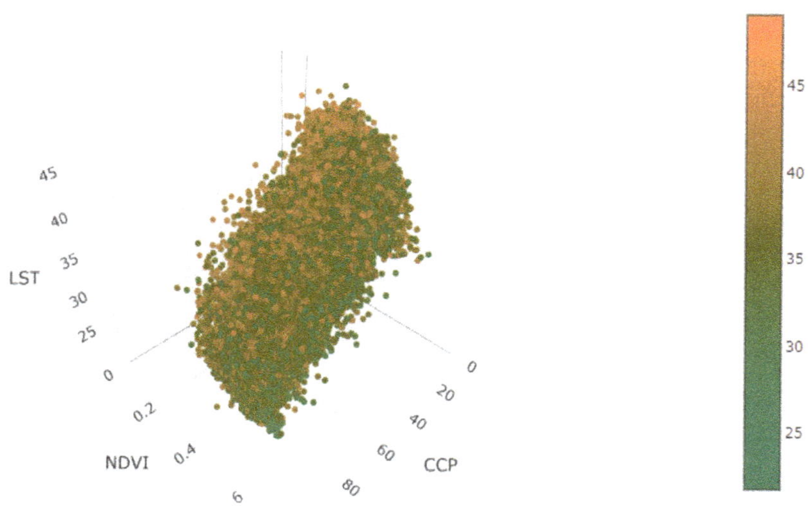

Figure 12. Co-effect of CCP and NDVI on LST (note: unit for LST is °C).

By comparing Figures 10 and 13, the joint effect of canopy coverage, vegetation density, and health on LST with road–land type interaction can be assessed. For example, the combination of a relatively high CCP (46.3%) and NDVI (0.40) in the context of a local road and residential land use yields one of the lowest mean LST values, at 36.8 °C. Similar patterns are observed for local roads with parkland and educational land use, as well as for state roads with medical land use. However, in contrast, the combination of a local road with residential land use also produces the highest recorded maximum LST, at 49.1 °C. The "hot spots" in these areas are attributed to lower canopy cover. In cases where both CCP (9.2%) and NDVI (0.22) are low, such as in state roads with parkland, the highest mean LST (41.2 °C) is observed. This indicates that grassy areas with limited canopy shading and a higher level of brown vegetation rather than green contribute significantly to the urban linear heat effect (UHI_{ULI}). Interestingly, combinations with relatively high CCP (45%) but lower NDVI (0.30)—such as state roads with educational land use—result in a lower mean LST (36.4 °C) compared to state roads with commercial land use, which have slightly lower CCP but higher NDVI (37.3 °C). Further comparison of the NDVI values in the state road and parkland combination and the regional road and commercial combination, both of which share a CCP of 9.2%, reveals that despite having a lower NDVI, the regional–commercial combination exhibits lower mean and maximum LST values than the state road–parkland combination.

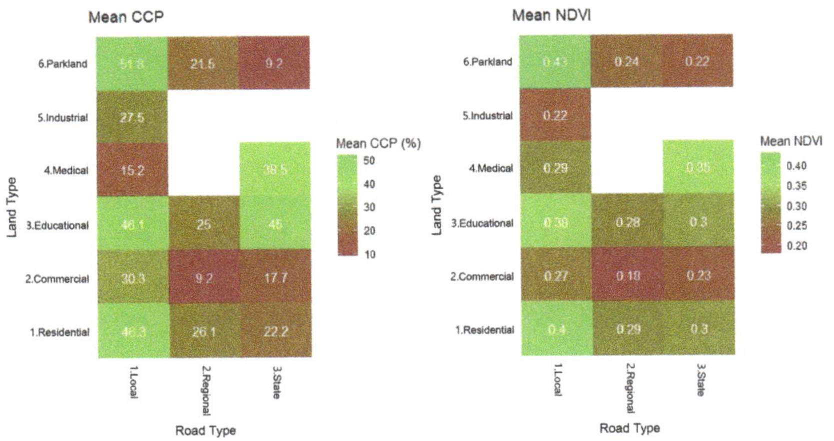

Figure 13. Distribution of CCP and NDVI over road typology and land use type (note: 1. white spaces mean no data; 2. scale is different for each heatmap).

4. Discussion

4.1. Road and Land Type

The results of this study report that as roads become wider, there is a greater sky view factor (SVF) and a lower aspect ratio (AR); LST declines and becomes cooler when other factors are constant. Regional roads (type 2) have an average LST 0.62 °C lower and state roads (type 3) 0.76 °C lower than local roads (type 1) per 100 m^2. This result is inconsistent with the study by Junsik Kim et al. [89] who reported that an increase of 0.08 in SVF causes the urban canyon surface to receive a greater level of radiation resulting in 0.64 °C higher average LST. However, other studies [90,91] report that as urban canyons become narrower, multi vortices appear, causing more complicated airflow and weaker ventilation and heat removal at near-ground level and leading to higher LST. In this research, decreased LST may be due to the proximity of road typologies to different land use types or predominant orientation. Some sections of state roads in the northern part of the study area are adjacent to parkland or regional roads and primarily follow north–south orientation, while local roads exhibit more varied alignments.

In terms of lowest to highest LST values, land types can be ordered as commercial (−0.37 °C), medical (−0.15 °C), educational (−0.13 °C), residential (0), industrial (0.05 °C), and parkland (0.11 °C). Other studies [91] have reported that the residential land use type has a lower LST than commercial, medical, and educational land uses. Marco Morabito et al. [92] also reported 0.5 °C cooler LST in residential areas than parkland areas. These studies reported that the variation in LST adjacent to the residential land use type could be explained by the different percentage of pervious surfaces and green fragmentation integrated with this land use. This study, however, did not analyze the percentage of pervious or impervious surfaces, so we are uncertain whether this explanation holds for Adelaide.

Commercial land use is estimated to have the lowest LST in this area, at 0.37 °C lower than residential land use (Table 4). This result differs from other studies [93,94] that report elevated LSTs across commercial areas relative to residential land. However, lower LSTs have been reported in dense commercial land use in the USA and China [95–97]. They highlighted that high-rise commercial buildings and complexes can provide greater amenities like green spaces, walking and cycling tracks, and also good public transport access around them. This open space may decrease anthropogenic heat emissions which was not considered in our study. However, in this study area, commercial land use types are mostly adjacent to state and regional roads (which are cooler than local roads), particularly those sections near parkland areas.

Educational land use type experiences a cooler LST by 0.13 °C per 100 m^2 compared to residential land use (Table 4). Our results differ to those obtained by Haizhen Wen et al. [98], who note that educational land use contains various facilities such as air conditioning or mechanical ventilation systems producing heat emissions even in suburbs causing increased LST compared to the others. However, Xincheng He [99] reported that densely vegetated areas around schools in this land type could reduce average local LST by a maximum of 0.5 °C in varying districts of China. The researchers also pointed out that local climate zones and local urban morphology could apparently contribute to LST values in educational lands. For example, compact mid-rise, large low-rise, bare paved or soil areas, higher building and road density, greater floor area ratio, and artificial surfaces elevate LST, while dense or scattered-tree areas, and high SVF mitigate LST. These findings align with our study, as educational land uses in the study area are located in open spaces, primarily grassed, and surrounded by canopy cover.

Medical land type, consisting of hospitals of different sizes, reports a cooler LST by 0.15 °C compared to the residential land type. Chen et al. looked at this land use but amalgamated with schools, cultural, political, and sport facilities [94]. Their study revealed that mean LST over this land type was 0.64 °C less than the commercial type and 0.25 °C higher than the residential type because of different levels of surrounding green infrastructure. Figures 10 and 13 indicate that while the educational land type does not show the highest vegetation values, it remains one of the most well-vegetated areas across all road typologies in this study area. This likely explains its lower LST compared to residential land uses.

Industrial land types report an LST 0.05 °C higher than residential areas. This is the smallest temperature difference (compared to the baseline) when comparing land use types. In [93,100], the highest LST is reported in industrial zones, including suburban regions characterized by extensive, contiguous industrial areas. However, in [93], industrial zones with substantial vegetated areas exhibit a lower recorded LST. This is consistent with Figures 10 and 13, which show that although industrial land uses have a slightly lower mean NDVI than medical land uses by 0.07, its 12.3% higher CCP results in similar LST values. In this study, industrial land use occupies the significantly smallest area and has the least proximity to the road typologies compared to the others (Table 2). This limited spatial extent may contribute to its lower average LST compared to other land uses and parkland in particular.

Table 4 also shows that parkland is estimated to be the hottest land type over road typologies, showing 0.11 °C higher than residential LST. This contrasts with other studies [94,101–103] reporting parkland as offering a cooling island in urban areas with 1–10 °C lower LST. The results in [94] imply a cooler mean LST of parkland by 2.48, 2.1, and 3.26 °C compared to residential, public services (schools, hospitals, sport facilities), and commercial land use, respectively. However, these differences declined to 0.6, 0.85, and 1.98 °C for brownfield, which was defined as unused lands or wastelands. According to these studies [94,101–103], the cooling effect of parkland strongly depends on factors like their distance to the city center, CCP, NDVI, and intended use such as public gardens, local parks, or sport fields. In our study, the lands categorized as parkland that were adjacent to regional and state roads had little grass, often bare soil, and contained little or no canopy cover (mean CCP of 9 and median CCP of 0 for state roads, mean CCP of 21 and median CCP of 2.6 for regional roads). Further, it was likely that these parks had no or poor irrigation, which resulted in a low NDVI value (mean and median of 0.2 compared to NDVI for healthy and green vegetation being 0.66 and more [104]). These values indicate the importance of maintaining healthy vegetation and canopy trees to provide shading if parks are to be used as a mitigation strategy to lessen the UHI effect.

4.2. CCP and NDVI

GAM analysis shows if CCP increases by 1% on a road, LST can decrease by 0.05 °C. For example, by increasing 60% of CCP, a 3 °C reduction in LST would occur. This correlation is in line with another study on a Mediterranean climate conducted in Portugal where areas with CCP larger than 70% experience 3.42 °C cooler LST than those between 10 and 30% [52]. Other studies [105,106] also show that CCP is a reliable contributing factor to LST mitigation because it can significantly mitigate solar short-wave radiation penetrating deeply under canopy. Combined with the evapotranspiration of canopy trees, this leads to higher latent heat flux compared to the sensible heat experienced in non-canopy areas, ultimately lowering LST.

As reported in Table 4, every increase of 0.1 in NDVI value can cause a 0.72 °C decline in LST, which is one of the most significant predictors, as suggested by the *p*-value. This is consistent with other research [107] that reports that higher greenness leads to lower LST. Some studies indicate a stronger correlation between NDVI and LST, such as a 19% decrease in NDVI that can increase LST by 7.2 °C [78] or an increase in NDVI by 0.87 that leads to a decrease of 13.44 °C in LST [79]. This negative correlation between NDVI and LST tends to be stronger in summer, indicating that higher NDVI suggests well-irrigated green canopy cover and lower LST [78].

Values resulting from the interaction between CCP and NDVI indicate that this combination has a more significant cooling effect on LST with strong confidence across our data, meaning this reliably leads to a reduction in LST. This cooling effect may be influenced by other factors like urban morphology but always remains in a negative correlation with UHI_{ULI}, in contrast to road–land interaction, which changes from separated to combined analysis and from one combination to another even in the same area. Unfortunately to date, existing studies' focus has been on the effect of either canopy cover structure or NDVI separately on UHI because the multi-variate regression model considers multiple inputs, leading to duplicated contributions from these layers, and there is a paucity of knowledge that considers this co-effect on LST in urban systems and urban linear systems. Although CCP and NDVI are dependent, the canopy cover percentage (CCP) represents the proportion of tree crown coverage within each road block, thus capturing the vertical shading and physical coverage directly above road surfaces. In contrast, NDVI reflects overall vegetation greenness and photosynthetic activity, including ground-level vegetation (shrubs, grasses, lawns) in addition to trees, which are strongly related to evapotranspiration processes contributing to surface cooling. Including both variables allows the model to consider both the structural shading effects (CCP) and the vegetation density effects (NDVI), which together provide a more comprehensive representation of vegetation's influence on land surface temperature. Thus, including both CCP and NDVI allows us to capture complementary mechanisms of urban heat mitigation: structural shading and evapotranspirative cooling (explanation for CCP and NDVI dependency was added).

As this research reveals, the role of road–land type in UHI_{ULI} mitigation is complex because of a wide range of geographical and environmental factors. This complexity suggests that the application of green infrastructure or nature-integrated-based solutions to achieve urban cooling may not necessarily achieve the desired temperature reduction outcomes [108]. For example, according to Zheng et al. [109], LST in local–residential use type is impacted by both the height of buildings and vegetation coverage along roadways, while Chen et al. [94] revealed that in parkland, industrial, and commercial land use in Berlin, Germany, LST is more affected by built environment factors like shading, materials, or anthropogenic heat, and in other land uses, NDVI is a dominant contributing factor to varying LST. This study, however, did not consider road typology and other relevant characteristics such as SVF for the land type under research in Berlin.

Our research provides a statistical model tested for the analysis of the key factors influencing UHI_{ULI} mitigation, integrating built types (road characteristics and land use) and land cover types (canopy cover quantity and quality, elevation, and coordinates) to explore correlations between these factors and UHI magnitude, measured by LST. We found that the impact of road typology and land type on LST varies due to the complexity of urban systems and services. However, the cooling effect of canopy cover, as part of natural systems, is consistently significant, mitigating the thermal effects of built and human systems [94]. Furthermore, from a practical perspective, urban green strategies can be applied to roads more easily when compared to other approaches such as changing the surface albedo or the design of the road [4–6]. From [57], Adelaide tree canopy coverage increased by approximately 4% between 2019 and 2022, particularly along local streets. However, localized areas of vegetation loss persist, primarily where construction demand is high and along road typologies undergoing infrastructure upgrades. This duality underscores the complex challenges of balancing urban development with environmental sustainability and the urban energy budget, especially in the face of intensifying heat waves.

To optimize heat mitigation policies and improve financial management, it is essential to consider land use, land cover, and road typology. This research has the following limitations. The available data on linear infrastructure remains generic and could not fully capture the uniqueness of all urban sites, limiting the precision of road typology classification. Additionally, there remains a significant knowledge gap in heat mitigation strategies for linear infrastructure, complicating direct comparisons of our findings with previous studies. Nonetheless, this paper addresses the most critical factors and contributes to advancement of knowledge in urban heat mitigation.

Further, the study relied solely on the Adelaide database, which is a single-day thermal image (under consistent meteorological conditions, using a high-resolution thermal dataset collected on a cloud-free day under peak summer heat), and constrained the analysis to the availability of data, including LST, NDVI, and CCP (explanation about thermal image limitation was added). This limitation may have increased the uncertainty of the results, as some significant factors, such as canopy tree species, leaf area index (LAI), pavement materials, and local air temperature records, could not be considered. For instance, NDVI values can reflect prior weather conditions like rainy weather, potentially affecting their accuracy as indicators. In this dataset also, NDVI is calculated from vegetation with all heights but CCP reflects only above-3m canopy trees, resulting in some discrepancies such as high CCP and low NDVI. The research employed a remote sensing approach to examine UHI_{ULI} effects in Adelaide's road networks during summer, relying on aerial imagery. One of the core contributions of this research is to propose a method for estimating the urban heat island effect specifically along road networks (UHI_{ULI}), which remains an unexplored component of the broader UHI phenomenon. The intention is not to generalize across all urban surfaces, but rather to offer a replicable framework for analyzing and mitigating heat exposure along linear infrastructure, where cooling interventions can have direct spatial and functional relevance. While this method was effective for large-scale analysis, more significant findings may emerge from similar studies conducted in experimental contexts. Future research could include logging local air temperatures along roads to enable surface–air temperature correlations and exploring new approaches or research directions that provide deeper insights into the effectiveness of urban heat mitigation strategies—particularly regarding tree canopy attributes such as crown width, height, species, and other types of vegetation cover. Additionally, with access to comparable databases, the methods and results of this study could be tested in cities with similar or contrasting climates, potentially enhancing their applicability and robustness.

5. Conclusions

In addressing three main questions, this research investigates how roads, crucial elements of urban linear infrastructure, can be converted into cool corridors. It examines the role of canopy cover percentage (CCP) in reducing land surface temperature (LST), explores the impact of additional contributing factors, and assesses the possibilities for optimizing CCP to enhance cooling effects. To explore these questions, remote sensing and statistical analysis approaches were applied. We first categorized roads into three types—local, regional, and state roads—and identified key factors such as canopy cover (percentage and NDVI) and land type (residential, commercial, educational, medical, industrial, and parkland). Then, we examined how LST varied under the combined influences of CCP, NDVI, and land type across different road typology and analyzed correlations between these variables and LST.

Unlike previous studies that examined canopy cover and NDVI separately, this research assesses their combined effect on LST. The findings highlight that interactions among contributing factors complicate the determination of optimal cooling values. While CCP and NDVI consistently show a negative correlation with LST, the combined effect of land type and road typology is more variable. These results underscore the complex interplay between built and natural environments, emphasizing the need for a multi-factor analysis rather than isolated assessments. While the absolute differences in LST between road types and land uses may appear numerically modest, even small reductions in surface temperature can have meaningful impacts on local heat dynamics. For example, a local road surface fully exposed to sunlight can reach 49 °C compared to a surface shaded 100% by tree canopy (28°C), representing a more than 10 degree difference, as shading reduces both incoming shortwave radiation and outgoing longwave re-radiation. Since roads constitute a large proportion of urban impervious surfaces, even marginal reductions in their surface temperature contribute cumulatively to lowering ambient air temperature, reducing heat stress, and improving human thermal comfort, particularly during heat waves (explanation for small difference was added).

Effective heat mitigation strategies require a holistic approach, where canopy cover along roads—the most influential factor—must be complemented by additional mitigation measures in adjacent land uses such as proper irrigation to support healthy green cover. Land use significantly affects LST, with commercial and medical areas exhibiting lower-than-expected temperatures due to green spaces and urban amenities. The interaction between road typology and land use further complicates temperature variations, as cooling effects depend on urban morphology and vegetation cover. While specific canopy cover percentages may provide significant cooling in some land uses, different configurations may be necessary in others. Therefore, the quantity of the canopy cover (CCP), its quality or health (NDVI), and strategic placement (relative to land type and road typology) of canopy trees are key determinants in mitigating UHI effect.

From a policy-making standpoint, urban linear heat is a built–natural environment phenomenon that is complex. Thus, coordination across various professional disciplines that have leverage on how roads, land use, and greening are planned and maintained is needed if urban cooling is to be achieved. We also note the need for additional research to further understand the effectiveness of urban heat mitigation strategies, particularly in relation to how the tree canopy (crown width, height, and species) and other vegetation cover can support UHI_{ULI} mitigation strategies.

Author Contributions: Conceptualization, E.M. and P.D.; methodology, E.M., P.D., M.C. and G.S.; software, E.M., M.C. and G.S.; validation, E.M. and M.C.; formal analysis, E.M., M.C. and G.S.; investigation, E.M., P.D., M.C. and G.S.; data curation, E.M. and M.C.; writing—original draft preparation, E.M.; writing—review and editing, P.D., M.C., G.S. and E.M.; visualization, E.M.; supervision, P.D., M.C. and G.S. All authors have read and agreed to the published version of the manuscript.

Funding: This research received no external funding.

Institutional Review Board Statement: Not applicable.

Informed Consent Statement: Not applicable.

Data Availability Statement: No new data were created in this study.

Conflicts of Interest: The authors declare no conflicts of interest.

Abbreviations

The following abbreviations are used in this manuscript:

UHI	Urban Heat Island
ULI	Urban Linear Infrastructure
UHI_{ULI}	Urban heat within linear infrastructure
LST	Land Surface Temperature
CCP	Canopy Cover Percentage
NDVI	Normalized Difference Vegetation Index
Ta	Air Temperature

References

1. Phelan, P.E.; Kaloush, K.; Miner, M.; Golden, J.; Phelan, B.; Silva III, H.; Taylor, R.A. Urban heat island: Mechanisms, implications, and possible remedies. *Annu. Rev. Environ. Resour.* **2015**, *40*, 285–307. [CrossRef]
2. Mirabi, E.; Davies, P.J. A systematic review investigating linear infrastructure effects on Urban Heat Island (UHIULI) and its interaction with UHI typologies. *Urban Clim.* **2022**, *45*, 101261. [CrossRef]
3. Oke, T.R.; Mills, G.; Christen, A.; Voogt, J.A. *Urban Climates*; Cambridge University Press: Cambridge, UK, 2017. [CrossRef]
4. Coutts, A.; Tapper, N. Trees for a Cool City: Guidelines for optimised tree placement. Melbourne Australia: Cooperative Research Centre for Water Sensitive Cities. *CRC Water Sensitive Cities Sch. Earth Atmos. Environ. Monash Univ.* **2017** . Available online: https://watersensitivecities.org.au/wp-content/uploads/2017/11/Trees-for-a-cool-city_Guidelines-for-optimised-tree-placement.pdf (accessed on 1 July 2025).
5. Santamouris, M.; Ding, L.; Fiorito, F.; Oldfield, P.; Osmond, P.; Paolini, R.; Prasad, D.; Synnefa, A. Passive and active cooling for the outdoor built environment–Analysis and assessment of the cooling potential of mitigation technologies using performance data from 220 large scale projects. *Sol. Energy* **2017**, *154*, 14–33. [CrossRef]
6. Wong, N.H.; Jusuf, S.K.; La Win, A.A.; Thu, H.K.; Negara, T.S.; Xuchao, W. Environmental study of the impact of greenery in an institutional campus in the tropics. *Build. Environ.* **2007**, *42*, 2949–2970. [CrossRef]
7. Droste, A.M.; Steeneveld, G.J.; Holtslag, A.A. Introducing the urban wind island effect. *Environ. Res. Lett.* **2018**, *13*, 094007. [CrossRef]
8. Wei, G.; He, B.J.; Sun, P.; Liu, Y.; Li, R.; Ouyang, X.; Luo, K.; Li, S. Evolutionary trends of urban expansion and its sustainable development: Evidence from 80 representative cities in the belt and road initiative region. *Cities* **2023**, *138*, 104353. [CrossRef]
9. Yu, Z.; Zhao, P. The factors in residents' mobility in rural towns of China: Car ownership, road infrastructure and public transport services. *J. Transp. Geogr.* **2021**, *91*, 102950. [CrossRef]
10. Oke, T.R. Canyon geometry and the nocturnal urban heat island: Comparison of scale model and field observations. *J. Climatol.* **1981**, *1*, 237–254. [CrossRef]
11. Zhou, D.; Zhao, S.; Liu, S.; Zhang, L. Spatiotemporal trends of terrestrial vegetation activity along the urban development intensity gradient in China's 32 major cities. *Sci. Total Environ.* **2014**, *488*, 136–145. [CrossRef]
12. Irmak, M.A.; Yilmaz, S.; Dursun, D. Effect of different pavements on human thermal comfort conditions. *Atmósfera* **2017**, *30*, 355–366. [CrossRef]
13. Gui, J.; Phelan, P.E.; Kaloush, K.E.; Golden, J.S. Impact of pavement thermophysical properties on surface temperatures. *J. Mater. Civ. Eng.* **2007**, *19*, 683–690. [CrossRef]
14. Gedzelman, S.; Austin, S.; Cermak, R.; Stefano, N.; Partridge, S.; Quesenberry, S.; Robinson, D. Mesoscale aspects of the urban heat island around New York City. *Theor. Appl. Climatol.* **2003**, *75*, 29–42. [CrossRef]
15. Kim, K.H.; Jeon, S.E.; Kim, J.K.; Yang, S. An experimental study on thermal conductivity of concrete. *Cem. Concr. Res.* **2003**, *33*, 363–371. [CrossRef]
16. Gungor, S.; Cetin, M.; Adiguzel, F. Calculation of comfortable thermal conditions for Mersin urban city planning in Turkey. *Air Qual. Atmos. Health* **2021**, *14*, 515–522. [CrossRef]
17. Oke, T.R. The energetic basis of the urban heat island. *Q. J. R. Meteorol. Soc.* **1982**, *108*, 1–24. [CrossRef]
18. Cetin, M.; Adiguzel, F.; Zeren Cetin, I. Determination of the effect of urban forests and other green areas on surface temperature in Antalya. In *Concepts and Applications of Remote Sensing in Forestry*; Springer: Berlin/Heidelberg, Germany, 2023; pp. 319–336. [CrossRef]
19. Rabl, A.; De Nazelle, A. Benefits of shift from car to active transport. *Transp. Policy* **2012**, *19*, 121–131. [CrossRef]
20. Kilicoglu, C.; Cetin, M.; Aricak, B.; Sevik, H. Site selection by using the multi-criteria technique—A case study of Bafra, Turkey. *Environ. Monit. Assess.* **2020**, *192*, 1–12. [CrossRef]
21. Cetin, M. The changing of important factors in the landscape planning occur due to global climate change in temperature, Rain and climate types: A case study of Mersin City. *Turk. J. Agric.-Food Sci. Technol.* **2020**, *8*, 2695–2701.
22. Al-Wreikat, Y.; Serrano, C.; Sodré, J.R. Effects of ambient temperature and trip characteristics on the energy consumption of an electric vehicle. *Energy* **2022**, *238*, 122028. [CrossRef]
23. Millard-Ball, A. The width and value of residential streets. *J. Am. Plan. Assoc.* **2022**, *88*, 30–43. [CrossRef]
24. Doulos, L.; Santamouris, M.; Livada, I. Passive cooling of outdoor urban spaces. The role of materials. *Sol. Energy* **2004**, *77*, 231–249. [CrossRef]
25. Feyisa, G.L.; Dons, K.; Meilby, H. Efficiency of parks in mitigating urban heat island effect: An example from Addis Ababa. *Landsc. Urban Plan.* **2014**, *123*, 87–95. [CrossRef]
26. Zeren Cetin, I.; Ozel, H.B.; Varol, T. Integrating of settlement area in urban and forest area of Bartin with climatic condition decision for managements. *Air Qual. Atmos. Health* **2020**, *13*, 1013–1022. [CrossRef]

27. Zeren Cetin, I.; Sevik, H. Investigation of the relationship between bioclimatic comfort and land use by using GIS and RS techniques in Trabzon. *Environ. Monit. Assess.* **2020**, *192*, 1–14. [CrossRef] [PubMed]
28. Cetin, M. Climate comfort depending on different altitudes and land use in the urban areas in Kahramanmaras City. *Air Qual. Atmos. Health* **2020**, *13*, 991–999. [CrossRef]
29. Cetin, M. Determination of bioclimatic comfort areas in landscape planning: A case study of Cide Coastline. *AGRIS—Int. Syst. Agric. Sci. Technol.* **2016**, *4*, 800–804.
30. Fahmy, M.; Sharples, S.; Yahiya, M. LAI based trees selection for mid latitude urban developments: A microclimatic study in Cairo, Egypt. *Build. Environ.* **2010**, *45*, 345–357. [CrossRef]
31. Gallay, I.; Olah, B.; Murtinová, V.; Gallayová, Z. Quantification of the cooling effect and cooling distance of urban green spaces based on their vegetation structure and size as a basis for management tools for mitigating urban climate. *Sustainability* **2023**, *15*, 3705. [CrossRef]
32. Leuzinger, S.; Vogt, R.; Körner, C. Tree surface temperature in an urban environment. *Agric. For. Meteorol.* **2010**, *150*, 56–62. [CrossRef]
33. Gachkar, D.; Taghvaei, S.H.; Norouzian-Maleki, S. Outdoor thermal comfort enhancement using various vegetation species and materials (case study: Delgosha Garden, Iran). *Sustain. Cities Soc.* **2021**, *75*, 103309. [CrossRef]
34. Norton, B.A.; Coutts, A.M.; Livesley, S.J.; Harris, R.J.; Hunter, A.M.; Williams, N.S. Planning for cooler cities: A framework to prioritise green infrastructure to mitigate high temperatures in urban landscapes. *Landsc. Urban Plan.* **2015**, *134*, 127–138. [CrossRef]
35. Abdollahzadeh, N.; Biloria, N. Outdoor thermal comfort: Analyzing the impact of urban configurations on the thermal performance of street canyons in the humid subtropical climate of Sydney. *Front. Archit. Res.* **2021**, *10*, 394–409. [CrossRef]
36. Wang, L.; Su, J.; Gu, Z.; Tang, L. Numerical study on flow field and pollutant dispersion in an ideal street canyon within a real tree model at different wind velocities. *Comput. Math. Appl.* **2021**, *81*, 679–692. [CrossRef]
37. Feldhake, C.; Danielson, R.; Butler, J. Turfgrass evapotranspiration. I. Factors influencing rate in urban environments 1. *Agron. J.* **1983**, *75*, 824–830. [CrossRef]
38. Tan, C.L.; Wong, N.H.; Tan, P.Y.; Jusuf, S.K.; Chiam, Z.Q. Impact of plant evapotranspiration rate and shrub albedo on temperature reduction in the tropical outdoor environment. *Build. Environ.* **2015**, *94*, 206–217. [CrossRef]
39. Dimoudi, A.; Nikolopoulou, M. Vegetation in the urban environment: Microclimatic analysis and benefits. *Energy Build.* **2003**, *35*, 69–76. [CrossRef]
40. Monteith, J.; Unsworth, M. *Principles of Environmental Physics: Plants, Animals, and the Atmosphere*; Academic Press: Cambridge, MA, USA, 2013.
41. Pauleit, S. Urban street tree plantings: Identifying the key requirements. In *Proceedings of the Institution of Civil Engineers-Municipal Engineer*; Thomas Telford Ltd.: London, UK, 2003; Volume 156, pp. 43–50. [CrossRef]
42. Wong, N.H.; Yu, C. Study of green areas and urban heat island in a tropical city. *Habitat Int.* **2005**, *29*, 547–558. [CrossRef]
43. Grace, J. 3. Plant response to wind. *Agric. Ecosyst. Environ.* **1988**, *22*, 71–88. [CrossRef]
44. Mirabi, E.; Davies, P.J. Mitigating urban heat along roadways; systematic review of impact and practicability. *Urban Clim.* **2024**, *58*, 102207. [CrossRef]
45. Sawka, M.; Millward, A.A.; Mckay, J.; Sarkovich, M. Growing summer energy conservation through residential tree planting. *Landsc. Urban Plan.* **2013**, *113*, 1–9. [CrossRef]
46. Vailshery, L.S.; Jaganmohan, M.; Nagendra, H. Effect of street trees on microclimate and air pollution in a tropical city. *Urban For. Urban Green.* **2013**, *12*, 408–415. [CrossRef]
47. Srivanit, M.; Hokao, K. Evaluating the cooling effects of greening for improving the outdoor thermal environment at an institutional campus in the summer. *Build. Environ.* **2013**, *66*, 158–172. [CrossRef]
48. Souch, C.; Souch, C. The effect of trees on summertime below canopy urban climates: A case study Bloomington, Indiana. *J. Arboric* **1993**, *19*, 303–312. [CrossRef]
49. Department, H.K.P. HKGBC Guidebook for Sustainable Built Environment. In *Hong Kong Planning Standards and Guidelines*; Chapter 11: Urban Design Guidelines; 2015. Available online: https://hkgbc.org.hk/eng/resources/publications/HKGBC-Publication/Guidebook/HKGBC-Guidebook-for-Sustainable-Built-Environment.pdf (accessed on 22 August 2024).
50. Ryliškė, J.; Puodžiukas, V.; Žilionienė, D. *New Approach to the Lithuanian Road Classification Based on Worldwide Experience*; VGTU Press: Vilnius, Lithuania, 2017. Available online: https://etalpykla.vilniustech.lt/handle/123456789/154999 (accessed on 22 August 2024).
51. Stewart, I.D.; Oke, T.R. Local climate zones for urban temperature studies. *Bull. Am. Meteorol. Soc.* **2012**, *93*, 1879–1900. [CrossRef]
52. Godinho, S.; Gil, A.; Guiomar, N.; Costa, M.J.; Neves, N. Assessing the role of Mediterranean evergreen oaks canopy cover in land surface albedo and temperature using a remote sensing-based approach. *Appl. Geogr.* **2016**, *74*, 84–94. [CrossRef]

53. Gorgani, S.A.; Panahi, M.; Rezaie, F. The Relationship between NDVI and LST in the urban area of Mashhad, Iran. In Proceedings of the International Conference on Civil Engineering Architecture & Urban Sustainable Development, Tabriz, Iran, 27–28 November 2013; Volume 2013.
54. Sun, D.; Kafatos, M. Note on the NDVI-LST relationship and the use of temperature-related drought indices over North America. *Geophys. Res. Lett.* **2007**, *34*, 1485. [CrossRef]
55. Marzban, F.; Sodoudi, S.; Preusker, R. The influence of land-cover type on the relationship between NDVI–LST and LST-T air. *Int. J. Remote Sens.* **2018**, *39*, 1377–1398. [CrossRef]
56. Khandelwal, S.; Goyal, R.; Kaul, N.; Mathew, A. Assessment of land surface temperature variation due to change in elevation of area surrounding Jaipur, India. *Egypt. J. Remote Sens. Space Sci.* **2018**, *21*, 87–94. [CrossRef]
57. Green Adelaide. Urban Heat and Tree Canopy Mapping. 2024. Available online: https://www.greenadelaide.sa.gov.au/projects/urban-heat-and-tree-canopy-mapping (accessed on 22 August 2024).
58. BOM. Adelaide: Geographic Information. 2017. Available online: http://www.bom.gov.au/water/nwa/2017/adelaide/regiondescription/geographicinformation.shtml#Gdescription (accessed on 22 August 2024).
59. DEW. Urban Heat and Tree Mapping of Adelaide. 2024. Available online: http://spatialwebapps.environment.sa.gov.au/urbanheat/?viewer=urbanheat (accessed on 22 August 2024).
60. Peel, M.C.; Finlayson, B.L.; McMahon, T.A. Updated world map of the Köppen-Geiger climate classification. *Hydrol. Earth Syst. Sci.* **2007**, *11*, 1633–1644. [CrossRef]
61. BOM. Seasonal Climate Summary for Greater Adelaide, Product code IDCKGC23L0. In *Greater Adelaide in Summer 2018–19: Warmer and Drier Than Average* ; Technical Report, Common Wealth; 2019. Available online: http://www.bom.gov.au/climate/current/season/sa/archive/201902.adelaide.shtml (accessed on 1 July 2025).
62. BOM. Greater Adelaide in Summer 2022–23: Drier Than Average; Generally Warmer Days. 2023. Available online: http://www.bom.gov.au/climate/current/season/sa/archive/202302.adelaide.shtml (accessed on 26 August 2024).
63. BOM. Monthly Climate Statistics. 2009. Available online: http://www.bom.gov.au/climate/averages/tables/cw_023090.shtml (accessed on 23 August 2024).
64. Ossola, A.; Jenerette, G.D.; McGrath, A.; Chow, W.; Hughes, L.; Leishman, M.R. Small vegetated patches greatly reduce urban surface temperature during a summer heatwave in Adelaide, Australia. *Landsc. Urban Plan.* **2021**, *209*, 104046. [CrossRef]
65. Cowan, T.; Purich, A.; Perkins, S.; Pezza, A.; Boschat, G.; Sadler, K. More frequent, longer, and hotter heat waves for Australia in the twenty-first century. *J. Clim.* **2014**, *27*, 5851–5871. [CrossRef]
66. Pezza, A.B.; Van Rensch, P.; Cai, W. Severe heat waves in Southern Australia: Synoptic climatology and large scale connections. *Clim. Dyn.* **2012**, *38*, 209–224. [CrossRef]
67. Tryhorn, L.; Risbey, J. On the distribution of heat waves over the Australian region. *Aust. Meteorol. Mag.* **2006**, *55*, 169–182.
68. Parker, T.J.; Berry, G.J.; Reeder, M.J. The influence of tropical cyclones on heat waves in Southeastern Australia. *Geophys. Res. Lett.* **2013**, *40*, 6264–6270. [CrossRef]
69. Guan, H.; Bennett, J.; Ewenz, C.; Benger, S.; Zhu, S.; Clay, R.; Soebarto, V. *Characterisation, Interpretation and Implications of the Adelaide Urban Heat Island*; Flinders University: Adelaide, Australia, 2013.
70. Erell, E.; Williamson, T. Intra-urban differences in canopy layer air temperature at a mid-latitude city. *Int. J. Climatol. A J. R. Meteorol. Soc.* **2007**, *27*, 1243–1255. [CrossRef]
71. Rogers, C.D.; Gallant, A.J.; Tapper, N.J. Is the urban heat island exacerbated during heatwaves in southern Australian cities? *Theor. Appl. Climatol.* **2019**, *137*, 441–457. [CrossRef]
72. Carlos, B.K.; Veronica, S.; Scott, H.; Ehsan, S. The Potential for Urban Canopy Cover to Reduce Heat-Related Mortality in Adelaide. In *Urban Overheating: Heat Mitigation and the Impact on Health*; Springer: Berlin/Heidelberg, Germany, 2022; pp. 249–273. [CrossRef]
73. Data.SA. Metropolitan Adelaide Tree Canopy and Built Environment 2018/2019 tp 2022. 2024. Available online: https://data.sa.gov.au/data/dataset/metropolitan-adelaide-tree-canopy-and-built-environment-change-detection-2018-2019-to-2022 (accessed on 22 August 2024).
74. Seed Consulting Services Western Adelaide Urban Heat Mapping Project. Technical report, Seed Consulting Services, Airborne Research Seed Consulting Services, Adelaide, South Australia 2017. Available online: https://www.adaptwest.com.au/sites/adaptwest/media/pdf/western_adelaide_urban_heat_mapping_report-(2).pdf (accessed on 23 August 2024).
75. ABS. Census Geography Glossary, Understanding Census Geography, Descriptions of All Geographies Available for 2021 Census Data. 2022. Available online: https://www.abs.gov.au/census/guide-census-data/geography/census-geography-glossary (accessed on 23 August 2024).

76. ABS. Mesh Blocks, Australian Statistical Geography Standard (ASGS) Edition 3. 2021. Available online: https://www.abs.gov.au/statistics/standards/australian-statistical-geography-standard-asgs-edition-3/jul2021-jun2026/main-structure-and-greater-capital-city-statistical-areas/mesh-blocks (accessed on 23 August 2024).
77. OSM. OpenStreetMaps. 2023. Available online : https://www.openstreetmap.org/edit#map=16/-33.77084/151.10624 (accessed on 23 August 2024).
78. Fatemi, M.; Narangifard, M. Monitoring LULC changes and its impact on the LST and NDVI in District 1 of Shiraz City. *Arab. J. Geosci.* **2019**, *12*, 127. [CrossRef]
79. Anbazhagan, S.; Paramasivam, C. Statistical correlation between land surface temperature (LST) and vegetation index (NDVI) using multi-temporal landsat TM data. *Int. J. Adv. Earth Sci. Eng.* **2016**, *5*, 333–346. [CrossRef]
80. Hastie, T.J. Generalized additive models. In *Statistical Models in S*; Routledge: London, UK, 2017; pp. 249–307. Available online: https://www.taylorfrancis.com/chapters/edit/10.1201/9780203738535-7/generalized-additive-models-trevor-hastie (accessed on 22 August 2024).
81. Wood, S.; Goude, Y.; Fasiolo, M. Interpretability in generalized additive models. In *Interpretability for Industry 4.0: Statistical and Machine Learning Approaches*; Springer: Berlin/Heidelberg, Germany, 2022; pp. 85–123. [CrossRef]
82. Wood, S.N. *Generalized Additive Models: An Introduction with R*; Chapman and Hall/CRC: Boca Raton, FL, USA, 2017. [CrossRef]
83. Wiley, M.; Wiley, J.F.; Wiley, M.; Wiley, J.F. GAMs. In *Advanced R Statistical Programming and Data Models: Analysis, Machine Learning, and Visualization*; Apress: New York, NY, USA, 2019; pp. 165–224. Available online: https://link.springer.com/book/10.1007/978-1-4842-2872-2 (accessed on 22 August 2024).
84. Phan, T.N.; Kappas, M.; Tran, T.P. Land surface temperature variation due to changes in elevation in northwest Vietnam. *Climate* **2018**, *6*, 28. [CrossRef]
85. Abubakar, B.A.; Abubakar, S.A. Effect of elevation on land surface temperature (LST) variation in Jos and Environs, Nigeria. *Intercont. Geoinf. Days* **2021**, *2*, 151–154.
86. Vasyl Cherlinka. All You Need To Know About Index. 2024. Available online: https://www.eos.com/blog/ndvi-faq-all-you-need-to-know-about-ndvi/?utm_source=chatgpt.com (accessed on 22 August 2024).
87. Vilcins, D.; Sly, P.D.; Scarth, P.; Mavoa, S. Green space in health research: An overview of common indicators of greenness. *Rev. Environ. Health* **2024**, *39*, 221–231. [CrossRef]
88. Rhew, I.C.; Vander Stoep, A.; Kearney, A.; Smith, N.L.; Dunbar, M.D. Validation of the normalized difference vegetation index as a measure of neighborhood greenness. *Ann. Epidemiol.* **2011**, *21*, 946–952. [CrossRef]
89. Kim, J.; Lee, D.K.; Brown, R.D.; Kim, S.; Kim, J.H.; Sung, S. The effect of extremely low sky view factor on land surface temperatures in urban residential areas. *Sustain. Cities Soc.* **2022**, *80*, 103799. [CrossRef]
90. Yang, H.; Chen, G.; Wang, D.; Hang, J.; Li, Q.; Wang, Q. Influences of street aspect ratios and realistic solar heating on convective heat transfer and ventilation in full-scale 2D street canyons. *Build. Environ.* **2021**, *204*, 108125. [CrossRef]
91. Chen, Q.; Cheng, Q.; Chen, Y.; Li, K.; Wang, D.; Cao, S. The influence of sky view factor on daytime and nighttime urban land surface temperature in different spatial-temporal scales: A case study of Beijing. *Remote Sens.* **2021**, *13*, 4117. [CrossRef]
92. Morabito, M.; Crisci, A.; Georgiadis, T.; Orlandini, S.; Munafò, M.; Congedo, L.; Rota, P.; Zazzi, M. Urban imperviousness effects on summer surface temperatures nearby residential buildings in different urban zones of Parma. *Remote Sens.* **2017**, *10*, 26. [CrossRef]
93. Lu, Y.; Yue, W.; Huang, Y. Effects of land use on land surface temperature: A case study of Wuhan, China. *Int. J. Environ. Res. Public Health* **2021**, *18*, 9987. [CrossRef] [PubMed]
94. Chen, S.; Haase, D.; Qureshi, S.; Firozjaei, M.K. Integrated land use and urban function impacts on land surface temperature: Implications on urban heat mitigation in Berlin with eight-type spaces. *Sustain. Cities Soc.* **2022**, *83*, 103944. [CrossRef]
95. Kim, J.I.; Jun, M.J.; Yeo, C.H.; Kwon, K.H.; Hyun, J.Y. The effects of land use zoning and densification on changes in land surface temperature in Seoul. *Sustainability* **2019**, *11*, 7056. [CrossRef]
96. Stone, B.; Hess, J.J.; Frumkin, H. Urban form and extreme heat events: Are sprawling cities more vulnerable to climate change than compact cities? *Environ. Health Perspect.* **2010**, *118*, 1425–1428. [CrossRef]
97. Chen, L.; Jiang, R.; Xiang, W.N. Surface heat island in Shanghai and its relationship with urban development from 1989 to 2013. *Adv. Meteorol.* **2016**, *2016*, 9782686. [CrossRef]
98. Wen, H.; Xiao, Y.; Zhang, L. School district, education quality, and housing price: Evidence from a natural experiment in Hangzhou, China. *Cities* **2017**, *66*, 72–80. [CrossRef]
99. He, X.; Gao, W.; Wang, R. Effect of urban structure on land surface temperature around elementary schools in Hangzhou based on local climate zones. *Sustain. Cities Soc.* **2024**, *114*, 105724. [CrossRef]
100. Su, W.; Gu, C.; Yang, G. Assessing the impact of land use/land cover on urban heat island pattern in Nanjing City, China. *J. Urban Plan. Dev.* **2010**, *136*, 365–372. [CrossRef]

101. Algretawee, H.; Rayburg, S.; Neave, M. Estimating the effect of park proximity to the central of Melbourne city on Urban Heat Island (UHI) relative to Land Surface Temperature (LST). *Ecol. Eng.* **2019**, *138*, 374–390. [CrossRef]
102. Feng, Y.; Zhang, K.; Li, A.; Zhang, Y.; Wang, K.; Guo, N.; Wan, H.Y.; Tan, X.; Dong, N.; Xu, X.; et al. Spatial and Seasonal Variation and the Driving Mechanism of the Thermal Effects of Urban Park Green Spaces in Zhengzhou, China. *Land* **2024**, *13*, 1474. [CrossRef]
103. Wang, X.; Cheng, H.; Xi, J.; Yang, G.; Zhao, Y. Relationship between park composition, vegetation characteristics and cool island effect. *Sustainability* **2018**, *10*, 587. [CrossRef]
104. Jena, J.; Misra, S.R.; Tripathi, K.P. Normalized difference vegetation index (NDVI) and its role in agriculture. *Agric. Food E-Newsl.* **2019**, *1*, 387–389.
105. Mildrexler, D.J.; Zhao, M.; Running, S.W. A global comparison between station air temperatures and MODIS land surface temperatures reveals the cooling role of forests. *J. Geophys. Res. Biogeosciences* **2011**, *116* , 1486. [CrossRef]
106. Huband, N.; Monteith, J. Radiative surface temperature and energy balance of a wheat canopy: I. Comparison of radiative and aerodynamic canopy temperature. *Bound.-Layer Meteorol.* **1986**, *36*, 1–17. [CrossRef]
107. Roberts, D.A.; Dennison, P.E.; Roth, K.L.; Dudley, K.; Hulley, G. Relationships between dominant plant species, fractional cover and land surface temperature in a Mediterranean ecosystem. *Remote Sens. Environ.* **2015**, *167*, 152–167. [CrossRef]
108. Liu, L.; Zhang, Y. Urban heat island analysis using the Landsat TM data and ASTER data: A case study in Hong Kong. *Remote Sens.* **2011**, *3*, 1535–1552. [CrossRef]
109. Zheng, Z.; Zhou, W.; Yan, J.; Qian, Y.; Wang, J.; Li, W. The higher, the cooler? Effects of building height on land surface temperatures in residential areas of Beijing. *Phys. Chem. Earth Parts A/B/C* **2019**, *110*, 149–156. [CrossRef]

atmosphere

Article

Summer Diurnal LST Variability Across Local Climate Zones Using ECOSTRESS Data in Lecce and Milan

Gianluca Pappaccogli [1], Antonio Esposito [1,2,*] and Riccardo Buccolieri [1]

[1] Dipartimento di Scienze e Tecnologie Biologiche ed Ambientali, University of Salento, S.P. 6 Lecce-Monteroni, 73100 Lecce, Italy; gianluca.pappaccogli@unisalento.it (G.P.); riccardo.buccolieri@unisalento.it (R.B.)
[2] Dipartimento di Matematica e Fisica, University of Salento, Via per Arnesano, snc, 73100 Lecce, Italy
* Correspondence: antonio.esposito@unisalento.it

Abstract: This study assesses the accuracy of Local Climate Zone (LCZ) classification and its impact on land surface temperature (LST) analysis in Mediterranean cities using high-resolution ECOSTRESS data. Two classification methods were compared: a Geographic Information System (GIS)-based approach integrating high-resolution geospatial data and an LCZ map derived from WUDAPT. Discrepancies in LCZ classification influenced the spatial distribution of urban forms, with WUDAPT overestimating LCZ 6 (open low-rise) and LCZ 8 (large low-rise) while underrepresenting more compact urban types. LST analysis revealed distinct thermal responses between Milan and Lecce, underscoring the influence of urban morphology and local climate. Densely built zones (LCZ 2, LCZ 5) exhibited the highest temperatures, especially at night, while LCZ 8 also retained significant heat. Milan's dense urban areas experienced pronounced nighttime overheating, whereas Lecce showed a clear daytime temperature gradient, with historic districts (LCZ 2) maintaining lower LST the light-colored and high thermal capacity of building materials. A Kruskal–Wallis test confirmed significant differences between the GIS-based and WUDAPT-derived LCZ maps, highlighting the impact of classification methodology and spatial resolution on LST analysis. These findings emphasize the need for multi-scale approaches to urban climate adaptation and mitigation, providing valuable advice for urban planners and policymakers in development of sustainable and climate-resilient cities. This research is also among the first to integrate ECOSTRESS data with LCZ maps to examine LST variations across spatial and temporal scales.

Keywords: land surface temperature; WUDAPT; Local Climate Zones; ECOSTRESS; urban overheating; remote sensing

1. Introduction

Urban planning and sustainable development are essential for enhancing the resilience and well-being of cities, particularly in the face of climate change and rapid urbanization. A comprehensive understanding of the interactions between climatic conditions and the urban microclimate is crucial for improving air quality and thermal environments [1]. The Local Climate Zone (LCZ) classification system plays a key role in this process by systematically categorizing urban areas based on their climatic and physical attributes [2,3]. The LCZ methodology provides a structured framework for urban planners to analyze distinct climate profiles within cities. Through LCZ classification, areas vulnerable to extreme temperatures, air pollution, or other environmental stressors can be identified, enabling targeted interventions to enhance climate resilience and urban living [4].

Analyzing local climate patterns has broader implications for sustainable urban development. Detailed climate data help identify opportunities to integrate green infrastructure, such as urban forests, green roofs, and permeable surfaces, which mitigate the Urban Heat Island (UHI) effect, improve air quality, enhance biodiversity, and promote the well-being of urban populations [5].

Additionally, the LCZ classification system facilitates intercity and interregional comparisons, fostering knowledge exchange among cities to address common urban challenges, such as rising temperatures, air pollution, and energy consumption. LCZ classifications serve as a foundation for implementing effective mitigation and adaptation strategies. Cities with similar LCZ patterns can share best practices in green space management or urban energy efficiency, accelerating progress toward urban sustainability [6].

In this context, this study aims to evaluate the accuracy and applicability of LCZ classifications in urban climate research. Specifically, it presents a comparative analysis of LCZs using geographic information systems (GIS) and morphological data to classify LCZs across several Italian cities located in the southern, central, and northern regions. A key component of this research is the comparison between two LCZ mapping approaches: a GIS-based LCZ map (hereinafter GIS-based map) and the European LCZ Level 0 map (hereinafter WUDAPT European LCZ map) developed by Demuzere et al. [6] is available as a downloadable GeoTIFF file from the World Urban Database and Access Portal Tools (WUDAPT) from which the city area of interest has been cropped. (https://figshare.com/articles/dataset/European_LCZ_map/13322450, last accessed 2 May 2025). According to their methodology, the validation was carried out through a comprehensive accuracy assessment, which involved a bootstrap cross-validation technique. In this process, the dataset was repeatedly sampled to evaluate the model's performance. Additionally, thematic benchmarks for 150 selected functional urban areas were established using independent global and open-source data on surface cover, surface imperviousness, building height, and anthropogenic heat [6,7]

The spatial distribution of heat in urban areas has traditionally been assessed through fixed-point air temperature measurements, satellite-based land surface temperature (LST) data [8–11], and numerical modeling. Extensive research has explored urban heat and evaluated the effectiveness of various mitigation strategies across different cities. Many studies utilize satellite-derived land surface temperature (LST) to analyze urban heat patterns from a broad, top-down perspective. The widespread use of LST data in heat mitigation studies is driven by its ability to provide consistent, global-scale observations of land surface temperatures through satellite monitoring [12,13]

Despite the availability of multiple remote sensing datasets, capturing diurnal variations in LST or Surface Urban Heat Island (SUHI) at a high spatial resolution remains a challenge. While satellites such as MODIS and Sentinel-3 provide frequent daily LST data [14–16], their relatively coarse spatial resolution (1 km^2) limits their effectiveness for urban microclimate studies. Similar studies such as Wu et al. [17] have characterized the spatial heterogeneity of the urban thermal environment at the city level using LSTs retrieved from Landsat-8. Higher-resolution satellites like Landsat or ASTER enable more precise spatial analyses but suffer from temporal limitations. However, ASTER provides only a limited number of images per city, while Landsat acquires data at fixed intervals and lacks nighttime observations [18,19].

To address these limitations, this study integrates high-resolution thermal data from the Ecosystem Spaceborne Thermal Radiometer Experiment on Space Station (ECOSTRESS) into urban temperature analyses.

This study assesses surface temperature patterns across different cities using ECOSTRESS satellite data. The analysis builds upon the previous objective by utilizing the two LCZ classifications described earlier (WUDAPT European LCZ and GIS-based maps). Integrating these datasets enables an evaluation of how different LCZ classification methods influence the assessment of the urban thermal environment. Additionally, a detailed examination of thermal variations across LCZs serves as a key indicator of the reliability and representativeness of the chosen classification approach.

The potential of ECOSTRESS for capturing diurnal LST variability has been highlighted in studies such as Chang et al. [20], Chang et al. [21], Hulley et al. [22], and Wei et al. [23]. Launched aboard the International Space Station (ISS) on 29 June 2018, ECOSTRESS provides LST observations with a high spatial resolution of 70 × 70 m, capturing thermal data at different times of day and night based on local time (LT), improving the temporal characterization of urban thermal environments [24].

The novelty of this study lies in its medium-to-long-term analysis of ECOSTRESS data, distinguishing it from prior research that primarily focused on short-term periods. For example, Chang et al. [20] examined ECOSTRESS LST data only between 1 June and 30 September 2019. In contrast, this study utilizes ECOSTRESS LST data for the summer months (June, July, and August, hereinafter JJA) from 2018 to 2024, combining them with LCZ maps to analyze diurnal LST dynamics in Lecce (a small city in southern Italy) and Milan (a metropolis in northern Italy). By leveraging ECOSTRESS's unique ability to capture LST variations at different times of the day over multiple years, this study provides a comprehensive assessment of the diurnal thermal behavior based on robust statistical analysis.

Advanced statistical techniques are employed to assess the significance of the data and to identify similarities among different LCZs in Lecce and Milan across various time periods. The findings contribute to a deeper understanding of the relationship between urban form and thermal dynamics, supporting urban planners and policymakers in developing climate adaptation and mitigation strategies.

This research provides essential knowledge into the role of LCZ mapping in shaping future urban growth and regeneration efforts. As urban areas continue to expand, integrating climate-sensitive planning approaches is critical to mitigating environmental stressors. LCZ data facilitate the identification of priority areas for climate adaptation measures, including increasing vegetation cover, improving energy efficiency, and optimizing road networks to enhance air circulation [25]. Ultimately, this study underscores the importance of incorporating high-resolution LST data into urban climate research to support sustainable development and urban resilience [26].

2. Methodology

2.1. Study Areas

This study focuses on five Italian cities—Bari, Naples, Rome, Milan, and Lecce. The selection includes four major metropolitan cities (Bari, Naples, Rome, and Milan) representing different regions of Italy (south, center, and north). In addition, Lecce was included because of its association with the University of Salento.

The selected cities include different climatic conditions according to Köppen's classification. Naples, Rome, and Lecce share a Mediterranean climate (Csa), characterized by mild winters and hot, dry summers. Bari, on the other hand, has a humid subtropical climate (Cfa), characterized by relatively high temperatures and evenly distributed precipitation throughout the year. Milan, located in northern Italy, falls into the category of west coast marine climate (Cfb), characterized by moderate temperature variations and consistent precipitation throughout the seasons (source: https://koeppen-geiger.vu-wien.ac.at/, accessed 19 December 2024).

Regarding urban development, Milan is the largest among these cities, exhibiting a radial expansion pattern with new districts emerging outward from the historical core. Rome similarly follows a radial growth model but features multiple urban centers interspersed with significant green areas. Lecce also expands radially, with its urban growth directed northwest toward the industrial sector, east toward the coast, and south in predominantly residential zones. As coastal cities, Naples and Bari have distinct development patterns: Naples has expanded along the shoreline, influenced by the presence of Mount Vesuvius, whereas Bari has grown further inland (Figure 1).

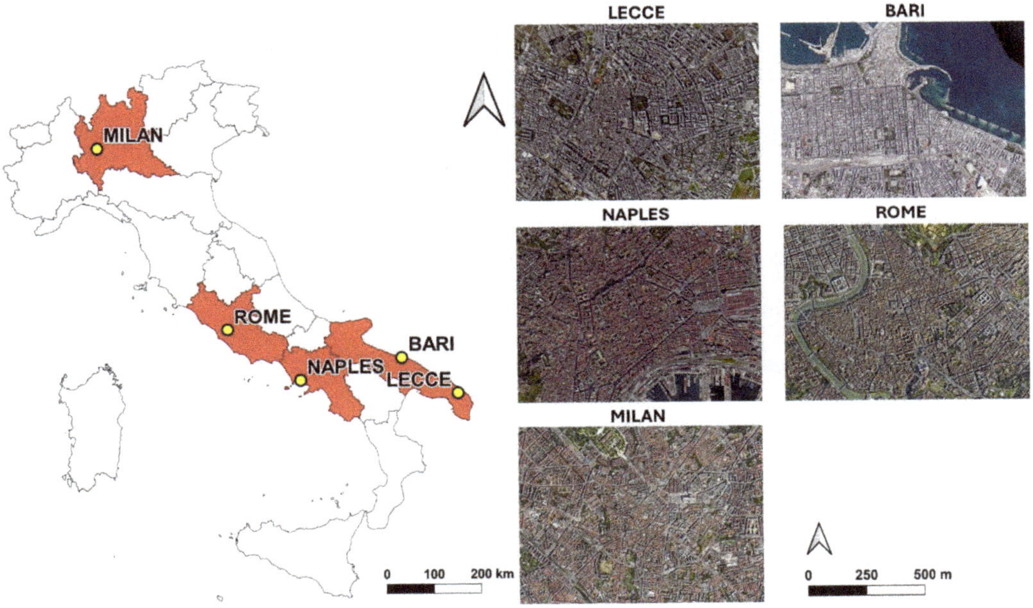

Figure 1. Location of selected cities (their regions colored). Satellite images of the Italian cities investigated (maps from Google Earth Pro).

2.2. Collecting Morphological Data

Morphological and land use data were collected and analyzed using shapefiles for all cities. The data were obtained from the following institutional sources:

- for Lecce and Bari, SIT Puglia (http://www.sit.puglia.it, last accessed on 14 April 2022);
- for Naples, Geoportale Nazionale (http://wms.pcn.minambiente.it, last accessed on 18 May 2022);
- for Rome Open Data Lazio (https://geoportale.regione.lazio.it, last accessed on 20 October 2022);
- for Milan, Milano Geoportale (https://geoportale.comune.milano.it, last accessed on 17 June 2022).

2.3. Estimation of Morphological Parameters Using GIS

A morphological analysis was performed using QGIS software (version: 3.22.1, https://www.qgis.org/en/site, accessed on 10 December 2024). A region of interest (ROI) was defined for each selected city, with its dimensions adjusted based on the city's size. Specifically, a 10 km × 10 km area was designated for Lecce and Bari (hereafter referred to as the "Lecce region" and "Bari region"), while a larger 20 km × 20 km area was allocated for Naples, Rome, and Milan (hereafter the "Naples region", and so forth).

The analysis included key urban canopy parameters (UCPs) such as mean building height, sky view factor, aspect ratio, building surface fraction, and the fractions of impervious and permeable surfaces. These parameters were computed at a spatial resolution of 100 m × 100 m following the classification framework proposed by Stewart and Oke [2]. For a comprehensive explanation of the calculation methodology, refer to Esposito et al. [27]. Regarding the building surface fraction (BSF), fraction of impervious surface (ISF) and permeable surface (PSF), Stewart and Oke [2] describe them respectively as the ratio of building plan area, the ratio of impervious surface (paved, rocks, roads, etc.) to total area (%) and as the ratio of permeable surface (bare soil, vegetation, water, green space, agricultural land, etc.) to total area (%).

In calculating these parameters, the sum of ISF, PSF, and BSF values in a single grid cell are equal to 1, so in the absence or lack of permeable elements for PSF calculation, a simple subtraction was performed to obtain this parameter:

$$\mathrm{PSF} = 1 - (\mathrm{BSF} + \mathrm{ISF}) \quad (1)$$

Figure 2 shows the maps of the PSF, ISF, and BSF parameters of all cities at a spatial resolution of 100 × 100.

All the cities show an urban center with prevalence of areas with both high BSF values (generally between 0.70 and 0.95) and high ISF values, underscoring the prevalence of impervious areas (roads, sidewalks and squares) and low PSF values, which instead shows a tendency to increase moving outward from the cities.

2.4. Construction of the GIS-Based Maps

The creation of GIS-based maps involved calculating the actual morphological parameters of a city and applying fuzzy logic to improve classification accuracy. This approach integrates geometric and surface coverage properties, including sky view factor (SVF), BSF, ISF, aspect ratio (AR), and average building height (HM), to classify urban and rural areas into distinct LCZs [28] based on Stewart and Oke's criteria [2]. The map was constructed by assigning an LCZ category to each grid cell based on the morphological parameters calculated for each cell and applying the fuzzy logic. Fuzzy logic assigns membership percentages to each grid cell, quantifying the degree of correspondence to specific LCZ types. Rather than employing a binary classification, this method evaluates how well a cell's features align with predefined parameter ranges for each LCZ, resulting in a fuzzy, continuous representation of spatial transitions between urban and rural areas. A linear membership function was applied for fuzzy membership assignment. Pixels with values within the specified range were assigned a membership value of 1, while those outside the range received membership values based on their proximity to the range, gradually decreasing to 0 at ± the range specified by the edge. A fuzzy minimum overlap method was then used to generate the fuzzy levels for all urban LCZ types. This approach captures the uncertainties in classification and allows the identification of the three most likely LCZs for each 100 × 100 m grid cell [29]. In this process, BSF was used to distinguish urban from nonurban areas, with a threshold value of 1. However, this threshold introduced a potential source of inaccuracy. Due to the lack of specific land cover information for some

LCZ types (e.g., LCZ A-G), these categories were combined into a single class called "No Built Area" [27].

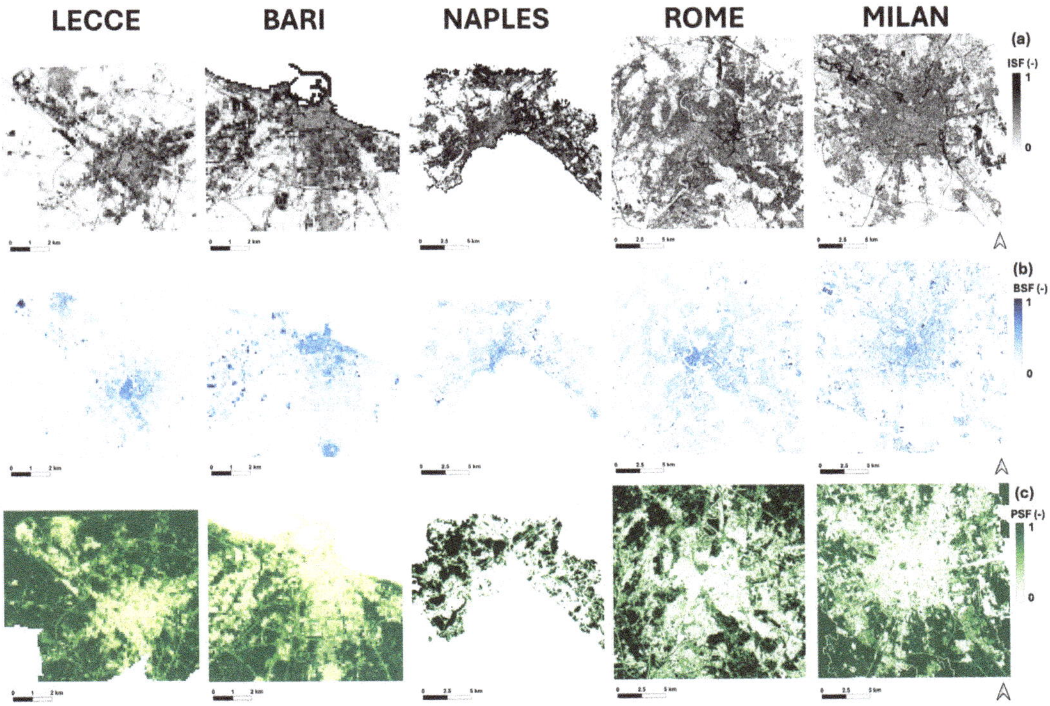

Figure 2. Maps of ISF (**a**), BSF (**b**), and PSF (**c**) for the cities of Lecce, Bari, Naples, Rome, and Milan.

2.5. Collecting ECOSTRESS LST Data

LSTs have been analyzed for two cities (Milan and Lecce). Data were derived from the ECOSTRESS Level-2 land surface temperature and emissivity (LSTE) product, available through the NASA Land Processes Distributed Active Archive Center (NASA-LPDAAC) (https://lpdaac.usgs.gov/, accessed on 6 February 2025). The ECOSTRESS mission, managed by NASA's Jet Propulsion Laboratory, provides several layers of data products, with increasing levels incorporating additional auxiliary information. The ECOSTRESS LST Level-2 (ECO2LSTE) product was specifically selected for its high spatial resolution and validated accuracy.

The ECOSTRESS LST product is generated using a physics-based temperature emissivity separation (TES) algorithm that dynamically retrieves LST from five thermal infrared bands with wavelengths between 8 and 12.5 μm [30]. The final product has a spatial resolution of 70 m × 70 m, resampled from the original pixel size of 38 × 68 m, and each image covers a wide band of about 400 km, allowing for a broad regional analysis.

Validation studies have demonstrated the high accuracy and reliability of the ECOSTRESS LST product. Results from the Algorithm Theoretical Basis Document indicate consistent accuracy to within 1 K over different types of land surfaces [24]. Additional validation efforts conducted by the ECOSTRESS LST science team reported uncertainties of <1 K [30]. A more recent validation study further confirmed the product's performance, reporting a root mean square error (RMSE) of 1.07 K, a mean absolute error (MAE) of 0.40 K, and an R^2 value greater than 0.99 at global validation sites [31].

For this study, ECOSTRESS LST data were analyzed to assess diurnal LST dynamics during JJA. This approach addresses the need to investigate the peak thermal response of urban areas, as summer represents the period of highest LST variability. Analyzing LST patterns during this season is essential for evaluating the influence of urban morphology on thermal anomalies, identifying areas most exposed to urban warming, and determining where adaptation strategies should be implemented to mitigate heat stress for residents.

All available images from 2018 to 2024 were collected, along with the corresponding cloud masks and quality control layers. A quality check was then performed to exclude measurement errors caused by cloud contamination. The dataset includes 189 images for Lecce and 300 images for Milan for the summer season, collected between 2018 and 2024. These images include both daytime and nighttime LST observations. For Lecce, the dataset includes 13 images from 2018, 20 from 2019, 24 from 2020, 38 from 2021, 33 from 2022, 31 from 2023, and 30 from 2024. For Milan, it comprises 22 images from 2018, 43 from 2019, 48 from 2020, 47 from 2021, 45 from 2022, 56 from 2023, and 39 from 2024.

To analyze the diurnal variability of LST in detail, the dataset was categorized into six time periods: 03:00–06:00, 07:00–10:00, 11:00–14:00, 15:00–18:00, 19:00–22:00, and 23:00–02:00. This segmentation ensured a balanced distribution of images across different periods of the day, to provide a robust statistical analysis of the data.

2.6. Kruskal–Wallis Test Analysis

To detect significant differences in LST across various LCZs throughout the daily cycle, the Kruskal–Wallis test was performed. As described by Ostertagova et al. [32], this nonparametric statistical method assesses whether significant differences exist among three or more independent groups. It serves as an alternative to one-way analysis of variance ANOVA when the assumptions of normality and homogeneity of variances are not satisfied. This test is rank-based and determines whether the samples originate from the same distribution.

The Kruskal–Wallis test is applicable under the following conditions:

- The observations are independent;
- The dependent variable is at least ordinal;
- The populations have similar distribution shapes, differing only in location (median).

The test examines the null hypothesis (H_0) that all groups originate from the same distribution:

$$H_0 : F_1(x) = F_2(x) = \cdots = F_k(x)\ \forall x \quad (2)$$

The alternative hypothesis (H_1) states that at least one group distribution differs:

$$H_1 : \exists_{i,j} \quad (3)$$

such that:

$$F_i(x) \neq F_j(x) \quad (4)$$

The test procedure involves ranking all observations across groups and computing the test statistic (H):

$$H = \frac{12}{N(N+1)} \sum_{i=1}^{k} \frac{R_i^2}{n_i} - 3(N+1) \quad (5)$$

where:

- N is the total number of observations.

- R_i is the sum of ranks for group i.
- n_i is the sample size of group i.

If tied ranks exist, a correction factor is applied to adjust H.

The test statistic follows an approximate chi-square (X^2) distribution with $k-1$ degrees of freedom. The null hypothesis is rejected if:

$$H > X^2_{\alpha,k-1} \quad (6)$$

where $X^2_{\alpha,k-1}$ is the critical value at a chosen significance level α.

If the *p*-value associated with H is less than a predetermined threshold (usually 0.05), the null hypothesis is rejected, suggesting that at least one group differs from the others. In this research, the Kruskal–Wallis test is used to test whether significant differences exist between multiple groups based on LST values.

3. Results and Discussion

3.1. Development of LCZ Maps

The evaluation of LCZ classification accuracy was performed using a confusion matrix, comparing the GIS-based map with the WUDAPT European LCZ map [6,7] (Figure 3). The rows of the matrix represent the WUDAPT LCZ classes, while the columns correspond to the GIS-based map. The diagonal of the matrix, highlighted in pink, indicates the number of pixels classified identically by both maps. Overall accuracy (OA) values were calculated by dividing the sum of correctly classified pixels (diagonal sum) by the total number of pixels, providing an overall percentage for classification accuracy [33]. The OA results revealed varying classification accuracy across cities: 73% for Lecce, 53% for Bari, 68% for Naples, 51% for Rome, and 48% for Milan. These values were obtained using a strict criterion that required exact pixel conformity to all UCPs tested, which might suggest that the actual percentage of correctly classified pixels might be higher in cases where tolerance or weighting adjustments were applied.

The OA values presented here are in line with the results of other studies, such as Hidalgo et al. [34], which also found discrepancies between the actual LCZ classifications and UCP value ranges. These deviations can be attributed to the inherent limitations of classifying LCZs based on predefined UCP thresholds, especially when morphologically and microclimatically similar LCZs are misclassified. In addition, the shape and size of spatial units in maps can affect the accuracy of classification. As noted by Wellinger et al. [35], LCZs typically do not conform to square shapes, resulting in mixed pixels along zone boundaries in gridded maps, which contributes to inconsistencies in UCP values.

Compared with the WUDAPT-based method, the GIS-based mapping approach offers greater accuracy and adaptability to local contexts. By utilizing high-resolution geospatial data from sources such as OpenStreetMap and digital elevation models (DEMs), along with complex morphological parameters like SVF and AR, GIS-based mapping provides a more detailed representation of LCZs. This is particularly advantageous in cities with intricate urban layouts, where WUDAPT, relying on satellite imagery and machine learning algorithms like Random Forest, struggles to capture vertical variability and distinguish heterogeneous urban forms [29,35]. Additionally, GIS methods allow for the adjustment of LCZ parameters based on city-specific characteristics, enhancing classification accuracy and relevance [35].

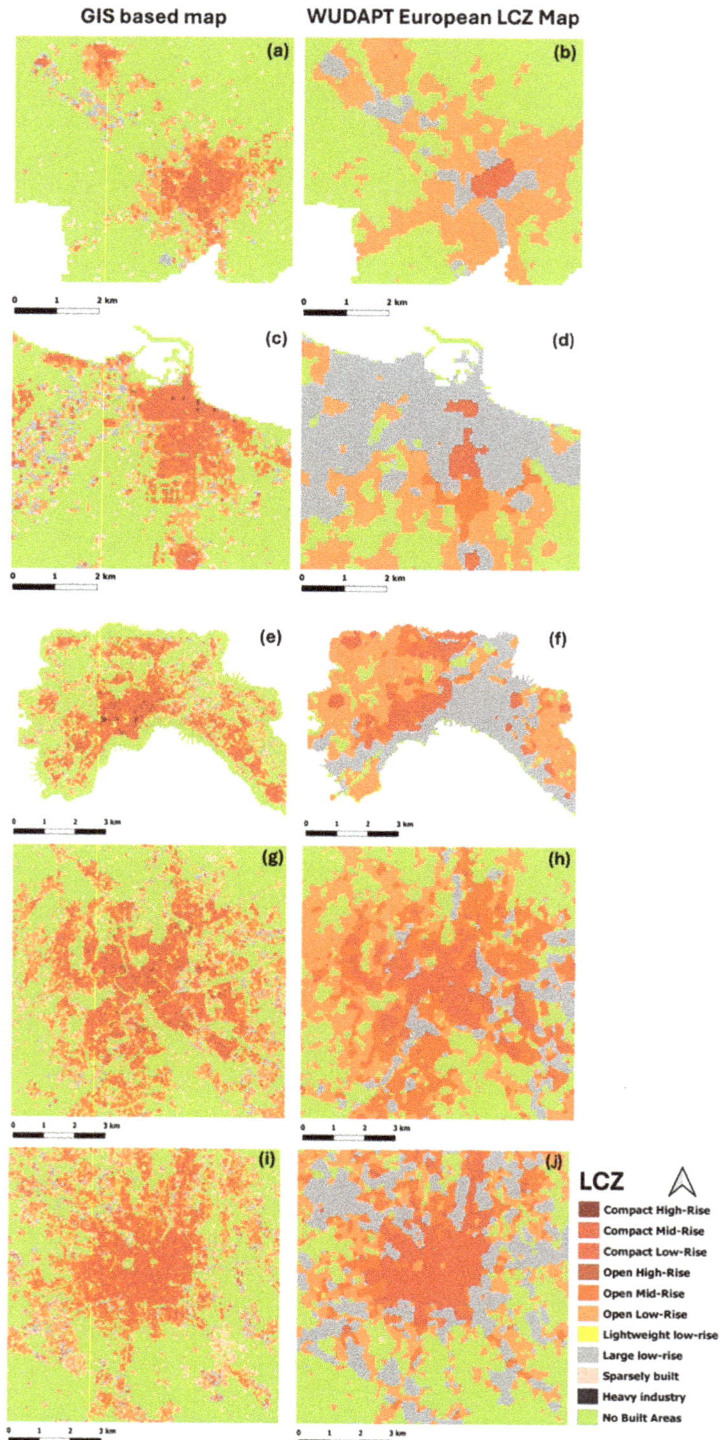

Figure 3. Comparison between GIS-based map (left panel) and WUDAPT European LCZ map (right panel) for Lecce (**a**,**b**), Bari (**c**,**d**), Naples (**e**,**f**), Rome (**g**,**h**), and Milan (**i**,**j**).

3.2. GIS-Based and WUDAPT European LCZ Map

In this section, the different distributions of LCZ classes from the two different maps were identified and quantified for all cities (Figure 4). The comparison of GIS and WUDAPT data shows significant discrepancies in the representation of LCZs in Italian cities ("No Built Areas" are not considered in the following analysis as it is idealized to make up for the lack of information of non-urban LCZs as explained in Section 2.4).

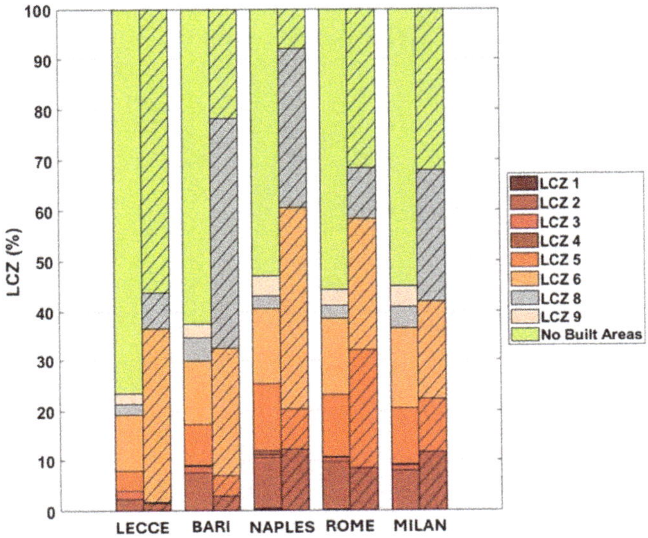

Figure 4. Percentage distribution of LCZs for all investigated ROIs: GIS-based map (left columns not lined) and the WUDAPT European LCZ Map (right columns lined).

GIS data tend to present a more uniform distribution of LCZs, with a consistent presence of LCZ 2 (compact midrise), LCZ 5 (open midrise), and LCZ 6 (open low-rise) in all cities. For example, GIS data for Naples assign 10% to LCZ 2 and 13% to LCZ 5, while Rome similarly presents 9% for LCZ 2 and 12% for LCZ 5. In contrast, WUDAPT data show a more heterogeneous distribution, often attributing higher percentages to LCZ 6 (35% in Lecce and 40% in Naples) and LCZ 8 (46% in Bari and 32% in Naples).

Notably, LCZ 1 (compact high-density) and LCZ 4 (open high-density) are absent or minimally represented in both datasets, suggesting a low prevalence of these urban forms in the cities. LCZ 3 (compact low-density) also shows consistently low percentages, with GIS reporting a maximum of 2% in Lecce and WUDAPT recording almost negligible values in all cities. On the other hand, LCZ 9 (sparsely built-up industrial areas) shows strong differences, with GIS identifying significant contributions in Milan (4%), while WUDAPT reports values close to zero.

These discrepancies underscore the methodological differences in classification approaches between the GIS and WUDAPT datasets. GIS captures a finer granularity in urban morphology, while WUDAPT tends to emphasize broader classifications, with a focus on zones such as LCZ 6 and LCZ 8.

3.3. Diurnal Variability of LST in Different LCZs

This section compares the variability of LST across different LCZs, using both GIS and European LCZ map, for the cities of Lecce and Milan.

Statistical analysis with the nonparametric Kruskal–Wallis test identifies significant differences between the different LCZs, represented by the letters above each boxplot. The letters were assigned based on the *p*-values. Specifically, the first boxplot in each figure is always assigned the first letter of the sequence (i.e., "a/A"). Lowercase/uppercase characters are used to distinguish Lecce and Milan, respectively, while maintaining the same statistical significance. If the *p*-value for the second boxplot is less than 0.05, indicating a significant difference, it is assigned the next letter in the sequence. If the *p*-value is greater than 0.05, the second boxplot retains the same letter as the first, as no significant difference is detected. This method is applied throughout the entire sequence, ensuring a consistent assignment of letters based on statistical significance. In the case of mixed behavior (as in the case of Figure 5 for the WUDAPT European LCZ Map in Lecce), the box is assigned two letters, again based on the *p*-values. This letter assignment system clearly highlights the similarities and differences between the LCZs based on the statistical test performed.

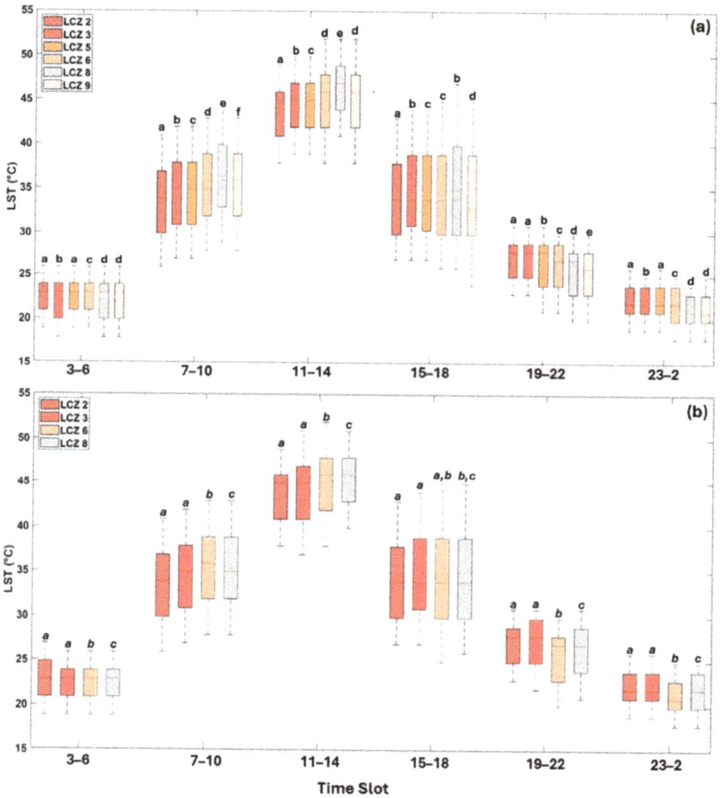

Figure 5. Diurnal variability of LST for various LCZs in the city of Lecce. GIS-based map (**a**) and the WUDAPT European LCZ Map (**b**). The red line present in the boxes indicates the median. The boxes illustrate the interquartile range, spanning from the 25th to the 75th percentile. The whiskers extend to ±2.7 standard deviations, encompassing 99.3% of the data. Different lowercase letters indicate whether there are significant differences between LCZs based on Kruskal–Wallis test, *p*-value 0.05. The classification starts with the letter "a" and refers to the first LCZ present.

3.3.1. Lecce

Figure 5a shows the diurnal cycle of LST for Lecce based on the GIS-based map, while Figure 5b presents the corresponding results using the WUDAPT map. Both figures show a characteristic diurnal trend, with the LST reaching its minimum values during the early morning hours and peaking between 11:00 and 14:00. There is a sharp rise in temperature in the morning, followed by a gradual decline in the afternoon and evening. These results validate the ability of ECOSTRESS satellite data to effectively capture the diurnal temperature cycle, accurately reflecting urban thermal dynamics.

Examining the differences between LCZs between 03:00 and 06:00, despite the general similarity of the median LST, specific patterns emerge. The classes of compact (LCZ2) and open (LCZ5) midrise have comparable LST values, as do the large low-rise (LCZ8) and sparsely built (LCZ9). However, significant differences are observed between the two low-rise building classes (LCZ3 and LCZ6), indicating that radiative cooling effects are more pronounced in these areas.

During the 07:00–11:00 period, substantial temperature differences emerge among all LCZs, underscoring the spatial heterogeneity of the urban thermal environment. This trend persists between 11:00 and 14:00, except for LCZ6 and LCZ9, which show no significant differences. LCZ2 records the lowest LST (43.6 °C), while LCZ8 reaches the highest value (45.8 °C), confirming that compact urban areas tend to experience slightly lower daytime peak temperatures than low-density areas. The urban core consistently records lower LST values than surrounding areas, a pattern consistent with Esposito et al. [14], who reported negative SUHI intensities in the Lecce core based on low-resolution satellite imagery. This urban cooling effect is particularly evident during the JJA season, probably due to reduced soil moisture in the suburban areas.

Between 15:00 and 18:00, the LST values of the compact (LCZ3) and large low-rise areas (LCZ8) converge (40.5 °C and 40.3 °C, respectively), while the low (LCZ5) and medium (LCZ6) open classes show similar temperatures (39.2 °C and 39.0 °C, respectively). During the evening transition (19:00 to 22:00), the two compacts midrise (LCZ2) and low-rise (LCZ3) areas follow comparable cooling rates, with average LSTs of 26.3 °C and 26.1 °C, respectively, while the other classes show more divergent thermal behaviors. This suggests that the compact zones experience prolonged heat retention due to accumulated energy during the day.

At night (23:00–02:00), LST distributions resemble those of the 03:00-06:00 period, albeit with slightly higher values, especially in denser urban areas. The highest LSTs are recorded in LCZ2 and LCZ3 (24.8 °C and 24.5 °C, respectively), while LCZ9 has the lowest values (23.2 °C).

Trend analysis of the WUDAPT-derived LSTs reveals similar temporal dynamics, except for the period 15:00–18:00, where LCZ6 aligns with LCZ3 instead of LCZ8. Throughout the day, statistical analysis confirms that the compact classes (LCZ2 and LCZ3) exhibit comparable LST trends, while the other LCZs show distinct thermal behaviors. During the 07:00-14:00 interval, compact areas maintain lower LSTs than openLCZs due to their higher heat storage capacity and increased shading due to dense building configurations. In contrast, nighttime conditions favor high LST in compact areas, as these areas release stored heat more gradually, potentially exacerbating nighttime overheating under certain environmental conditions. Overall, Figure 5a,b highlight the diurnal thermal response of different LCZs in Lecce, reinforcing the role of urban morphology in modulating LST dynamics. Compact urban areas generally have lower daytime temperature peaks but retain heat more efficiently at night, contributing to pronounced SUHI effects. The presence of vegetation further influences these patterns, with LCZ9 consistently showing lower temperatures due to evapotranspiration processes. In addition, coastal breezes from the Adriatic Sea likely enhance nighttime cooling, further modulating thermal contrasts between LCZs. These results underscore the complex interplay between urban structure, thermal properties, and microclimatic influences in shaping diurnal LST variations in Mediterranean cities.

3.3.2. Milan

Figure 6 shows the diurnal variation of LST in the different LCZs of Milan. The GIS-based map (Figure 6a) shows that during the early morning period (03:00–06:00), LCZ1 and LCZ2 have similar LST values, averaging about 21.00 °C. In contrast, LCZ9 records the lowest LST of 20.12 °C, highlighting the greater heat retention capacity of the compact urban areas (LCZ1 and LCZ2) compared to the stronger radiative cooling experienced in the less dense areas such as LCZ9. This difference can be attributed to the absence of high-rise buildings and narrow street canyons in LCZ9, which promote more efficient nighttime cooling.

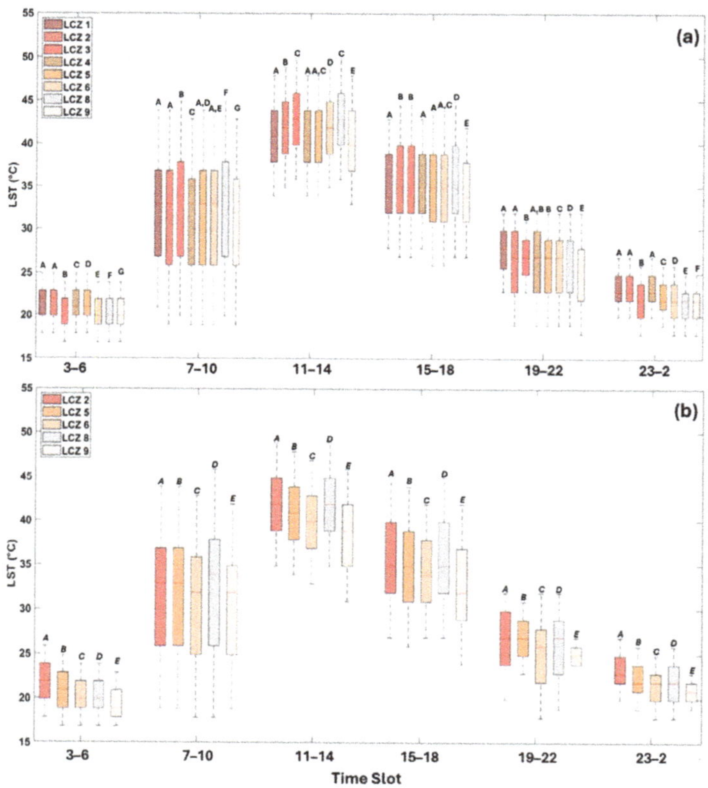

Figure 6. As in Figure 5 but for the city of Milan.

During the morning period (07:00–10:00), LCZ1 and LCZ2 remain thermally similar, while LCZ3 and LCZ4 show distinct temperature trends. LCZ3 reaches the highest LST at 32.67 °C, followed closely by LCZ8 at 32.58 °C. LCZ9 remains the coolest, with an average temperature of 31.08 °C. In the midday hours (11:00 to 14:00), the disparities in LST between LCZs become more pronounced, with LCZ3 and LCZ8 recording the highest temperatures, with about 42.50 °C. In contrast, LCZ9 continues to record the lowest temperature, at 40.08 °C, underscoring the role of vegetated areas in mitigating temperature extremes during periods of peak warming. These results are in line with previous studies, which have shown that compact urban and industrial areas experience rapid temperature increases due to a high fraction of impervious surfaces [20,23].

The middle hours of the day present the greatest thermal heterogeneity, reflecting the different heating rates of different urban structures. In the afternoon period (15:00–18:00), LST values begin to decrease, although LCZ8 maintains the highest average temperature at 35.55 °C, while LCZ9 remains the coolest at 34.25 °C. As the evening progresses (19:00–22:00), LCZ1 and LCZ2 maintain high LST values, with recorded temperatures of 27.18 °C and 26.88 °C, respectively, while LCZ9 remains the coolest at 25.24 °C. Notably, a similar thermal pattern emerges between the compact and open LCZs, except for LCZ6 (open low-rise), suggesting that despite the variations in urban fabric, similar heat release occurs during this period.

During the nighttime period (23:00–02:00), differences in LST persist, with LCZ1 and LCZ2 having the highest temperatures (22.90 °C). This trend is very similar to that observed in the early morning hours (03:00–06:00), except for LCZ4, which shows a different cooling rate than the more compact zones. These observations are in line with literature findings indicating that denser LCZs retain heat more efficiently at night, while suburban and vegetated areas cool faster due to enhanced radiative cooling [17].

Figure 6b illustrates the distinct thermal behavior of LCZs, as captured by the European LCZ map, revealing pronounced temperature differences, particularly in the morning and midday hours. While the WUDAPT map effectively characterizes temperature variations among LCZs, the diurnal LST cycle shows substantial spatial heterogeneity. Overall, the most densely built-up classes (LCZ2 and LCZ5) show higher temperatures than the other LCZs, although these differences become less pronounced in the morning hours. LCZ8 consistently records high LST values during this period.

Comparative analysis of the results of the Kruskal–Wallis test between the GIS-based map and the European map of LCZs indicates variations in statistical differentiation and classification. The European LCZ map assigns greater variability to LCZs, suggesting greater statistical significance in LST distributions. This discrepancy is likely attributable to differences in classification methodologies and spatial resolution, which influence the statistical interpretation of temperature variations between LCZs.

Milan shows diurnal LST trends comparable to those observed in Lecce; however, the magnitude of the temperature variation between LCZs is more pronounced. Unlike Lecce, the composition of Milan's land cover exacerbates temperature disparities, with the widespread use of concrete and asphalt—materials with high thermal inertia—contributing to prolonged heat retention, particularly in LCZ1 and LCZ3.

During the peak heating hours (11:00–14:00), the differences in LST become increasingly evident. The LCZ3 and LCZ8 reach the highest temperatures (42.50 °C), confirming that dense urban environments with minimal vegetation experience the most extreme warming. In contrast, LCZ9 remains cooler (40.08 °C), emphasizing the mitigating influence of green spaces during periods of intense solar radiation. These results are consistent with previous studies documenting rapid temperature escalation in urban environments due to extensive coverage of impervious surfaces [17,20].

In the afternoon cooling phase (15:00–18:00), LSTs decrease throughout the city, but differences between LCZs persist. LCZ8 maintains the highest mean temperature at 35.55 °C, while LCZ9 remains the coolest at 34.25 °C, highlighting the role of evapotranspiration in reducing heat retention. As the evening approaches (19:00–22:00), the compact urban LCZs (LCZ1 and LCZ2) continue to show high LSTs, indicative of prolonged heat accumulation and gradual nocturnal release by the impermeable surfaces. This trend extends into the early morning hours (03:00–06:00), where LCZ1 and LCZ2 show comparable LSTs, reinforcing their efficiency in retaining heat. Meanwhile, LCZ9, characterized by low-density development and substantial vegetation, experiences greater radiative cooling, resulting in lower temperatures in the absence of significant building structures.

3.3.3. Inter-City Comparison

Comparative analysis of the thermal responses of different Local Climate Zones (LCZs) in Lecce and Milan reveals distinct patterns, particularly in nighttime overheating. In Milan, denser LCZs (LCZ1, LCZ2, LCZ4, and LCZ5) experience higher nighttime temperatures, indicative of more pronounced urban heat retention. This trend contrasts with Lecce, where temperature variations among LCZs are less marked, except in the WUDAPT-generated map, probably due to the different distribution of LCZs. In Lecce, a well-defined temperature gradient is observed during the daytime period (07:00–14:00), with temperatures progressively increasing from denser to less dense LCZs. This pattern is less evident in Milan, where higher temperatures are primarily concentrated in high-rise areas (LCZ8 and LCZ3). The historic core of Lecce, mainly classified as LCZ2, maintains lower daytime temperatures due to the high albedo and thermal capacity of its historic buildings and paved roads, which limit heat peaks, consistent with findings in other Mediterranean cities [14,36]. In contrast, Milan's larger size and higher urban density contribute to greater overheating, particularly during nighttime hours, intensifying the SUHI effect.

Despite sharing the same LCZ classifications, the thermal responses of corresponding LCZs differ between the two cities, underscoring the role of urban morphology, land cover composition, and climate in modulating LST variability. This aligns with findings from previous research in cities such as Valencia, Spain [23], and Xi'an, China [20], which highlight the influence of these factors on intra-LCZ temperature variations.

During the evening (19:00–22:00) and night (23:00–02:00) hours, both cities exhibit a progressive cooling trend, with urban LCZs retaining higher temperatures than suburban areas. However, this effect is more pronounced in Milan, likely due to its larger urban extent and higher building density in the high-midrise LCZs. The observed discrepancies between Milan and Lecce reinforce the importance of localized urban characteristics in shaping thermal responses, even within standardized LCZ classifications.

3.3.4. Contributions and Limitations

The findings contribute to a deeper understanding of SUHI dynamics in Mediterranean urban environments and emphasize the necessity of city-specific adaptation strategies to mitigate urban overheating effects. This research contributes to the broader understanding of urban climate adaptation by demonstrating the potential of high-resolution remote sensing for urban thermal assessments. The contributions gained can support data-driven policymaking and urban planning strategies aimed at mitigating the impacts of extreme heat in Mediterranean cities. The study also emphasizes the necessity of incorporating high-resolution thermal monitoring into climate adaptation policies, particularly in Mediterranean cities facing intensifying heat stress.

Despite the study's contributions, several limitations must be acknowledged. While ECOSTRESS provides high-resolution thermal data, its irregular acquisition times limit temporal continuity, particularly during transitional periods such as early morning and late evening. Additionally, while the GIS-based approach offers improved classification accuracy, its scalability is constrained by the availability of high-resolution morphological data, limiting its applicability in regions with less detailed geospatial coverage. Furthermore, this study did not explicitly account for anthropogenic heat emissions, wind dynamics, and surface moisture, which are known to influence urban thermal behavior.

4. Conclusions

This study analyzed LST variability in Mediterranean cities using high-resolution ECOSTRESS data and two distinct LCZ classification methods: a GIS-based approach and the WUDAPT European LCZ map. This represents one of the first applications of ECOSTRESS data for diurnal thermal analysis in combination with LCZ classifications, offering a comprehensive assessment of urban heat dynamics during summer. The high spatial resolution and frequent temporal coverage of ECOSTRESS enable a more detailed analysis of urban thermal patterns compared to traditional in-situ measurements, which often lack representativeness in heterogeneous urban environments.

The primary objective was to evaluate classification accuracy and its impact on urban thermal analysis in Milan and Lecce. The GIS-based method, integrating high-resolution geospatial datasets, provided superior classification accuracy, particularly in cities with complex urban structures. In contrast, WUDAPT, which relies on satellite imagery and machine learning, exhibited broader classifications that sometimes misrepresented vertical variability, leading to lower accuracy in cities like Milan. The classification discrepancies between the two methods influenced the spatial distribution of LCZs: the GIS-based approach provided a more detailed representation, particularly for LCZ2, LCZ5, and LCZ6, whereas WUDAPT tended to overestimate LCZ6 and LCZ8. These differences directly affected the interpretation of LST patterns, with WUDAPT struggling to differentiate LCZs with similar urban morphologies.

The LST analysis revealed distinct thermal behaviors in Milan and Lecce, driven by differences in climate, urban form, and material properties. While LCZ classifications provided a useful framework for identifying thermal trends, results showed that identical LCZs can exhibit varying temperature responses depending on local conditions. This underscores the need to integrate LCZ classifications with localized climatic and morphological data for a more precise assessment of urban heat dynamics.

From a practical perspective, the findings offer important insights for urban heat mitigation strategies. The integration of high-resolution satellite data with detailed geospatial datasets allows for targeted identification of overheating zones, supporting climate adaptation planning. Urban policymakers can leverage this information to implement heat resilience strategies such as increasing vegetation cover, optimizing building materials, and enhancing urban ventilation.

Future research should consider additional variables related to the thermal properties of urban surfaces (e.g., roofs, walls, and roads) as well as meteorological forcings such as humidity, wind field, and anthropogenic heat fluxes to refine urban heat assessments. The greater spatial detail of remote sensing data should be enhanced by detailed land use mapping. Combining ECOSTRESS data within situ temperature measurements, reanalysis data (e.g., ERA5), and numerical modeling approaches would improve the robustness and generalizability of the results.

Author Contributions: Conceptualization, A.E., G.P. and R.B.; methodology, A.E. and G.P.; software, A.E. and G.P.; validation, A.E.; formal analysis, A.E.; investigation, A.E. and G.P.; resources, R.B.; data curation, A.E.; writing—original draft preparation, A.E. and G.P.; writing—review and editing, G.P. and R.B.; visualization, A.E. and G.P.; supervision, R.B.; project administration, R.B.; funding acquisition, R.B. All authors have read and agreed to the published version of the manuscript.

Funding: A.E. acknowledges the PhD financial support of the Italian Ministry of University and Research (MUR) by the PON "Ricerca e Innovazione 2014-2020—Asse IV"—PhD course in "Scienze e Tecnologie Biologiche ed Ambientali"–XXXVII cycle-University of Salento.

Institutional Review Board Statement: Not applicable.

Informed Consent Statement: Not applicable.

Data Availability Statement: Derived data supporting the findings of this study are available from the corresponding author on request. The data derived from the GIS-based approach is available on the open Zenodo repository https://doi.org/10.5281/zenodo.15075278.

Conflicts of Interest: The authors declare no conflicts of interest.

Abbreviations

LST (°C)	Land Surface Temperature
WUDAPT	World Urban Database and Access Portal Tools
LCZs	Local Climate Zones
SUHI (°C)	Surface Urban Heat Island
ECOSTRESS	Ecosystem Spaceborne Thermal Radiometer Experiment on Space Station
UCPs	Urban Canopy Parameters
BSF (-)	Building Surface Fraction
ISF (-)	Impervious Surface Fraction
PSF (-)	Pervious Surface Fraction
SVF (-)	Sky View Factor
AR (-)	Aspect Ratio
HM (m)	Average Building Height
LSTE	Land Surface Temperature and Emissivity
ECO2LSTE	ECOSTRESS LST Level-2
RMSE (-)	Root Mean Square Error
MAE (-)	Mean Absolute Error
H_0 (-)	Null Hypothesis
H_1 (-)	Alternative Hypothesis
OA (%)	Overall Accucary
DEMs	Digital Elevation Models

References

1. Donateo, A.; Dinoi, A.; Pappaccogli, G. Impact on Ultrafine Particles Concentration and Turbulent Fluxes of SARS-CoV-2 Lockdown in a Suburban Area in Italy. *Atmosphere* **2021**, *12*, 407. [CrossRef]
2. Stewart, I.D.; Oke, T.R. Local climate zones for urban temperature studies. *Bull. Am. Meteorol. Soc.* **2012**, *93*, 1879–1900.
3. Feng, W.; Liu, J. A Literature Survey of Local Climate Zone Classification: Status, Application, and Prospect. *Buildings* **2022**, *12*, 1693. [CrossRef]
4. Huang, F.; Jiang, S.; Zhan, W.; Bechtel, B.; Liu, Z.; Demuzere, M.; Huang, Y.; Xu, Y.; Ma, L.; Xia, W.; et al. Mapping local climate zones for cities: A large review. *Remote Sens. Environ.* **2023**, *292*, 113573.
5. Gill, S.E.; Handley, J.F.; Ennos, A.R.; Pauleit, S. Adapting cities for climate change: The role of the green infrastructure. *Built Environ.* **2007**, *33*, 115–133.
6. Demuzere, M.; Bechtel, B.; Middel, A.; Mills, G. Mapping Europe into local climate zones. *PLoS ONE* **2019**, *14*, e0214474.
7. Demuzere, M.; Kittner, J.; Martilli, A.; Mills, G.; Moede, C.; Stewart, I.D.; van Vliet, J.; Bechtel, B. A global map of local climate zones to support earth system modelling and urban-scale environmental science. *Earth Syst. Sci. Data* **2022**, *14*, 3835–3873.
8. Leconte, F.; Bouyer, J.; Claverie, R.; Pétrissans, M. Using local climate zone scheme for UHI assessment: Evaluation of the method using mobile measurements. *Build. Environ.* **2015**, *83*, 39–49.
9. Lin, Y.; Jim, C.Y.; Deng, J.; Wang, Z. Urbanization effect on spatiotemporal thermal patterns and changes in Hangzhou (China). *Build. Environ.* **2018**, *145*, 166–176.

10. Yan, C.; Guo, Q.; Li, H.; Li, L.; Qiu, G.Y. Quantifying the cooling effect of urban vegetation by mobile traverse method: A local-scale urban heat island study in a subtropical megacity. *Build. Environ.* **2020**, *169*, 106541.
11. Martilli, A.; Sánchez, B.; Santiago, J.L.; Rasilla, D.; Pappaccogli, G.; Allende, F.; Martín, F.; Roman-Cascón, C.; Yagüe, C.; Fernández, F. Simulating the Pollutant Dispersion during Persistent Wintertime Thermal Inversions over Urban Areas: The Case of Madrid. *Atmos. Res.* **2022**, *270*, 106058.
12. Zhou, D.; Xiao, J.; Bonafoni, S.; Berger, C.; Deilami, K.; Zhou, Y.; Frolking, S.; Yao, R.; Qiao, Z.; Sobrino, J.A. Satellite Remote Sensing of Surface Urban Heat Islands: Progress, Challenges, and Perspectives. *Remote Sens.* **2019**, *11*, 48.
13. Naserikia, M.; Hart, M.A.; Nazarian, N.; Bechtel, B.; Lipson, M.; Nice, K.A. Land Surface and Air Temperature Dynamics: The Role of Urban Form and Seasonality. *Sci. Total Environ.* **2023**, *905*, 167306. [PubMed]
14. Esposito, A.; Pappaccogli, G.; Donateo, A.; Salizzoni, P.; Maffeis, G.; Semeraro, T.; Santiago, J.L.; Buccolieri, R. Urban Morphology and Surface Urban Heat Island Relationship During Heat Waves: A Study of Milan and Lecce (Italy). *Remote Sens.* **2024**, *16*, 4496. [CrossRef]
15. Chang, Y.; Xiao, J.; Li, X.; Frolking, S.; Zhou, D.; Schneider, A.; Weng, Q.; Yu, P.; Wang, X.; Li, X.; et al. Exploring diurnal cycles of surface urban heat island intensity in Boston with land surface temperature data derived from GOES-R geostationary satellites. *Sci. Total Environ.* **2021**, *763*, 144224. [CrossRef]
16. Pinker, R.T.; Ma, Y.; Chen, W.; Hulley, G.; Borbas, E.; Islam, T.; Hain, C.; Cawse-Nicholson, K.; Hook, S.; Basara, J. Towards a unified and coherent land surface temperature earth system data record from geostationary satellites. *Remote Sens.* **2019**, *11*, 1399. [CrossRef]
17. Wu, Z.; Yao, L.; Ren, Y. Characterizing the spatial heterogeneity and controlling factors of land surface temperature clusters: A case study in Beijing. *Build. Environ.* **2020**, *169*, 106598.
18. Geletič, J.; Lehnert, M.; Savić, S.; Milošević, D. Inter−/intra-zonal seasonal variability of the surface urban heat island based on local climate zones in three central European cities. *Build. Environ.* **2019**, *156*, 21–32.
19. Wang, C.; Middel, A.; Myint, S.W.; Kaplan, S.; Brazel, A.J.; Lukasczyk, J. Assessing local climate zones in arid cities: The case of Phoenix, Arizona and Las Vegas, Nevada. *ISPRS J. Photogramm. Remote Sens.* **2018**, *141*, 59–71.
20. Chang, Y.; Xiao, J.; Li, X.; Middel, A.; Zhang, Y.; Gu, Z.; He, S. Exploring diurnal thermal variations in urban local climate zones with ECOSTRESS land surface temperature data. *Remote Sens. Environ.* **2021**, *263*, 112544. [CrossRef]
21. Chang, Y.; Xiao, J.; Li, X.; Weng, Q. Monitoring diurnal dynamics of surface urban heat island for urban agglomerations using ECOSTRESS land surface temperature observations. *Sustain. Cities Soc.* **2023**, *98*, 104833.
22. Hulley, G.; Shivers, S.; Wetherley, E.; Cudd, R. New ECOSTRESS and MODIS Land Surface Temperature Data Reveal Fine-Scale Heat Vulnerability in Cities: A Case Study for Los Angeles County, California. *Remote Sens.* **2019**, *11*, 2136. [CrossRef]
23. Wei, L.; Sobrino, J.A. Surface urban heat island analysis based on local climate zones using ECOSTRESS and Landsat data: A case study of Valencia city (Spain). *Int. J. Appl. Earth Obs. Geoinf.* **2024**, *130*, 103875. [CrossRef]
24. Hulley, G. *ECOsystem Spaceborne Thermal Radiometer Experiment on Space Station (ECOSTRESS) Mission–Level 2 Product User Guide*; LP DAAC: Sioux Falls, SD, USA, 2019.
25. Akbari, H.; Konopacki, S. Calculating energy-saving potentials of heat-island reduction strategies. *Energy Policy* **2005**, *33*, 721–756.
26. Emmanuel, R.; Krüger, E. Urban heat island and its impact on climate change resilience in a shrinking city: The case of Glasgow, UK. *Build. Environ.* **2012**, *53*, 137–149. [CrossRef]
27. Esposito, A.; Grulois, M.; Pappaccogli, G.; Palusci, O.; Donateo, A.; Salizzoni, P.; Santiago, J.L.; Martilli, A.; Maffeis, G.; Buccolieri, R. On the Calculation of Urban Morphological Parameters Using GIS: An Application to Italian Cities. *Atmosphere* **2023**, *14*, 329. [CrossRef]
28. Muhammad, F.; Xie, C.; Vogel, J.; Afshari, A. Inference of Local Climate Zones from GIS Data, and Comparison to WUDAPT Classification and Custom-Fit Clusters. *Land* **2022**, *11*, 747. [CrossRef]
29. Estacio, I.; Babaan, J.; Pecson, N.J.; Blanco, A.C.; Escoto, J.E.; Alcantara, C.K. GIS-based mapping of local climate zones using fuzzy logic and cellular automata. *Int. Arch. Photogramm. Remote Sens. Spat. Inf. Sci.* **2019**, *XLII-4/W19*, 199–206. [CrossRef]
30. Hook, S.J.; Cawse-Nicholson, K.; Barsi, J.; Radocinski, R.; Hulley, G.C.; Johnson, W.R.; Rivera, G.; Markham, B. In-flight validation of the ECOSTRESS, Landsats 7 and 8 thermal infrared spectral channels using the Lake Tahoe CA/NV and Salton Sea CA automated validation sites. *IEEE Trans. Geosci. Remote Sens.* **2020**, *58*, 1294–1302.
31. Hulley, G.C.; Gottsche, F.M.; Rivera, G.; Hook, S.J.; Freepartner, R.J.; Martin, M.A.; Cawse-Nicholson, K.; Johnson, W.R. Validation and Quality Assessment of the ECOSTRESS Level-2 Land Surface Temperature and Emissivity Product. *IEEE Trans. Geosci. Remote Sens.* **2022**, *60*, 5000523.
32. Ostertagova, E.; Ostertag, O.; Kováč, J. Methodology and application of the Kruskal-Wallis test. *Appl. Mech. Mater.* **2014**, *611*, 115–120.

33. Sun, Y.; Zhang, N.; Miao, S.; Kong, F.; Zhang, Y.; Li, N. Urban morphological parameters of the main cities in China and their application in the WRF model. *J. Adv. Model. Earth Syst.* **2021**, *13*, e2020MS002382.
34. Hidalgo, J.; Dumas, G.; Masson, V.; Petit, G.; Bechtel, B.; Bocher, E.; Foley, M.; Schoetter, R.; Mills, G. Comparison between local climate zones maps derived from administrative datasets and satellite observations. *Urban Clim.* **2019**, *27*, 64–89.
35. Wellinger, N.; Gubler, M.; Müller, F.; Brönnimann, S. GIS-Based Revision of a WUDAPT Local Climate Zones Map of Bern, Switzerland. *City Environ. Interact.* **2024**, *21*, 100135.
36. Donateo, A.; Palusci, O.; Pappaccogli, G.; Esposito, A.; Martilli, A.; Santiago, J.L.; Buccolieri, R. Analysis of Urban Heat Island and Human Thermal Comfort in a Mediterranean City: A Case Study of Lecce (Italy). *Sustain. Cities Soc.* **2023**, *98*, 104849.

atmosphere

Article

Development of Visualization Tools for Sharing Climate Cooling Strategies with Impacted Urban Communities

Linda Powers Tomasso [1,*], Kachina Studer [2], David Bloniarz [3,4], Dillon Escandon [5] and John D. Spengler [1]

1 Department of Environmental Health, Harvard T.H. Chan School of Public Health, Boston, MA 02115, USA; spengler@hsph.harvard.edu
2 Visualization Research and Teaching Laboratory, Department of Earth and Planetary Science, Harvard University, Cambridge, MA 02138, USA; kachinastuder@gmail.com
3 Department of Environmental Conservation, University of Massachusetts, Amherst, MA 02003, USA; bloniarz@umass.edu
4 USDA Forest Service Northern Research Station, Amherst, MA 01003, USA
5 Graduate School of Design, Harvard University, Cambridge, MA 02138, USA; dillon_escandon@gsd.harvard.edu
* Correspondence: tomasso@hsph.harvard.edu

Abstract: Intensifying heat from warming climates regularly concentrates in urban areas lacking green infrastructure in the form of green space, vegetation, and ample tree canopy cover. Nature-based interventions in older U.S. city cores can help minimize the urban heat island effect, yet neighborhoods targeted for cooling interventions may remain outside the decisional processes through which change affects their communities. This translational research seeks to address health disparities originating from the absence of neighborhood-level vegetation in core urban areas, with a focus on tree canopy cover to mitigate human susceptibility to extreme heat exposure. The development of LiDAR-based imagery enables communities to visualize the proposed greening over time and across seasons of actual neighborhood streets, thus becoming an effective communications tool in community-engaged research. These tools serve as an example of how visualization strategies can initiate unbiased discussion of proposed interventions, serve as an educational vehicle around the health impacts of climate change, and invite distributional and participatory equity for residents of low-income, nature-poor neighborhoods.

Keywords: nature-based solutions; urban heat islands (UHIs); street trees; community-engaged research; visualization technologies; augmented reality; climate change; environmental equity

1. Introduction

Identifying inequities between population exposures and outcomes is the bread and butter of environmental health. The associations our research generates provide evidence for public health policy or regulatory action to protect vulnerable citizens against harmful exposures. Still, the temptation exists to create even more nuanced evidence of environmental risk while falling short of true research policy translation, interventions, or other actions that stem from recommendations suggested by the evidence. Here we aim instead at translational research that has applications for addressing health disparities originating from the absence of neighborhood-level vegetation, whether owing to systemic underinvestment or benign neglect, with a focus on tree canopy cover to mitigate human susceptibility to extreme heat exposure.

Changing climate has intensified heat and the patterns of damaging precipitation events. However, less riveting harms regularly occur in urban areas lacking green infrastructure, specifically those with minimal urban green space, sparse vegetation, and low rates of tree canopy cover, in the form of extreme heat exposure. Solar radiation concentrates in denser built environments, where extended areas of asphalt, concrete building mass, and impermeable surfaces like sidewalks absorb and reflect infrared solar radiation back to the ground-level atmosphere, compounding these underlying disparities. Historic municipal underinvestment in core city neighborhoods—and residential poverty among those left behind by suburban out-migration—have amplified health disparities rooted in the built environment. These include failure to provide green scaping to reduce temperatures, maintain shade tree canopies that invite physical activity and socialization, construct sidewalks and bike lanes for safer mobility, and invest in basic neighborhood amenities such as benches or seating, traffic crossings, and park repair and maintenance.

Areas of disproportionate heat gain are known as urban heat islands (UHIs) due to their concentrated building mass that absorbs high levels of solar radiation. With few trees providing shade and deflecting incoming radiation, UHIs can harbor daytime temperatures 1–7 °F higher and nighttime temperatures 2–5 °F higher than suburbs with more abundant vegetation [1]. UHIs grow more acute with denser thermal massing, impermeable surface enlargement, and less nighttime cool-off, all typical of aging urban city cores.

UHIs pose health harms to vulnerable subgroups, particularly elderly and low-income residents, resulting in increased heat stroke, ischemic heart disease, hospitalization, and cognitive loss [2–4]. Heat gain that accompanies climate change aggravates existing health disparities that frequently track along racial lines. Excessive ambient temperatures in urban neighborhoods add heat-related health risks to city-dwellers already beset by inequitable burdens of poverty, many traced to HOLC redlining [5,6]. The direct benefits of enhanced thermal comfort these interventions create extend to lower mortality rates and reduced energy costs, in addition to greater physical activity and socialization. Studies have shown a range of health benefits from residential proximity to urban nature. Living near urban green space can lower the risk for health problems such as heat stroke [7,8], cancer [9,10], heart disease [11,12], hypertension [13,14], obesity [15,16], diabetes [17,18], pre-term birth [19,20], and asthma [21,22] and can improve mental health [23–25].

Nature-based interventions across the U.S.'s older city cores can help minimize the UHI effect. The presence of urban vegetation increases urban albedo via higher reflectivity of sunlight off of surfaces like foliage and the expansion of evaporative ground cover [26]. Simulations of urban climates support the conclusion that increasing albedo, evapotranspiration, and vegetative cover effectively counteracts urban heating. Vegetation can lower temperature by provisioning shade, inducing individuals outside to socialize, itself a mediator of health [27,28], and prompting more engagement in greater physical activity than they might in hotter conditions [29,30]. The cooling effects of urban greening have been amply documented, proven through direct measurements as well as computer simulations [31]. Zardo et al. (2017) has estimated how tree canopy coverage and soil cover as gauged by size and shape constitute adaptation measures that enhance cooling capacity in urban environments [32]. Larger parks of 10 hectares or more and urban forests can lower daytime temperatures by 0.94 °C [33]. Street trees, green roofs [34], and green façades--smaller forms of green infrastructure—also are effective at reducing ambient temperatures [35].

Afforestation strategies hold enormous potential to address the health challenges UHIs pose in low-income neighborhoods like Springfield, Massachusetts'. Street tree plantings can directly reduce radiated heat contributing to UHIs: the presence of tree canopies can reduce the physiological equivalent temperature (PET) by 4 °C to 11 °C of densely built environments [36], as compared to vegetated surfaces receiving full sun exposure. Given that 37% of recorded heat-related deaths since 1991 are attributable to global warming [37], afforestation to counter UHIs is a central tenet of urban temperature-reducing strategies. In fact, the literature is replete with modeled estimates of UHI reductions enabled by tree-based solutions [38–41]. The numerical and spatial models in these appraisals, however, rarely translate into implementable designs. The few that do fail to consult the communities targeted for these interventions.

This study addresses a proposed intervention to redress climate change-related heat gain that disproportionately impacts poor, nature-deficient residential neighborhoods. Springfield, Massachusetts, serves as a pilot case to display street tree planting as a nature-based intervention helping cities achieve their long-term goal of mitigating urban heat differentials, while also promoting the underlying health and wellbeing of community residents by providing them with a more comfortable, temperate neighborhood environment for pedestrian mobility, exercise, and socialization, as well as lower building energy needs.

This pilot project builds on the integrated design work led by landscape architect Martha Schwartz to develop Springfield, Massachusetts' linear urban forest (LUF), a research initiative supported by a 2021 Harvard Climate Change Grant. The interdisciplinary research team consisted of Harvard Graduate School of Design landscape designers, Harvard T.H. Chan School of Public Health researchers, US Forest Service biologists, and municipal policy practitioners who collaborated on the original LUF design. The LUF was designed to encompass all streets and run throughout the city. As a living ecological green infrastructure, LUF provisions many eco-system services that address environmental and social challenges, among them city cooling, buffering adjacent river flooding, better management of stormwater runoff, human health protection, urban biodiversity enhancement, decreased energy bills, increased quality of life, enhanced walkability, reduced greenhouse gas emissions, intensified carbon sequestration, and climate equity.

The pilot LUF launched the next step for project realization by supporting the development of dynamic visualization modeling and design prototyping to bring a digitalized green solution to community residents. In this way, community members living in heat-stressed neighborhoods could participate in the decision-making process for reimagining their streets in ways that co-introduce climate mitigation, health, regenerative nature, and equity. This study therefore made use of a community-engaged research (CER) strategy as the strategic framework for developing tools to engage local citizens as co-generators of knowledge regarding neighborhood transformation. CER, or its broader umbrella term, community academic partnerships (CAP) [42], is distinguished by a commitment to partnering with communities to co-create knowledge arising from both researcher and user expertise [43]. By engaging local users in the production of knowledge, researchers aspire to social impact benefiting community members for whom the interventions under discussion are targeted. Collaboration should result in an attenuated "know–do" lag, meaning a shortening of the cycle between research findings and implementation [43].

Community-based research partnerships recognize parity among interventionists and impacted stakeholders in holding seats at the table, different from a strictly research-centered practice where effected community participants are historically excluded. CERs also address procedural equity whereby decisional tools are made available to facilitate full participation in knowledge-creation sessions. Many aspects of community-based participatory research (CBPR) strategies were employed in this collaboration, enabled through the researchers' partnership with ReGreen Springfield, a trusted community organization whose facilitation of the design charettes had the result of placing community at the center of these sessions, a central tenet of CBPR [38]. Like CBPR, community equity was an explicit goal of the selected research process; moreover, ReGreen Springfield, with its expertise regarding tree presence in these neighborhoods, made for a strong community partner since the design stage of the research. Knowledge co-creation by community members was limited to feedback from visualization of neighborhood intervention, however. CAP and CBPR can assist in translating research knowledge where a lack of prior stakeholder engagement helped sustain disparities in communities now targeted for tree planting [44].

Some examples exist where communities participated in processes that co-produced risk mitigation strategies when faced with climate change. One example of community-based research was a flood vulnerability assessment and risk mitigation in Milwaukee, Wisconsin, with a similar focus on green infrastructure as a partial solution [45]. Visualization tools were applied to graph epistemic disciplines of green and blue infrastructure—tools under scrutiny to counteract urban heat island effects and human health with its increasing vulnerability to climate change risks, including that of extreme heat [46]. However, in these two cases, visualization tools were presented as static mapping without the dynamic visualization of our methods. In another study, Koulouris et al. [47] used augmented reality that takes advantage of locational data to create interactive, real-world scenarios placing humans in the center of urban environments. However, the captured locational data mimicked actual surroundings rather than projecting alternative scenarios, making this a tool of gamification rather than projection. We know of no other study that attempted to visualize nature-based interventions affecting climate risk or human health by interactive means. There were three aims of this research:

1. Develop visualization tools in collaboration with the Visualization Research and Teaching Laboratory in Harvard's Department of Earth and Planetary Sciences, then pilot these visualization technologies with residents of Springfield's impacted neighborhoods where the LUF is proposed.
2. Apply the InVEST Natural Capital Project's accounting model to estimate the reduction in urban temperature expected from expanding the urban street canopy.
3. Conduct design charettes to introduce a new form of pedagogy in community participatory research, including environmental literacy in connecting the health benefits of the presence of nature to reduce heat and asthma burdens and promote physical activity.

This research does not, however, attempt to put forth cost–benefit analysis of this project as engineering, maintenance, or quantified health gains, nor does it measure improvements in air quality or the reduction in CO_2 that may result from full implementation of the project. It does not compare the potential impacts of green infrastructure in Springfield to other cities that have implemented them. However, it does seek to demonstrate how public adoption of nature-based solutions that include climate–health and nature–health implementation goals can be accelerated through integrated assessment methodologies, dynamic visualization modeling, and design prototyping for presenting and soliciting feedback from locally impacted communities.

2. Materials and Methods

Springfield, Massachusetts, an aging industrial city with a legacy of air pollution, was selected as the design canvas for a linear urban forest due to its public health challenges and relatable population size. Springfield ranks number one among U.S. cities for asthma prevalence: the city has the highest number of asthma-related emergency room visits per capita in the U.S., as well as a high rate of asthma prevalence overall. High pollen counts aggravate the susceptibility of Springfield's residents to asthma [48]. Springfield also ranks tenth among the poorest cities per capita in the U.S. with populations over 100,000, having numerous environmental justice neighborhoods [49]. The city's residential population, at 21% black and 46% Hispanic [50], has the lowest life expectancy of all cities in Massachusetts [51]. We focused our visualization efforts on the neighborhood known as Upper Hill, which has a high percentage of Hispanic and black households, older citizens, and individuals under the poverty rate, with many of these categories overlapping. The physical environment in the Upper Hill neighborhood is densely populated with aging multi-family homes and limited recreational space.

The study population consisted of approximately 40 individuals who attended one of two Springfield design charette sessions. These individuals ranged in age from 16 to senior citizens. Demographics were equally balanced between males and females, with many couples in attendance. Participants represented a range of racial and economic diversity, although no non-native English speakers attended the sessions. The audience included a mixture of representatives of the neighborhoods selected for greening, members of our non-profit research partner ReGreen Springfield, young adults working in the tree-planting arm of that organization, and representatives of city government. Some charette participants did not live in the exact neighborhoods targeted for tree planting but rather came from the adjacent Forest Park neighborhood, with its mature tree canopy and ample shade that exemplifies the goals of urban greening. A few individuals from other tree-poor neighborhoods attended to see if proposals might transfer to their area. Most attendees actively participated with commentary, providing a representative opinion sample from which to draw data. A survey of participant preferences for specific street tree scenarios was informally conducted through a show of hands and voluntary comments.

2.1. Technology Applications

This study utilized advanced LiDAR scanning technology to capture and digitize the three-dimensional structure of impacted neighborhoods, enabling the creation of highly accurate digital models of streets, landscaping, and existing tree planting conditions. The integration of LiDAR scanning in urban street tree prototyping introduced a novel approach to environmental planning and participatory urban design.

2.1.1. Data Collection and Processing

LiDAR scans were conducted in the Upper Hill neighborhood, where street tree planting prototypes were implemented. Specialists from the Visualization Research and Teaching Laboratory at Harvard University's Department of Earth and Planetary Sciences collected on the streets of Springfield geolocated point cloud datasets using a VLX LiDAR scanner, manufactured by NavViz, Munich, Germany. This mobile, shoulder-mounted unit integrates LiDAR sensors, cameras, and GPS, allowing for efficient on-foot data acquisition. The system outputs a point cloud density of approximately 8000 points per square meter, achieving a resolution of around 5 mm accuracy. By measuring the two-way travel time of laser pulses and processing calculations based on the speed of light, the scanner generated dense, spatially accurate 3D point clouds. These datasets provided a precise digital representation of neighborhood infrastructure and served as the foundation for subsequent modeling and visualization.

2.1.2. 3D Modeling and Visualization

The collected point cloud data were imported into Blender, where a base mesh was generated to accurately represent buildings, existing tree plantings, and urban infrastructure. To improve the realism of the visualizations, Google Street View imagery was integrated as

360-degree background textures, ensuring that the surrounding neighborhood homes and structures were depicted with a high level of visual detail. The 360 images were precisely matched to the models by aligning their locations with the geospatially referenced LiDAR scans, allowing for an accurate representation of each site. A structured library of digital assets was developed to support scenario visualization. This included a catalog of tree models representing various species at different growth stages, as well as ground vegetation and hardscape elements such as sidewalks, benches, and street fixtures. These assets were integrated into the models to reflect proposed urban forest plans and growth projections.

Four viewpoints were selected for detailed visualization, each serving as a representative section of the study area. For each viewpoint, six distinct planting scenarios were developed, and each scenario was further rendered across four seasonal conditions. This resulted in a total of 96 high-fidelity 360-degree images, which were used in interactive presentations and public engagement sessions. The depicted scenarios ranged from current plantings maturing 50 years into the future to new plantings at spacing intervals of 40, 20, and 15 feet. This last scenario entailed converting the two-way street to one direction and foregoing a row of street parking to accommodate a wider sidewalk median for planting a double row of street trees. A final scenario added vegetation and smaller plantings as an understory to develop a ground-level ecological community around the trees.

2.1.3. Interactive Application Development

To facilitate public engagement and feedback, an interactive application was developed using the Unity game engine. The application featured a user-friendly interface that allowed participants to explore the 360-degree views at the designated viewpoints and toggle between different planting scenarios and seasonal conditions. The interface was designed to provide an intuitive navigation experience, enabling users to compare environmental changes dynamically. Beyond tree planting variations, the application also allowed exploration of additional urban design considerations, such as road layout modifications, linear forest programming, and potential community engagement features. Options included micro parks, small public gardens, and exercise areas integrated along the linear forests, encouraging users to consider broader urban greening strategies. Additionally, the application provided data on the heat island reduction associated with each scenario. By selecting different configurations of tree planting densities and green space allocations, users could observe potential temperature mitigation effects, reinforcing the ecological and social benefits of urban reforestation.

The interactive tool was hosted on AWS, ensuring accessibility for researchers, urban planners, and community members. The structured visualization of urban tree planting impacts across different timescales and environmental conditions provided a robust foundation for participatory decision-making and long-term urban forestry planning.

2.1.4. Species Selection

Tree selection for the visualization tool relied on ten species currently favored by the Springfield Parks Department for total tree composition in the city's master forest plan. Concerns with species resiliency in urban areas mandate that no more than ten percent of any one species be represented in the planting mix to avoid excessive mortality due to pest or disease. Relying on this guideline, the street tree visualization incorporated six common New England tree species already present in Springfield, with these being red maple, black tupelo, Eastern white pine, white oak, Northern red oak, and black oak (Table 1). Heights at maturity of these species were calculated for accurate representation in street tree canopies. These six species were interspersed in the visualizations, although no adjustments were made in species clustering across the different tree spacing scenarios.

Table 1. The six primary species used in Springfield's master forest plan for reforestation.

Type			Height (ft)		
Percent	Name	Scientific Name	New	Young	Mature
~10%	Red Maple	Acer rubrum	3–6	15–25	40–60
~10%	Black Tupelo/Gum	Nyssa sylvatica	3–6	15–25	30–50
~10%	Eastern White pine	Pinus strobus	2–4	15–30	50–80
~10%	White Oak	Quercus alba	3–6	15–25	60–100
~10%	Northern Red Oak	Quercus rubra	3–6	15–25	60–75
~10%	Black Oak	Quercus velutina	2–4	15–30	50–60

Additional consideration was given to species resiliency to increasing urban temperature due to climate change in planning for the linear urban forest, such that some species not endemic to New England might thrive in warming climate zones at future times. These new species were not represented in the visualization scenarios as their viability has not yet been established. Of note, saplings common to more southern climate zones will be incorporated into the test plots for the urban forest to observe their adaptability to warming northern climates zone such as Springfield's.

2.2. InVEST Urban Cooling Model Application

Concurrent modeling of estimated cooling potential projected from the original linear urban forest design was performed using the Natural Capital InVEST Model's Urban Cooling Module [52]. This model generates maps of the heat mitigation index (HMI) to project cooling capacities of urban green spaces based on the land use classes contained in the area under consideration. The InVEST model requires inputs of area land use maps where greening is present or designated and biophysical estimates of annual tree growth, evapotranspiration, and albedo surface reflectivity, measures derived from the scientific literature, as well as a percentage measure of building intensity vis-à-vis vegetated space. For Springfield land use we used MassGIS Data: 2016 Land Cover/Land Use, which combines land cover mapping from 2016 aerial imagery and land use derived from standardized assessor parcel information for Massachusetts at a 1 m spatial resolution [53].

Land maps inputted into the InVEST model typically reflect existing categories of land use. Here instead, the challenge was to model projections of street trees not captured in the LULC maps. The yet-unrealized greening required a work-around of streets designated for the linear urban forest to displace a lane of impervious pavement. This was carried out through a manual reassignment in ArcGIS of a percentage of street parcels from the impervious pavement category to mixed forest. These simulated green interventions appear as new entries in the LULC table as Roads2030, Roads2050, and Roads2070. To these road categories we assigned a decreasing building intensity of 0.86, 0.8, and 0.72, respectively, and a percentage increase in shade and albedo. In this way, we derived future-year estimates of modeled rather than realized cooling.

The remaining model inputs came from ArcGIS mapping or from values contained in the scientific literature. The evapotranspiration raster map used was the MODIS Evapotranspiration product, which uses data on earth land surface from satellite remote sensing obtained at a 1 km^2 land scale, with evapotranspiration coefficient Kc. Binary measures of greenspace and shading switches are values between 0 and 1, representing the proportion of tree cover (0 for no trees, 1 for full tree cover, with trees >2 m. The default value of 2000 m was applied for average air temperature radius. The reference temperature for a rural area 10 km or more outside of Springfield was 0.59 °C, resulting in a UHI magnitude, or the difference between the rural reference temperature and the maximum temperature observed in the city. The albedo assigned to the land cover map was 2.09 (Table 2).

Table 2. Inputs to land use categories used to model urban cooling in Springfield through the InVEST Natural Capital Model.

LU Code	Description	Shade [1]	Kc [2]	Albedo [3]	Green Area	Building [4] Intensity
0	Background	0	0	0	0	0
1	Unclassified (cloud, shadow)	0	0	0	0	0
2	High intensity (>80% IA)	0.05	0.37	0.18	0	0.95
3	Med intensity (50–80%)	0.18	0.55	0.19	0	0.3
4	Low intensity (20–50%)	0.33	0.76	0.18	0	0.1
5	Open space (<20%)	0.45	0.93	0.2	0	0.1
6	Cultivated land	0.3	0.7	0.2	1	0
7	Pasture/hay	0.3	1	0.2	1	0
8	Grassland	0.3	1	0.2	1	0
9	Deciduous forest	0.8	1	0.15	1	0
10	Evergreen forest	0.8	1	0.15	1	0
11	Mixed forest	0.8	1	0.15	1	0
12	Scrub/shrub	0.4	1	0.2	1	0
13	Palustrine forested wetland	0.6	1	0.11	1	0
14	Palustrine scrub/shrub wetland	0.4	1	0.11	1	0
15	Palustrine emergent wetland	0.4	1	0.11	1	0
16	Estuarine forested wetland	0.8	1	0.11	1	0
17	Estuarine scrub/shrub wetland	0.4	1	0.11	1	0
18	Estuarine emergent wetland	0.4	1	0.11	1	0
19	Unconsolidated shore	0	0	0	0	0
20	Bare land	0	0.3	0.2	0	0
21	Water	0.3	1	0.06	1	0
22	Palustrine aquatic bed	0.3	1	0.06	1	0
23	Estuarine aquatic bed	0.3	1	0.06	1	0
24	Tundra	0	1	0.2	1	0
25	Snow/ice	0	1	0.75	0	0
26	Roads20	0.05	0.37	0.18	0	0.95
27	Roads30	0.14	0.37	0.22	1	0.86
28	Roads50	0.2	0.37	0.28	1	0.8
29	Road70	0.28	0.37	0.36	1	0.72

[1] Tree Canopy | Multi-Resolution Land Characteristics (MRLC) Consortium, U.S. Geological Service [54]; [2] evapotranspiration coefficient (Kc)—[55,56]; [3] Taha et al., 1988 [26]; [4] a floating-point value between 0 and 1. This is calculated by dividing the floor area by the land area, normalized between 0 and 1.

The estimates of city-wide ambient cooling generated by the InVEST model provided expectations of temperature reduction if the urban forest were to be realized in full. We chose not to model the full linear urban forest as designed because many of the assumptions drawn to project urban streets circa 2070 embodied futuristic scenarios that we felt distracted from the main premise of urban cooling—for example, fully autonomous vehicles and driveway elimination. For this reason, we focused on street tree planting alone.

Several demonstration sessions to beta-test the technology were held prior to the in situ community design charettes in Springfield. Two demonstration pilot sessions were presented to students and professors at UMass Amherst Department of Environmental

Conservation, and two pilots were presented in the Harvard Visualization Laboratory Theater with students from Northeastern and Harvard Universities. Participants attending these demonstrations encouraged the use of infographics to enhance visualization by explaining the impact of changes within the street scenarios. From these sessions we added temperature reduction, seasonal changes to the tree canopy, and planned engineering designs for stormwater drainage as a transect below street level.

With the visualization tools ready, we held two community design forums with residents of neighborhoods identified for urban tree planting. Sessions were organized by our non-profit partner, ReGreen Springfield, led by Dr. David Bloniarz, in collaboration with the City of Springfield Public Works Department, which is responsible for implementing neighborhood street tree planting. A brief presentation highlighted the environmental and public health benefits of increased residential contact with nature: temperature and hydrology regulation, particulate filtration, stormwater infiltration, and carbon sequestration, as well as prospective health improvements for impacted residents. Following the visualization sequence, community members responded to the street tree scenarios introduced in their neighborhoods from which we collected feedback. Each participant received a gift card for their time.

3. Results

These digital design charettes demonstrated how climate change impacts of extreme heat and intensive precipitation can be mitigated by repurposing streets for tree planting.

Each of the two community charettes produced a de-identified recording or transcript from the community participatory research in which both the various designs for street tree densification across 2024–2070 were presented to community residents. From this we learned (1) preferences for form and degree of nature-based solutions for climate mitigation, and (2) feedback on the process of community research itself to reiterate and improve heat mitigation projects in other cities.

An initial rendering of the LiDAR-based street diagrams showing street tree densification we shared succeeded in providing an aerial overview of the tree canopy densification (Figure 1). However, the charette participants felt the aerial perspective lacked the immediacy of walking out one's front door into the street-level environment and experiencing the tree canopy overhead.

Similarly, the wide-angle LiDAR depictions of street-level views distorted the spatial arrangement of buildings, trees, and the street, altering the dimensions of each and rendering these views unusable for public presentation (Figure 2).

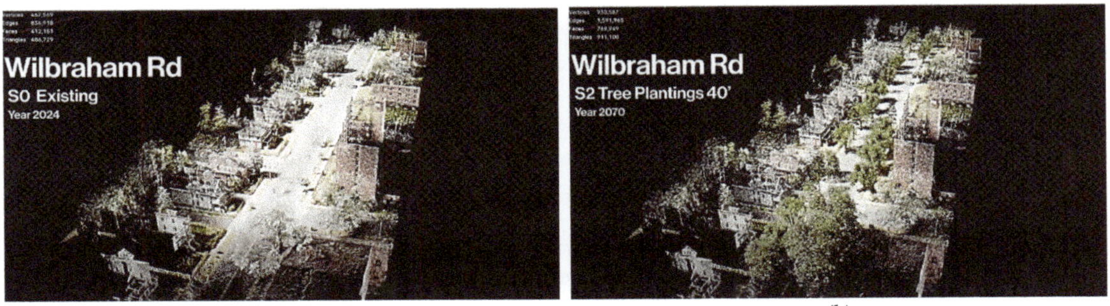

(a) (b)

Figure 1. *Cont.*

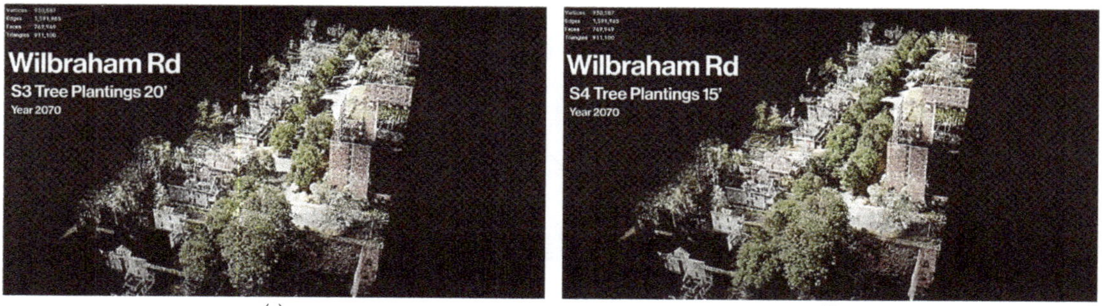

(**c**) (**d**)

Figure 1. LiDAR-based aerial diagrams of Wilbraham Road describing from the upper left (**a**) existing trees in the year 2024, (**b**) mature trees in the year 2070 planted at a 40′ distance, (**c**) mature trees in the year 2070 planted at a 20′ distance, and (**d**) mature trees in the year 2070 planted at a 15′ distance.

(**a**) (**b**)

Figure 2. LiDAR-based street-level diagrams of Wilbraham Road describing (**a**) existing trees for the year 2024 and (**b**) additional trees planted at 40 feet maturing in the year 2070.

The imaging specialists resorted to Google Street View superimposed with a series of street tree plantings that corresponded to scenarios of current plantings maturing 50 years into the future, plus tree plantings at 40-, 20-, and 15-feet spacings. These renderings were well received for their realism and ground-level relatability.

Of the four scenarios presented, charette participants favored street tree planting at 20 feet of distance to maximize the tree canopy without sacrificing a parking lane. This selection corresponds to a modeled 1.75 °F city-wide reduction in ambient temperature. The fact that participants were willing to accept a 0.75 °F increase in temperature is an indication that peri-urban culture remains vehicle-based, without the necessary transit infrastructure to permit alternative infrastructure types, even those such as green infrastructure, which would increase residential comfort and health.

Figure 3 presents still images of the street tree plantings spaced at 20 feet of distance for (a) Wilbraham Road and (b) Albemarle Street. Interactive representations of these scenarios are found in Videos S1 and S2.

The InVEST Urban Cooling Model generated temperature changes for Springfield based on models of a city-wide LUF. From this assumption, we see that the average cooling capacity rises from 0.46 °F under current landscape conditions to 0.54 °F in 2070 under conditions of maturing street trees. Average temperature variation falls slightly from 4.13 °F to 4.09 °F, and average annual temperatures from 3.07 °F to 3.04 °F (Table 3).

Figure 3. Before and after renderings of street trees planted on (**a**) Wilbraham Road, a main connector road in Springfield, and (**b**) Albemarle Street, a residential neighborhood street that intersects Wilbraham. The tree planting renderings feature the residents' preferred selection of 20 feet of planting distance between trees.

Table 3. Change in temperature output from InVEST Urban Cooling Model 3.87.

Parameter	InVEST Model Springfield-Wide UHI Profile TΔ °C			
	2020	**2030**	**2050**	**2070**
Square_Miles	33.08	33.08	33.08	33.08
avg_cooling capacity	0.254C/0.46F	0.281C/0.506F	0.285C/0.513F	0.298C/0.536F
avg_tmp_variation	1.9747	2.292C/4.13F	2.286/4.12F	2.269/4.09F
avg_tmp_annual	1.3847	1.702C/3.07F	1.696C/3.05F	1.690/3.04F
Temp_air_nomix	29.50–30.98	2.45 (0.61C–2.68C) (33.10F–36.82F)	2.45 (0.61C–2.68C) (33.10F–36.82F)	2.45 (0.61C–2.68C) (33.10F–36.82F)
Temp_air (high)	30.75	2.64 °C = 36.75 °F	2.64 °C = 36.75 °F	2.64 °C = 36.75 °F
Temp_air_(low)	30.62	2.18 °C = 35.92 °F	2.17 °C = 35.91 °F	2.15 °C = 35.87 °F
Heat mitigation	0–0.71	0–0.99053	0–0.99053	0–0.99053

4. Discussion

Digitally created street images showing how streets can be transformed in the face of rising heat offered community members the opportunity to express preferences and co-create neighborhood-level responses. Creating digital images of neighborhoods allowed them to envision future projections of their actual streets. In doing so, the research team created a new pedagogy for involving citizens and civic leaders, giving them a realistic sense of what adaptations to climate change might look like at the neighborhood scale, thus realizing the project's innovation potential. In addition to its technical contributions, this work played a vital role in community engagement and education. By combining state-of-the-art technology with a participatory approach, this research provides a replicable framework for urban planning that addresses environmental challenges while fostering community involvement.

Visualizing the impact of urban street trees on cooling cities facilitated on-the-ground collaboration with the Springfield community. Street tree plantings may affect changes to light, shading, sightlines, and resident security, even as they offer heat mitigation and other known physical and emotional health benefits associated with contact with urban nature. Bringing nature-based scenarios to the very residents of these targeted neighborhoods opened a dialogue that sensitized designers of green interventions to residents' concerns. Through the community design charettes the researchers solicited feedback to the different scenarios. Reactions were very positive to the projected neighborhood greening. No one appeared to favor the status quo after seeing the more tree-rich scenarios. In fact, the audience seemed to recoil upon viewing the seeming barrenness of the current street views that cycled around after the projected tree densifications.

Charette participants explained their voluntary attendance as interest in the design proposals but also as the opportunity to have a participatory voice in a municipal-driven process. Having a third party moderate the discussion also invited participation of individuals typically excluded from such procedures. The presence of city officials meant that feedback to the proposals concerning property security, additional maintenance, and traffic flow was heard directly and could influence final design. For example, participants raised questions about loss of sight lines impacting a sense of safety resulting from tree densification. Another safety concern of residents was sidewalk maintenance resulting from root protrusion, which presents tripping hazards for pedestrians, and the related burden of homeowner liability. The case of tree planting at 15 feet of distance projected onto a widened sidewalk median on residential streets would result in street narrowing. Charette participants roundly voted down this option, rejecting the loss of parking on one side of the street and the conversion of two-way streets to a single direction to accommodate wider tree planting.

The visualization charette elicited many reflective comments from attendees beyond approval and disapproval. One woman familiar with the green-lined streets of Brookline, Massachusetts, said, "When you drive through neighborhoods where there is a complete lack of vegetation, there is a sense of economic degradation. You don't feel that when you drive along streets filled with trees no matter their economic level." This comment challenges notions of green gentrification, whereby residents often resist neighborhood nature interventions for fear of increased rent, which greening can regrettably trigger.

Caution tempered some of the enthusiasm otherwise observed during the charettes. A young man spoke in favor of evaluating multiple changes being simultaneously proposed for his neighborhood. "Between solar panels, [tree plantings], the streets changes, there's a lot to consider. If you go straight to one of those things without double-checking, it could be two times worse than what it was." Another individual worried that opposition to raking could block such a project. A few individuals pointed out that tree roots would likely upheave the adjacent sidewalk blocks, creating tripping hazards, a frequent result of street tree plantings. Another participant simply attended because he wanted to see if trees might be restored in his neighborhood after their previous removal. When cities are planning infrastructure improvements, residents' concerns are often not adequately considered. For researchers, community involvement in poorer neighborhoods is often the missing component of translating evidence-based recommendations both as a commonsense matter of public governance and as a question of environmental justice.

Our community participatory research helped accelerate the climate–health and nature–health understanding of nature-based interventions via integrated assessment methodologies, dynamic visualization modeling, and design prototyping by presenting and soliciting feedback from locally impacted communities. The development of tangible state-of-the-art visualization tools permitted community residents to review proposed changes to their neighborhoods. In this way, visualization advanced design efforts beyond sheer prototyping and invited community participation of future interventions as an essential component of climate mitigation that embodied and not just spoke to environmental justice.

Tangible benefits also emerged from this project. First, the City of Springfield has planned three pilot sites to test the survivability rates of the ten tree species when planted along busy Springfield traffic arteries. Second, the visualizations engaged the City Parks and Recreations Department to allow employees to voice the need for increased time and budget dedicated to the street tree maintenance as well as training to deal with homeless individuals who often seek shelter in vegetated areas. Most importantly, the charettes gave community members the chance to be active partners in the design process.

This pilot study has the potential to grow in three areas of interest: (A) expanding the prototype for nature-based interventions, with actual measurements to be taken on both urban cooling and correlated ecosystem services related to urban afforestation. These interventions might include various types of green infrastructure from which to choose as well as offering various cost–benefit points. (B) Demonstrating proof of concept that reimagined urban neighborhoods, visualized through augmented reality and displayed for citizens and community leaders, can redress the social determinants of health and health inequities. (C) Working with city public works and parks departments to develop robust cost–benefit analysis for implementing heat-reducing green infrastructure. Visualizing the LUF may positively activate community engagement, prompt climate concern within vulnerable groups, and stake a claim for environmental justice. Opinions voiced during the digital community charettes and focus groups and received by civic decision-makers can enhance knowledge generated by the extensive modeling currently being developed and become translational. Equity-based urban health is already of interest to health-centered foundations and philanthropies as climate change exacerbates these inequities.

Having tested the pedagogies for engaging citizens, the next steps are to improve imagery to make it more interactive for participants. The technology required to scale this research to cities and neighborhoods beyond Springfield requires the skills of an able modeler and readily available software. Scalability makes this approach adaptable to differing socio-economic conditions, including linguistic barriers, given its visual imagery basis. One way of reaching larger audiences might be to conduct design charettes via Zoom or similar platforms to overcome participant inconvenience or reticence. Scalability to new cities and audiences might include not just street tree plantings but also other forms of green infrastructure such as green roofs, afforestation of open lots, and more creative interventions, ranking participant preference among various design options. The use of digital images can also assist with gauging cost benefits of green interventions within communities, e.g., assessing "willingness to pay" for climate adaptation with green infrastructure.

5. Conclusions

This pilot study tested technologies that invite communities to partake in decisional processes affecting neighborhood change related to a reduction in urban heat islands. The development of tools that invite and facilitate community-engaged research allows neighborhood dwellers to participate in discussions that affect their comfort and autonomy, while recognizing the importance of distributional equity that comes from having a voice in such discussions. Moreover, the impact of changes addressing temperature mitigation have downstream effects on individual health and wellbeing, with cross-generational benefits in communities such as Springfield's. We hope that this innovative pedagogical model can find traction for use in similar urban communities where nature-based solutions to decreasing heat can rectify decades of municipal underinvestment.

Supplementary Materials: Supplementary materials are available online. Video S1: Visualization of street tree planting for the year 2070 on Wilbraham Road (https://urldefense.proofpoint.com/v2/url?u=https-3A__drive.google.com_file_d_1B-2DOMCyMhQ4pmmaoqr7ggLMpd2PY53Prc_view-3Fusp-3Ddrive-5Flink&d=DwMFaQ&c=WO-RGvefibhHBZq3fL85hQ&r=ZMAzWaUd1AjmpNVdp0SFg-riY3p-8eaMS6853st7I9c&m=kYThjFw8Mbx_EE2Pyk92tSWJpXyjs04opg2ZHFZsnzU5HhzVywrPzGSNnJgtebzg&s=0Bd8pgFk6NRwP78NrUYFv6aWX6sFuA2EQ1wV0CUupAE&e=, accessed 10 November 2024). Video S2: Visualization of street tree planting for the year 2070 on Albemarle Street (https://urldefense.proofpoint.com/v2/url?u=https-3A__drive.google.com_file_d_1bKbT32qHPST2MlU-2D-5F8jb23-2DRAMUM3YcR_view-3Fusp-3Ddrive-5Flink&d=DwMFaQ&c=WO-RGvefibhHBZq3fL85hQ&r=ZMAzWaUd1AjmpNVdp0SFg-riY3p-8eaMS6853st7I9c&m=kYThjFw8Mbx_EE2Pyk92tSWJpXyjs04opg2ZHFZsnzU5HhzVywrPzGSNnJgtebzg&s=Y3H1CNylSH4X3nBkbslMxFk3UVavl72PxoYgDRZ7qLo&e=, accessed 10 November 2024).

Author Contributions: Conceptualization, L.P.T. and J.D.S.; methodology, L.P.T., K.S., D.B., D.E., and J.D.S.; software, K.S.; validation, L.P.T., K.S., D.B., D.E., and J.D.S.; investigation, L.P.T., K.S., D.B., D.E., and J.D.S.; resources, D.B. and J.D.S.; writing—original draft preparation, L.P.T. and K.S.; writing—review and editing, J.D.S.; visualization, K.S.; supervision, L.P.T., D.B., and J.D.S.; project administration, L.P.T. and J.D.S.; funding acquisition, L.P.T. and J.D.S. All authors have read and agreed to the published version of the manuscript.

Funding: This research was funded by the Harvard Chan NIEHS Center for Environmental Health, grant number P30ES000002.

Institutional Review Board Statement: Not applicable.

Informed Consent Statement: Informed consent was obtained from all subjects involved in the study.

Data Availability Statement: The original contributions presented in this study are included in the article/supplementary material. Data is unavailable due to privacy restrictions. Further inquiries can be directed to the corresponding author.

Acknowledgments: The authors gratefully acknowledge ReGreen Springfield and the City of Springfield, MA, Public Works Department for facilitating these community design charettes; the residents of the Springfield communities who attended them; and Martha Schwartz for her foundational landscape design work for Springfield's linear urban forest.

Conflicts of Interest: The authors declare no conflicts of interest. The funders had no role in the design of the study; in the collection, analyses, or interpretation of data; in the writing of the manuscript; or in the decision to publish the results.

Abbreviations

The following abbreviations are used in this manuscript:

UHI	Urban heat island
PET	Physiological equivalent temperature
EPA	U.S. Environmental Protection Agency
CDC	Centers for Disease Control
HOLC	Home Owners' Loan Corporation
LUF	Linear urban forest
CER	Community-engaged research
CAP	Community–academic partnerships
CBPR	Community-based participatory research

References

1. USEPA. Heat Island Effect. 2014. Available online: https://www.epa.gov/heatislands (accessed on 6 December 2024).
2. Aghamohammadi, N.; Chng, S.F.; Daliana, N.F.; Ramakreshnan, L.; Mohammadi, P.A. Heat and Mental Health in Cities. In *Urban Overheating: Heat Mitigation and the Impact on Health*; Aghamohammadi, N., Santamouris, M., Eds.; Springer Nature: Singapore, 2022; pp. 81–107. [CrossRef]
3. Cleland, S.E.; Steinhardt, W.; Neas, L.M.; West, J.J.; Rappold, A.G. Urban Heat Island Impacts on Heat-Related Cardiovascular Morbidity: A Time Series Analysis of Older Adults in US Metropolitan Areas. *Environ. Int.* **2023**, *178*, 108005. [CrossRef] [PubMed]
4. Ramly, N.; Hod, R.; Rohaizat Hassan, M.; Hasni Jaafar, M.; Isa, Z.; Ismail, R. Identifying Vulnerable Population in Urban Heat Island: A Literature Review. *Int. J. Public Health Res.* **2023**, *13*, 1678–1693. [CrossRef]
5. Hoffman, J.S.; Shandas, V.; Pendleton, N. The effects of historical housing policies on resident exposure to intra-urban heat: A study of 108 US urban areas. *Climate* **2020**, *8*, 12. [CrossRef]
6. Mosiniak, H. Redlining and Urban Heat Islands: An Analysis of Historic Housing Discrimination and Heat Exposure. Ph.D. Dissertation, University of Michigan, Ann Arbor, MI, USA, 2021.
7. Bao, J.; Cao, Z.; Huang, C.; Lei, L.; Yang, Y.; Peng, J.; Shi, X. Modification Effect of Urban Landscape Characteristics on the Association between Heat and Stroke Morbidity: A Small-Scale Intra-Urban Study in Shenzhen, China. *Sci. Total Environ.* **2021**, *786*, 147223. [CrossRef]
8. Whyte, M.; Douwes, J.; Ranta, A. Green Space and Stroke: A Scoping Review of the Evidence. *J. Neurol. Sci.* **2024**, *457*, 122870. [CrossRef]
9. Browning, M.H.E.M.; Rigolon, A.; McAnirlin, O.; Yoon, H.V. Where Greenspace Matters Most: A Systematic Review of Urbanicity, Greenspace, and Physical Health. *Landsc. Urban Plan* **2022**, *217*, 104233. [CrossRef]
10. Porcherie, M.; Linn, N.; Le Gall, A.R.; Thomas, M.-F.; Faure, E.; Rican, S.; Simos, J.; Cantoreggi, N.; Vaillant, Z.; Cambon, L.; et al. Relationship between Urban Green Spaces and Cancer: A Scoping Review. *Int. J. Environ. Res. Public Health* **2021**, *18*, 1751. [CrossRef] [PubMed]
11. Bikomeye, J.C.; Beyer, A.M.; Kwarteng, J.L.; Beyer, K.M.M. Greenspace, Inflammation, Cardiovascular Health, and Cancer: A Review and Conceptual Framework for Greenspace in Cardio-Oncology Research. *Int. J. Environ. Res. Public Health* **2022**, *19*, 2426. [CrossRef]
12. Dalton, A.M.; Jones, A.P. Residential Neighbourhood Greenspace Is Associated with Reduced Risk of Cardiovascular Disease: A Prospective Cohort Study" ed. David Meyre. *PLoS ONE* **2020**, *15*, e0226524. [CrossRef] [PubMed]
13. Astell-Burt, T.; Feng, X. Urban Green Space, Tree Canopy, and Prevention of Heart Disease, Hypertension, and Diabetes: A Longitudinal Study. *Lancet Planet. Health* **2019**, *3*, S16. [CrossRef]
14. Moreira, T.C.L.; Polizel, J.L.; de Souza Santos, I.; Silva Filho, D.F.; Bensenor, I.; Lotufo, P.A.; Mauad, T. Green Spaces, Land Cover, Street Trees and Hypertension in the Megacity of São Paulo. *Int. J. Environ. Res. Public Health* **2020**, *17*, 725. [CrossRef] [PubMed]
15. De la Fuente, F.; Saldías, M.A.; Cubillos, C.; Mery, G.; Carvajal, D.; Bowen, M.; Bertoglia, M.P. Green Space Exposure Association with Type 2 Diabetes Mellitus, Physical Activity, and Obesity: A Systematic Review. *Int. J. Environ. Res. Public Health* **2021**, *18*, 97. [CrossRef] [PubMed]
16. Lu, M.; Murphy, M.; Kim, A.; Lingwall, M.; Barr, E.A. The Relationship between Natural Environments and Obesity: A Systematic Review. *Int. J. Environ. Health Res.* **2024**, 1–21. [CrossRef] [PubMed]
17. Bodicoat, D.H.; O'Donovan, G.; Dalton, A.M.; Gray, L.J.; Yates, T.; Edwardson, C.; Hill, S.; Webb, D.R.; Khunti, K.; Davies, M.J.; et al. The Association between Neighbourhood Greenspace and Type 2 Diabetes in a Large Cross-Sectional Study. *BMJ Open* **2014**, *4*, e006076. [CrossRef]
18. Ccami-Bernal, F.; Soriano-Moreno, D.R.; Fernandez-Guzman, D.; Tuco, K.G.; Castro-Díaz, S.D.; Esparza-Varas, A.L.; Medina-Ramirez, S.A.; Caira-Chuquineyra, B.; Cortez-Soto, A.G.; Yovera-Aldana, M.; et al. Green Space Exposure and Type 2 Diabetes Mellitus Incidence: A Systematic Review. *Health Place* **2023**, *82*, 103045. [CrossRef] [PubMed]
19. Twohig-Bennett, C.; Jones, A. The Health Benefits of the Great Outdoors: A Systematic Review and Meta-Analysis of Greenspace Exposure and Health Outcomes. *Environ. Res.* **2018**, *166*, 628–637. [CrossRef] [PubMed]
20. Ye, T.; Guo, Y.; Huang, W.; Zhang, Y.; Abramson, M.J.; Li, S. Heat Exposure, Preterm Birth, and the Role of Greenness in Australia. *JAMA Pediatr.* **2024**, *178*, 376–383. [CrossRef] [PubMed]
21. Cilluffo, G.; Ferrante, G.; Fasola, S.; Malizia, V.; Montalbano, L.; Ranzi, A.; Badaloni, C.; Viegi, G.; La Grutta, S. Association between Asthma Control and Exposure to Greenness and Other Outdoor and Indoor Environmental Factors: A Longitudinal Study on a Cohort of Asthmatic Children. *Int. J. Environ. Res. Public Health* **2022**, *19*, 512. [CrossRef] [PubMed]
22. Hartley, K.; Ryan, P.; Brokamp, C.; Gillespie, G.L. Effect of Greenness on Asthma in Children: A Systematic Review. *Public Health Nurs.* **2020**, *37*, 453–460. [CrossRef] [PubMed]

23. Astell-Burt, T.; Feng, X. Association of Urban Green Space with Mental Health and General Health Among Adults in Australia. *JAMA NETW OPEN* **2019**, *2*, e198209. [CrossRef]
24. Van den Berg, M.; van Poppel, M.; van Kamp, I.; Andrusaityte, S.; Balseviciene, B.; Cirach, M.; Danileviciute, A.; Ellis, N.; Hurst, G.; Masterson, D.; et al. Visiting Green Space Is Associated with Mental Health and Vitality: A Cross-Sectional Study in Four European Cities. *Health Place* **2016**, *38*, 8–15. [CrossRef] [PubMed]
25. Bratman, G.N.; Anderson, C.B.; Berman, M.G.; Cochran, B.; de Vries, S.; Flanders, J.; Folke, C.; Frumkin, H.; Gross, J.J.; Hartig, T.; et al. Nature and Mental Health: An Ecosystem Service Perspective. *Sci. Adv.* **2019**, *5*, eaax0903. [CrossRef] [PubMed]
26. Taha, H.; Akbari, H.; Rosenfeld, A.; Huang, J. Residential cooling loads and the urban heat island—The effects of albedo. *Build. Environ.* **1988**, *23*, 271–283. [CrossRef]
27. Doan, S.; MacDonald, S.; Swaminathan, K. The Socialization of Positive Emotions: Implications for Physical Health and Psychological Adjustment. *Ment. Health Prev.* **2023**, *30*, 200272. [CrossRef]
28. Kubzansky, L.D.; Winning, A.; Kawachi, I. Affective States and Health. *Soc. Epidemiol.* **2014**, *2*, 320–364.
29. Coventry, P.A.; Brown, J.; Pervin, J.; Brabyn, S.; Pateman, R.; Breedvelt, J.; Gilbody, S.; Stancliffe, R.; McEachan, R.; White, P. Nature-Based Outdoor Activities for Mental and Physical Health: Systematic Review and Meta-Analysis. *SSMPopul. Health* **2021**, *16*, 100934. [CrossRef] [PubMed]
30. Shanahan, D.F.; Franco, L.; Lin, B.B.; Gaston, K.J.; Fuller, R.A. The Benefits of Natural Environments for Physical Activity. *Sports Med.* **2016**, *46*, 989–995. [CrossRef] [PubMed]
31. Haider, T. Urban climates and heat islands: Albedo, evapotranspiration, and anthropogenic heat. *Energy Build.* **1997**, *25*, 99–103. [CrossRef]
32. Zardo, L.; Geneletti, D.; Pérez-Soba, M.; Van Eupen, M. Estimating the cooling capacity of green infrastructures to support urban planning. *Ecosyst. Serv.* **2017**, *26*, 225–235. [CrossRef]
33. Armson, D.; Stringer, P.; Ennos, A.R. The effect of tree shade and grass on surface and globe temperatures in an urban area. *Urban For. Urban Green.* **2012**, *11*, 245–255. [CrossRef]
34. Li, D.; Bou-Zeid, E.; Oppenheimer, M. The effectiveness of cool and green roofs as urban heat island mitigation strategies. *Environ. Res. Lett.* **2014**, *9*, 055002. [CrossRef]
35. Aram, F.; García, E.H.; Solgi, E.; Mansournia, S. Urban green space cooling effect in cities. *Heliyon* **2019**, *5*, e01339. [CrossRef]
36. Balany, F.; Ng, A.W.M.; Muttil, N.; Muthukumaran, S.; Wong, M.S. Green Infrastructure as an Urban Heat Island Mitigation Strategy—A Review. *Water* **2020**, *12*, 3577. [CrossRef]
37. Vicedo-Cabrera, A.M.; Scovronick, N.; Sera, F.; Roye, D.; Schneider, R.; Tobias, A.; Gasparrini, A. The Burden of Heat-Related Mortality Attributable to Recent Human-Induced Climate Change. *Nat. Clim. Change* **2021**, *11*, 492–500. [CrossRef] [PubMed]
38. Elmes, A.; Rogan, J.; Williams, C.; Ratick, S.; Nowak, D.; Martin, D. Effects of Urban Tree Canopy Loss on Land Surface Temperature Magnitude and Timing. *ISPRS J. Photogramm. Remote Sens.* **2017**, *128*, 338–353. [CrossRef]
39. Estoque, R.C.; Murayama, Y.; Myint, S.W. Effects of Landscape Composition and Pattern on Land Surface Temperature: An Urban Heat Island Study in the Megacities of Southeast Asia. *Sci. Total Environ.* **2017**, *577*, 349–359. [CrossRef] [PubMed]
40. Rahman, M.A.; Stratopoulos, L.M.F.; Moser-Reischl, A.; Zölch, T.; Häberle, K.-H.; Rötzer, T.; Pretzsch, H.; Pauleit, S. Traits of Trees for Cooling Urban Heat Islands: A Meta-Analysis. *Build. Environ.* **2020**, *170*, 106606. [CrossRef]
41. Tan, Z.; Lau, K.K.; Ng, E. Urban Tree Design Approaches for Mitigating Daytime Urban Heat Island Effects in a High-Density Urban Environment. *Energy Build* **2016**, *114*, 265–274. [CrossRef]
42. Drahota, A.; Meza, R.D.; Brikho, B.; Naaf, M.; Estabillo, J.A.; Gomez, E.D.; Vejnoska, S.F.; Dufek, S.; Stahmer, A.C.; Aarons, G.A. Community-Academic Partnerships: A Systematic Review of the State of the Literature and Recommendations for Future Research. *Milbank Q.* **2016**, *94*, 163–214. [CrossRef]
43. Jull, J.; Giles, A.; Graham, I.D. Community-Based Participatory Research and Integrated Knowledge Translation: Advancing the Co-Creation of Knowledge. *Implement. Sci.* **2017**, *12*, 150. [CrossRef] [PubMed]
44. Collins, S.E.; Clifasefi, S.L.; Stanton, J.; Straits, K.J.; Gil-Kashiwabara, E.; Rodriguez Espinosa, P.; Nicasio, A.V.; Andrasik, M.P.; Hawes, S.M.; Miller, K.A.; et al. Community-Based Participatory Research (CBPR): Towards Equitable Involvement of Community in Psychology Research. *Am. Psychol.* **2018**, *73*, 884–898. [PubMed]
45. Herreros-Cantis, P.; Hoffman, L.; Kennedy, C.; Kim, Y.; Charles, J.; Gillet, V.; Getzin, A.; Littlefield, D.; Zielinski, A.; Bernstein, J. Co-Producing Research and Data Visualization for Environmental Justice Advocacy in Climate Change Adaptation: The Milwaukee Flood-Health Vulnerability Assessment. *Cities* **2024**, *155*, 105474. [CrossRef]
46. Capari, L.; Wilfing, H.; Exner, A.; Höflehner, T.; Haluza, D. Cooling the City? A Scientometric Study on Urban Green and Blue Infrastructure and Climate Change-Induced Public Health Effects. *Sustainability* **2022**, *14*, 4929. [CrossRef]
47. Koulouris, D.; Menychtas, A.; Maglogiannis, I. An IoT-enabled platform for the assessment of physical and mental activities utilizing augmented reality exergaming. *Sensors* **2022**, *22*, 3181. [CrossRef] [PubMed]

48. AAFA Community Services. Springfield, Massachusetts: Why It's the #1 Asthma Capital. Asthma and Allergy Foundation of America, 2019. Available online: https://community.aafa.org/blog/springfield-massachusetts-why-it-s-the-1-asthma-capital (accessed on 18 November 2024).
49. World Population Review. Poorest Cities in America 2024. Available online: https://worldpopulationreview.com/us-city-rankings/poorest-cities-in-america (accessed on 18 November 2024).
50. US Census Bureau. 2021 Releases. Census.gov. Available online: https://www.census.gov/data/what-is-data-census-gov/latest-releases/2021.html (accessed on 18 November 2024).
51. Centers for Disease Control. National Center for Health Statistics. 2024. Available online: https://www.cdc.gov/nchs/index.html (accessed on 18 November 2024).
52. The Natural Capital Project, Stanford University. Urban Cooling. Available online: https://naturalcapitalproject.stanford.edu/invest/urban-cooling (accessed on 12 December 2024).
53. Executive Office of Technology and Security Services, Commonwealth of Massachusetts. Available online: https://maps.massgis.digital.mass.gov/MassMapper/MassMapper.html?bl=MassGIS%20Basemap__100&l=massgis:GISDATA.LANDCOVER_USE_INDEX_POLY__GISDATA.LANDCOVER_USE_INDEX_POLY::Default__ON__100&b=-73.69628906250001,41.164181671865485,-69.74121093750001,42.958432 (accessed on 5 March 2024).
54. U.S. Geological Service. Multi-Resolution Land Characteristics (MRLC) Consortium, Tree Canopy. Available online: https://www.mrlc.gov/data/type/tree-canopy (accessed on 20 December 2024).
55. Allen, R.G.; Pereira, L.S.; Raes, D.; Smith, M. Crop evapotranspiration-Guidelines for computing crop water requirements-FAO Irrigation and drainage paper 56. *Fao Rome* **1998**, *300*, D05109.
56. Akbari, H. Cooling Our Communities. A Guidebook on Tree Planting and Light-Colored Surfacing. 2009. Available online: https://escholarship.org/content/qt98z8p10x/qt98z8p10x.pdf (accessed on 24 October 2023).

atmosphere

MDPI

Article

The Influence of Multisensory Perception on Student Outdoor Comfort in University Campus Design

Hichem Touhami [1], Djihed Berkouk [2,3,*], Tallal Abdel Karim Bouzir [3], Sara Khelil [3] and Mohammed M. Gomaa [2,4,*]

1 Laboratory of Design and Modelling of Architectural and Urban Forms and Ambiances (LACOMOFA), Department of Architecture, Biskra University, Biskra 07000, Algeria; hichem.touhami@univ-biskra.dz (H.T.)
2 Department of Architecture, School of Engineering, Computing & Design, Dar Al-Hekma University, Jeddah 22246, Saudi Arabia
3 Department of Architecture, Biskra University, Biskra 07000, Algeria; bouzirtallal@gmail.com (T.A.K.B.); sara.khelil@univ-biskra.dz (S.K.)
4 Department of Architectural Engineering, Faculty of Engineering, Aswan University, Aswan 81542, Egypt
* Correspondence: dberkouk@dah.edu.sa (D.B.); mgomaa@dah.edu.sa (M.M.G.)

Abstract: The user's experience is critical in spatial design, particularly in outdoor spaces like university campuses, where the physical environment significantly influences students' relaxation and stress relief. This study investigates the combined impact of thermal, luminous, and auditory environments on students' perceptions within recreational areas at Bordj Bou Arreridj University Campus. A mixed-method approach combined field surveys and on-site measurements across eleven locations within three distinct spatial configurations. The findings from this study indicate that the auditory environment had the most substantial influence on overall perceptions, surpassing luminous and thermal factors. The open courtyard (Area 1) was perceived as less comfortable due to excessive heat and noise exposure. The shaded zone (Area 2) was identified as the most vulnerable, experiencing significant thermal stress and noise disturbances. In contrast, the secluded patio (Area 3) achieved the highest comfort rating and was perceived as the most cheerful and suitable space. Correlation analysis revealed significant interrelationships between physical and perceptual dimensions, highlighting the critical role of factors such as wind velocity, sky view factor, and illuminance in shaping thermal, luminous, and acoustic perceptions. A fuzzy logic model was developed to predict user perceptions of comfort, suitability, and mood based on measured environmental parameters to address the complexity of multisensory interactions. This study highlights the importance of integrating multisensory evaluations into spatial design to optimize the quality of outdoor environments.

Keywords: multisensory perception; outdoor thermal comfort; soundscape; university campus design; student well-being; fuzzy logic model

1. Introduction

Outdoor environments foster relaxation, social interaction, and stress relief, particularly within university campuses, hubs of academic and social activity [1,2]. However, traditional architectural design often prioritizes aesthetics and structural form over the quality of outdoor spaces. As a result, the well-being of users in recreational areas is frequently treated as a secondary consideration [3,4]. This oversight underscores the need for a more user-centered approach to outdoor spatial design that integrates multisensory environmental factors to enhance comfort and satisfaction [5].

A combination of sensory stimuli, including thermal, luminous, and auditory environments, shapes perceptions of outdoor spaces. Extensive research has explored the impact of these factors individually [6–11]. For instance, studies on thermal comfort have highlighted its importance in outdoor usability, particularly in hot climates [7]. The thermal environment is a critical determinant of outdoor comfort, particularly in regions with extreme climatic conditions. Studies have demonstrated that thermal stress directly affects physiological and psychological well-being [8]. For example, shade, wind, and surface materials were emphasized in mitigating thermal discomfort in outdoor spaces [9–11]. However, most research focuses on temperature as an isolated factor, with limited consideration of its interaction with other environmental dimensions. In university campuses, thermal comfort is particularly vital, as prolonged exposure to heat can deter students from utilizing recreational spaces.

Similarly, research on soundscapes has emphasized the role of auditory environments in influencing emotional responses [12,13]. The auditory environment, often called the soundscape, significantly influences emotional responses and perceived comfort in outdoor spaces. Kang and Zhang (2010) demonstrated that natural sounds, such as birdsong or flowing water, enhance positive experiences, while noise pollution, such as traffic or construction, detracts from overall satisfaction [14]. Despite growing interest in soundscapes, the combined effect of auditory stimuli with thermal and luminous conditions remains underexplored, particularly in recreational areas within university campuses.

While luminous conditions influence visual comfort, mood, and spatial perception, research by [15,16] highlighted the significance of light intensity, shading, and contrast in shaping outdoor experiences. While studies on daylighting strategies are well-documented in architectural design [17,18], they primarily address indoor spaces, leaving a gap in the understanding of how luminous conditions interact with other sensory factors in outdoor settings. This gap is critical for designing functional and aesthetically pleasing spaces.

University campuses present unique research settings for exploring multisensory perceptions. These environments must balance functionality, aesthetics, and user well-being, particularly in recreational spaces where students seek respite from academic stress. Although some studies have examined campus design, most fail to address the multisensory interactions that shape users' experiences in outdoor areas. Despite evidence that well-designed campus landscapes contribute significantly to psychological restoration and overall well-being, natural environments positively influence mental health and reduce stress levels [19,20]. University recreational areas, critical for promoting relaxation, social interaction, and stress relief, remain underexplored in the scientific literature. These spaces provide an ideal context for investigating the influence of environmental factors on human perception. While existing studies have extensively explored individual environmental factors on university campuses, such as thermal comfort, noise levels, and lightscape experience, limited research has integrated these variables into a multisensory framework [16,21]. This gap is particularly evident in regions such as North Africa, where climatic conditions and cultural practices create distinct challenges and opportunities for outdoor design.

To address this knowledge gap, the present study investigates the combined impact of thermal, luminous, and auditory environments on students' perceptions of recreational spaces at the Faculty of Science and Technology, Bordj Bou Arreridj University. A mixed-methods approach was employed, integrating field surveys and on-site measurements of environmental parameters across distinct spatial configurations. This study identifies the factors contributing to comfort, suitability, and mood and introduces a fuzzy logic model to predict user perceptions based on measured environmental data. This approach represents a novel application of fuzzy logic in the context of outdoor recreational spaces.

This research contributes to the growing literature on multisensory design in architecture and urban planning by highlighting the interactions between thermal, luminous, and auditory factors. It underscores the importance of integrating sensory evaluations into spatial design processes to optimize the quality of outdoor environments. The findings of this study have broader implications for enhancing user well-being in recreational areas, particularly on university campuses, where outdoor spaces play a pivotal role in promoting relaxation and social connectivity.

2. Materials and Methods

Bordj Bou Arreridj city is located approximately 200 km southeast of Algiers, in a region classified as semi-arid according to the UNEP, De Matrone, and Emberger indices [22]. To investigate the effect of thermal, visual, and auditory environments on the students' perception, an investigation was conducted in the recreational areas surrounding and within the science and technology faculty at BBA University, situated around 3 km southeast of the city (Figure 1).

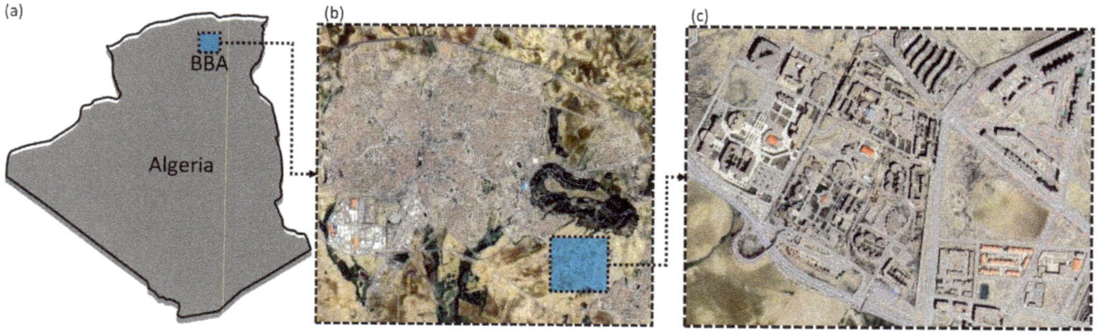

Figure 1. Case study area: (**a**) Bordj Bou Arreridj BBA localization, (**b**) university localization, (**c**) Bordj Bou Arreridj University.

2.1. Physical Context

The context of this study was divided into three areas based on the sky view factor (SVF). The first area is considered an open space due to the significant distance from the surrounding buildings with SVF > 0.9, and the second area, with 0.6 < SVF > 0.7, is a dense core positioned very near to the external walls of the faculty. The third area, a highly dense core with SVF < 0.6, represents the faculty's patio [23]. The mixed-method approach was utilized to collect quantitative and qualitative data simultaneously [24] at eleven points where the students gathered across the three areas (Figure 2).

The methodology of this research was based on the methodologies used in some previous studies [10,16], and the contribution of this study lies in adding a multisensory fuzzy logic model by using the MATLAB R2024a software (Figure 3). The aim of the in-situ measurement of the physical dimensions of thermal, visual, and auditory environments was performed when the participants were responding to the questionnaire handed out to ensure the direct relationship between participants' perception and physical dimensions, supporting real-time monitoring.

Figure 2. Distribution of measurement stations.

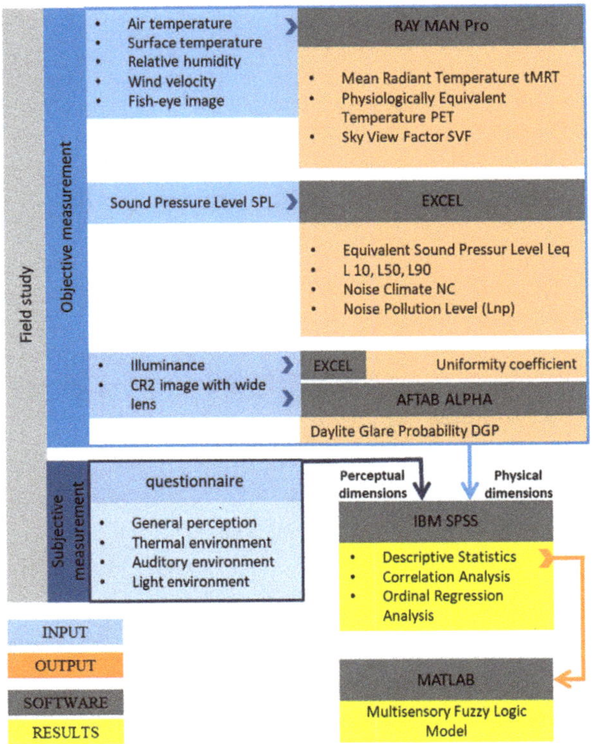

Figure 3. Methodological framework.

2.2. Objective Measurement and Material Used

The quantitative data collection operation was executed by measuring the different physical parameters of the environment (thermal, visual, and audible) through 11 measuring stations corresponding to the points identified as most valuable by the students (Figure 2). The first area covered stations one and two, stations three to eight were included in the second area, and the third area involved stations nine, ten, and eleven. The measurement was carried out systematically at every point for all the physical factors, as shown in Table 1, to ensure a rigorous and robust approach while achieving comparable results.

Table 1. Measurement stations information.

Areas	Stations	Image Type	Time Range	Recording Duration (min)	Factors Measured	Station Elevation (m)
Area 1 Area 2 Area 3	1, 2 3–8 9–11	-SVF Hemispherical Image -HDR image	12:30–12:40 12:50–13:40 13:50–14:10	3 min per station	Ts, Ta, RH, V, SPL, Illuminance	1.5 m above the ground

The sky view factor (SVF) was the primary dimension used to divide the recreational space into three areas depending on its variance, which ranged from 0.458 to 0.998. The method employed to calculate the SVF, which is based on fish-eye images and using the RAYMAN pro 1.2 software, has been adopted by numerous researchers [16,21,25,26].

A professional humidity meter was used to measure the ambient air temperature (Ta) and the relative humidity (RH) to assess the thermal environment. On the other hand, the soil surface temperature (Ts) was calculated using an infrared thermometer. In addition, an anemometer was employed to determine the wind velocity (Va). These parameters were used as inputs in the RAYMAN pro 1.2 software to calculate the mean radiant temperature (MRT) and the physiologically equivalent temperature (PET). Several studies have used this method considerably [16,21,27].

To evaluate the luminous environment, illuminance is regarded as one of the important indices in the field [28,29] and was quantified using a digital lux meter (MM-LM 01). Hence, CR2-format images were taken on site by a digital single-lens reflex (DSLR) camera (CANON EOS 5D Mark III, Canon, Japan) with a wide lens. This process aimed to calculate the daylight glare probability (DGP) with the support of the EVALGLAR plug-in incorporated with the AFTAB ALPHA 2.2 software [30,31].

Regarding the audible environment, the sound pressure level (SPL) measurement was carried out using a digital sound level meter. This was conducted to compute the equivalent sound pressure level (Leq), the percentile noise levels (L10, L50, L90), the noise climate (NC), and the noise pollution level (Lnp) due to their importance in the assessment of the impact of noise [21,32]. Furthermore, a stereo soundtrack was recorded at each station by a ZOOM H1 Sound Recorder [33,34].

2.3. Perceptual Dimensions and Questionnaire

Along with evaluating the physical dimensions and the measurement of the parameters, a subjective assessment was carried out to analyze the multisensory perceptual dimensions and determine the students' perceptions in the three recreational areas. For this reason, a questionnaire that included 12 semantic items was divided into four groups based on a seven-point bipolar rating scale assessment (Table 2). It was delivered to the participants after a detailed explanation to ensure clarity. The first group presented general perceptions, including comfort, suitability, and mood, which served as the dependent variables in the ordinal regression analysis. This group was adapted from a previous study that addressed multisensory interactions between thermal, auditory, and visual environments [17]. The second group covered the thermal environment, involving heat (cold–hot), used in the thermal sensation vote (TSV) [35,36], humidity (dry–humid), used in humidity sensation vote (HSV) [37], and wind conditions (windy–mild), which were used previously as not windy–windy [17] as well as in a study that mentioned wind conditions in the questionnaire as an influent parameter [38]. The third group introduced the luminous environment, consisting of glare (no glare–glare), lighting (dark–light), and uniformity (non-uniform–uniform), which have been largely used in lightscape assessment questionnaires [16,17,39,40]. The last group covered the audible environment, comprising sound excitation (monotonous–exciting), calmness (chaotic–calm), and sound variety (simple–varied), which were obtained from studies that explored an interesting number of soundscape attributes [14,41].

Table 2. Questionnaire used in the subjective assessment.

Environments		Very	Fairly	Little	Neutral	Little	Fairly	Very	
General perception	Uncomfortable	−3	−2	−1	0	1	2	3	Comfortable
	Not-suitable	−3	−2	−1	0	1	2	3	Suitable
	Depressive	−3	−2	−1	0	1	2	3	Cheerful
Thermal environment	Cold	−3	−2	−1	0	1	2	3	Hot
	Dry	−3	−2	−1	0	1	2	3	Humid
	Windy	−3	−2	−1	0	1	2	3	Mild
Luminous environment	No glare	−3	−2	−1	0	1	2	3	Glare
	Non-uniform	−3	−2	−1	0	1	2	3	Uniform
	dark	−3	−2	−1	0	1	2	3	light
Audible environment	Monotonous	−3	−2	−1	0	1	2	3	Exciting
	Chaotic	−3	−2	−1	0	1	2	3	Calm
	simple	−3	−2	−1	0	1	2	3	varied

2.4. Participants

This study included a sample size of 106 participants, comprising, 58 males (54.7%) and 48 females (45.3%), aged between 18 and 46 years, Table 3. The mean age of the participants was 21.85 years, with a standard deviation of 4.61 years, indicating a young group representative of university students. The participants were distributed across three areas, with 36 in the first area, 33 in the second area, and 37 in the third area. It is noteworthy that each respondent remained in the same position for 5 to 10 min before completing the questionnaire. It is essential to note that all the participants were in good health, and any medical condition or disease was registered.

Table 3. Descriptive statistics of participants by gender and age.

Statistic	Total	Males	Females
Number of Participants	106	58	48
Minimum Age	18	18	18
Maximum Age	46	46	45
Mean Age	21.85	22.36	21.25
Standard Deviation	4.61	4.59	4.606

2.5. Data Analysis

The statistical tests and data analysis were conducted utilizing the Statistical Package for the Social Science SPSS V.29 developed by the IBM Corporation (Armonk, NY, USA). The physical dimension data obtained from on-site measurements were processed by applying descriptive statistics. In terms of perceptual data, an ordinal regression analysis was executed to assess the effect of thermal, luminous, and audible perception variables on suitability, comfort, and mood. Finally, three correlation analyses were conducted to examine the relationships between the various dimensions. Pearson's correlation was used to assess relationships among the physical dimensions, Kendall's tau-b correlation was applied to evaluate associations among the perceptual dimensions, and Spearman's correlation was utilized to explore the relationships between perceptual and physical dimensions.

2.6. Fuzzy Logic Model

In urban planning, fuzzy logic, FL, has been deemed an effective tool in spatial evaluation. It can treat different memberships for complex uncertainty-related topics when addressing urban sustainability factors [39] and help make wise decisions [42]. The construction of the model encompasses four main steps: fuzzification, creating the rules, inference, and defuzzification [43]. In this study, the model was developed using the MATLAB software.

The fuzzification step represents the modeling of the input and output based on reference values obtained from previous studies. In light of this, the linguistic classification of the membership function of the equivalent sound pressure level (Leq) was taken from a study evaluating noise pollution's impact on people's health [44]. The physiologically equivalent temperature (PET) membership function was constructed based on a previous classification widely adopted in the literature [45]. The sky view factor (SVF) classification was inspired by research that explored the relationship between the sky view factor and land temperature [23]. Regarding illuminance, the linguistic classification was adapted from evaluating urban park lighting quality perception [29]. Finally, the mean radiant temperature (tMRT) was adjusted from a previous classification of the thermal comfort fuzzy logic model [46]. The membership functions of the input data were designed, as well as the three semantic items comfort, suitability, and mood that constituted the output variables, as shown in (Figure 4).

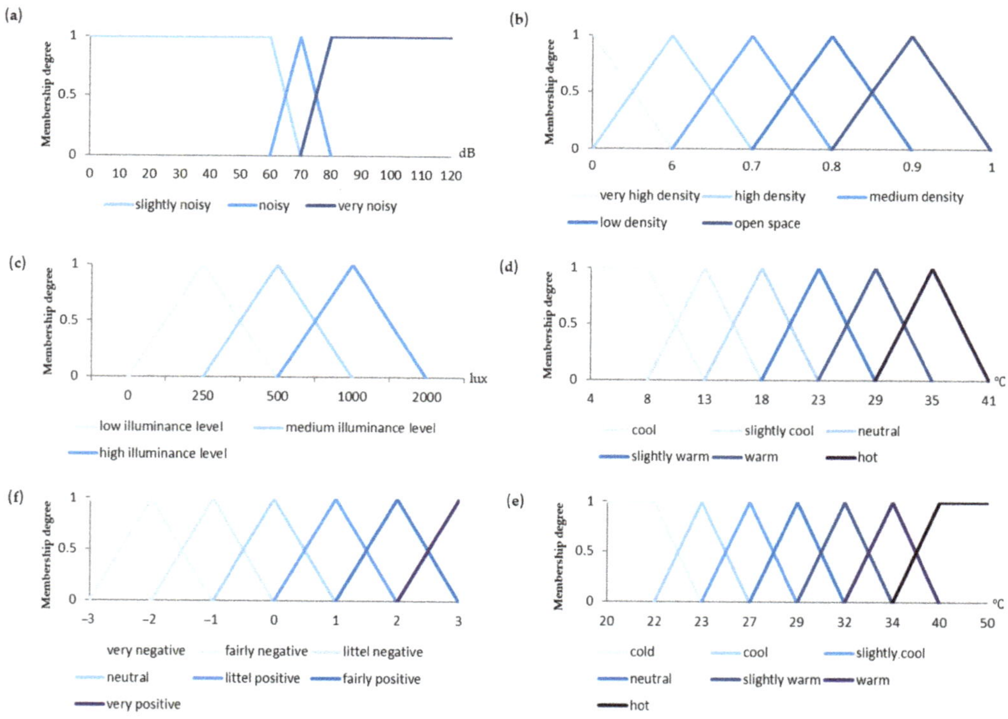

Figure 4. Membership functions: (**a**) Leq. (**b**) SVF. (**c**) Illuminance. (**d**) PET. (**e**) tMRT. (**f**) Participant responses (output).

Rules production provided the relationships among the linguistic variables for both inputs and outputs, which was composed of precondition (If part) and consequence (Then part) [46]. In this case, the number of rules corresponded to the number of variations in the variables' values, and the inference adopted was developed with the Mamdani's fuzzy method (Table 4). The defuzzification consisted of translating the fuzzy output to the crisp value. The method employed in this model was centroid defuzzification.

Table 4. If–then rules of the fuzzy logic model.

Rules		Illuminance	SVF	PET	Leq	tMRT		Suitability	Comfort	Mood
1	If	High	Open space	Slightly warm	Noisy	Hot	Then	Little not suitable	Little uncomfortable	Little depressive
2	If	High	Open space	Slightly warm	Noisy	Hot	Then	Fairly not suitable	Fairly uncomfortable	Little depressive
3	If	High	Medium density	Neutral	Noisy	Hot	Then	Little not suitable	Little uncomfortable	Little depressive
4	If	Low	Medium density	Slightly cool	Very noisy	Warm	Then	Fairly not suitable	Fairly uncomfortable	Little depressive
5	If	Medium	High density	Slightly warm	Noisy	Hot	Then	Little suitable	Little uncomfortable	Little depressive
6	If	Low	Medium density	Slightly warm	Noisy	Hot	Then	Little suitable	Little comfortable	Little cheerful
7	If	High	Open space	Slightly warm	Noisy	Hot	Then	Little not suitable	Very uncomfortable	Little depressive
8	If	Medium	Very high density	Warm	Noisy	Hot	Then	Fairly suitable	Little comfortable	Little cheerful
9	If	High	High density	Slightly warm	Noisy	Hot	Then	Fairly suitable	Fairly comfortable	Fairly cheerful
10	If	Medium	Very high density	Slightly warm	Noisy	Hot	Then	Fairly suitable	Little comfortable	Fairly cheerful
11	If	Medium	Very high density	Neutral	Noisy	Neutral	Then	Very suitable	Fairly comfortable	Little depressive
12	If	Low	High density	Neutral	Noisy	Neutral	Then	Little suitable	Little comfortable	Little cheerful
13	If	Low	High density	Slightly cool	Noisy	Neutral	Then	Little not suitable	Fairly comfortable	Neutral

3. Results

3.1. Physical Dimensions

Following the on-site measurements of thermal, luminous, and audible dimensions, descriptive statistics analyses were performed with the IBM SPSS V.29 software, and the results are presented in Table 5. The values of the SVF calculated at the different station measurements ranged from 0.458 to 0.998, with a mean of 0.71 and an SD of ±0.22. The notable difference in SVF values reflects the diversity in spatial morphology across the three areas. Concerning the thermal environment, the physiologically equivalent temperature (PET), varying between 20.7 °C and 32.2 °C, was generally considered as comfortable to slightly warm, with a mean of 27.8 °C and an SD of ±3.2 [45]. In addition, the mean radiant temperature (tMRT) ranged between 29.4 °C and 44.4 °C, with a mean value of 38.7 and an SD of ±3. The wide range of values in the thermal dimensions highlights the impact of morphology on thermal comfort. Additionally, orientation and tree cover can influence thermal sensation through direct sun exposure and shading. Regarding the luminous environment, the average illuminance measured was 1137.47 lux, with minimum and maximum values of 206 lux and 1880 lux, respectively, and an SD of ±627. This variation underscores the lack of light uniformity, which can adversely affect visual comfort [47]. The daylight glare probability (DGP) values ranged from 0.123 to 0.269, with a mean of 0.22 and a standard deviation (SD) of ±0.04. As the glare values did not exceed 0.4, the glare phenomenon may be not disturbing for users [16]. Regarding the audible environment, the sound equivalent level (Leq) values oscillated between 63.6 dB and 90.8 dB, with a mean of 71.8 and an SD of ±7.3. Based on the mean Leq, the sound environment was perceived as noisy, as its value exceeded 66 dB, which is the recommended limit for urban areas [48]. The noise climate (NC) values ranged from 8.5 dB and 21.4 dB, with a mean of 13.8 (±3.3). The notable values of the noise climate explain the gap between the background sound and the peak sound levels, suggesting that the peak levels may be considered as noise pollutants. Furthermore, the mean noise pollution level (Lnp) was 84.5 dB, with minimum and maximum values of 67.2 dB and 109 dB, respectively, and an SD of ±10.7. Considering the permissible limit for Lnp, which is 88 dB [32,49], the recorded values suggest that the sound environment was noisy, particularly given the maximum value of 109 dB, which significantly exceeded the allowable limit. Overall, the audible environment showed excessive values compared to the comfort thresholds, followed by the thermal environment. In contrast, the luminous environment values remained below the recommended limits, as illustrated in Figure 5.

Table 5. Descriptive statistics of the physical dimensions.

	Variables	Obs	Minimum	Maximum	Mean	SD
Thermal environment	SVF	11	0.458	0.998	0.71	0.22
	PET (°C)	11	20.7	32.2	27.8	3.2
	tMRT (°C)	11	29.4	44.4	38.7	3
Luminous environment	DGP	11	0.123	0.269	0.22	0.04
	Luminance (lux)	11	206	1880	1137.47	627
Audible environment	Leq (dB)	11	63.6	90.8	71.8	7.3
	NC (dB)	11	8.5	21.4	13.8	3.3
	Lnp (dB)	11	67.2	109	84.5	10.7

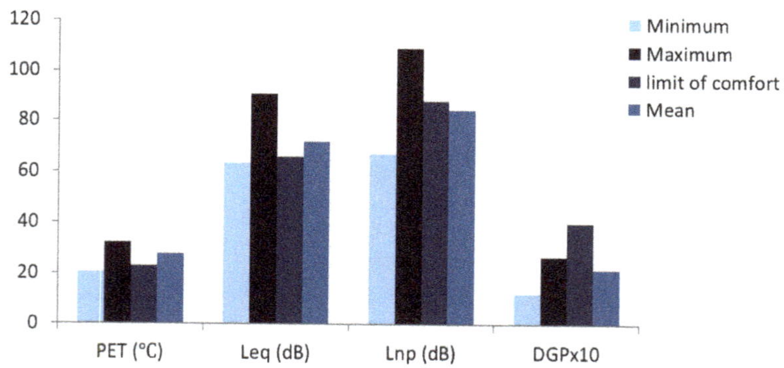

Figure 5. Minimum, maximum, and mean of PET, Leq, Lnp, and DGP values compared to the permissible limits.

3.2. *Perceptual Dimensions*

3.2.1. Overall Perception of the Space: Responses Analysis

To analyze the perceptual data, an analysis of the average scores and the percentage of the semantic difference was conducted after processing the descriptive statistics to examine the students' perceptions across the three areas, as shown in Figure 6. Concerning the general perception of suitability, comfort, and mood, the first and the second spaces were perceived closely alike, with a little non-suitable, a little uncomfortable, and neutral for the mood, and a little non-suitable, relatively uncomfortable, and a little depressive, respectively. However, the third area was perceived as reasonably suitable, comfortable, and cheerful. Regarding the thermal sensation, the first area was perceived as very hot compared to the second and third spaces, which were perceived as reasonably hot. Regarding the luminous environment, the first and the third areas were perceived as significantly glared and light, while the second spaces were perceived as relatively glared and reasonably light. Regarding audible perception, the first and the third areas were perceived as reasonably chaotic, whereas the second space was perceived as a little chaotic.

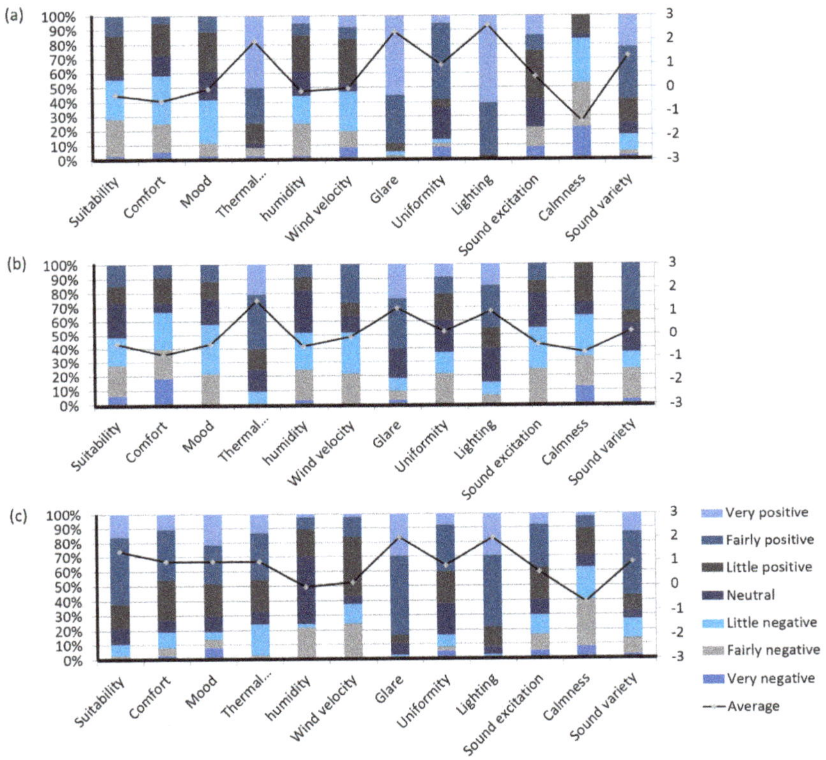

Figure 6. Average scores and percentage of semantic difference analysis. (**a**) First area, (**b**) second area, and (**c**) third area.

3.2.2. General Perception Dependence: Ordinal Regression Model

Ordinal regression was performed to identify which semantics of the thermal, luminous, and audible environment perception influenced the suitability, comfort, and mood through the three areas, yielding the results reported in Table 6.

Regarding zone 1, the results indicate the dominance of thermal and sound factors influencing the suitability of spaces. In this zone, thermal sensation ($p = 0.01$) and noise excitation ($p = 0.006$) were critical factors, showing that extreme temperatures and noise levels directly affected the usability perception of the space. This indicates that excessive heat caused significant discomfort, confirming Nikolopoulou and Steemers' previous study associating high temperatures with lower space usability [50]. In terms of noise levels, relaxation was affected, reducing the suitability of the space for recreational activities. Furthermore, the results show that thermal sensation was the only predictor of comfort ($p = 0.013$), underlining the preponderant influence of thermal stress in zone 1. This suggests that thermal conditions were the primary determinant of overall comfort and could be more important than other environmental factors. Emotional ambience, however, was influenced by the uniformity of light ($p = 0.011$) and sound excitation ($p = 0.000$). This confirms that non-uniform lighting probably caused visual discomfort, reducing positive emotional responses, in line with research associating uniform light with relaxation and safety [47]. Furthermore, the strong association between noise excitation and emotional ambience ($p = 0.000$) indicates that noise not only affected the space's suitability but also detracted from its emotional atmosphere.

Table 6. Ordinal regression results.

	Area 1			Area 2			Area 3		
Dependent Variables	**Predictors**	***p* Value**	**Model Info**	**Predictors**	***p* Value**	**Model Info**	**Predictors**	***p* Value**	**Model Info**
Suitability	Thermal sensation	0.01	Model Fitting sig = 0.000 $pR^2 = 1$ Parallel lines test = 1	thermal sensation uniformity	0.000 0.018	Model Fitting sig = 0.000 $pR^2 = 0.635$ Parallel lines test = 1	wind condition uniformity	0.03 0.01	Model Fitting sig = 0.05 $pR^2 = 0.397$ Parallel lines test = 0.105
	Sound excitation	0.006		lighting sound excitation	0.034 0.033				
				calmness	0.013				
Comfort	Thermal sensation	0.013	Model Fitting sig = 0.000 $pR^2 = 1$ Parallel lines test = 1	thermal sensation glare	0.006 0.013	Model Fitting sig = 0.012 $pR^2 = 0.491$ Parallel lines test = 0.99	calmness	0.001	Model Fitting sig = 0.003 $pR^2 = 0.519$ Parallel lines test = 0.998
				uniformity sound excitation	0.033 0.027				
				calmness	0.021				
Mood	Uniformity	0.011	Model Fitting sig = 0.000 $pR^2 = 0.939$ Parallel lines test = 0.68	glare	0.026	Model Fitting sig = 0.037 $pR^2 = 0.478$ Parallel lines test = 1	wind condition	0.026	Model Fitting sig = 0.033 $pR^2 = 0.4$ Parallel lines test = 0.925
	Sound excitation	0.000		calmness	0.012		sound excitation	0.026	

Regarding zone 2, the suitability and comfort in this area were influenced by a combination of thermal sensation, light uniformity, sound excitation, and calmness, with additional impacts from lighting conditions (dark/light), representing the contrast phenomena on suitability and glare on comfort. Direct sun exposure likely exacerbated heat stress, consistent with studies on the effects of solar radiation on outdoor comfort [51]. Light uniformity positively influenced perceptions, enhancing the sense of suitability, while glare detracted from comfort, emphasizing the need for balanced lighting solutions. The auditory environment showed a dual impact, where noise from nearby traffic or parking areas conflicted with moments of calmness provided by wind or quieter intervals, reflecting a complex soundscape. These diverse factors highlight the vulnerability of this space to environmental challenges because its open orientation and proximity to noise sources created a challenging environment for users.

Regarding zone 3, wind speed ($p = 0.03$) and light uniformity ($p = 0.01$) were identified as significant predictors of suitability, highlighting their positive contribution to the usability of the outdoor space. Perceived wind likely enhanced comfort and usability by providing natural cooling and masking unwanted noise, consistent with findings on the role of wind in enhancing outdoor experiences [52]. Similarly, uniform lighting enhances security and appropriateness, particularly in shaded areas. Comfort was strongly influenced by calmness ($p = 0.001$), underlining the essential role of a calm auditory environment in promoting comfort, as confirmed by the findings of Kang and Zhang [14]. The mood was influenced by wind speed ($p = 0.026$) and sound excitation ($p = 0.000$). The wind positively impacted mood by improving thermal and auditory conditions. At the same time, the strong association with sound excitation underlined the direct influence of auditory stimulation, whether pleasant or intrusive, on emotional responses. These results underline the importance of integrating the elements of wind and light while minimizing auditory disturbance to create more comfortable, emotionally supportive outdoor spaces.

3.3. Correlation Analysis

Figure 7 illustrates the interrelationships network, highlighting significant correlations between physical and perceptual dimensions across Kendall's tau-b, Pearson's, and Spearman's analyses.

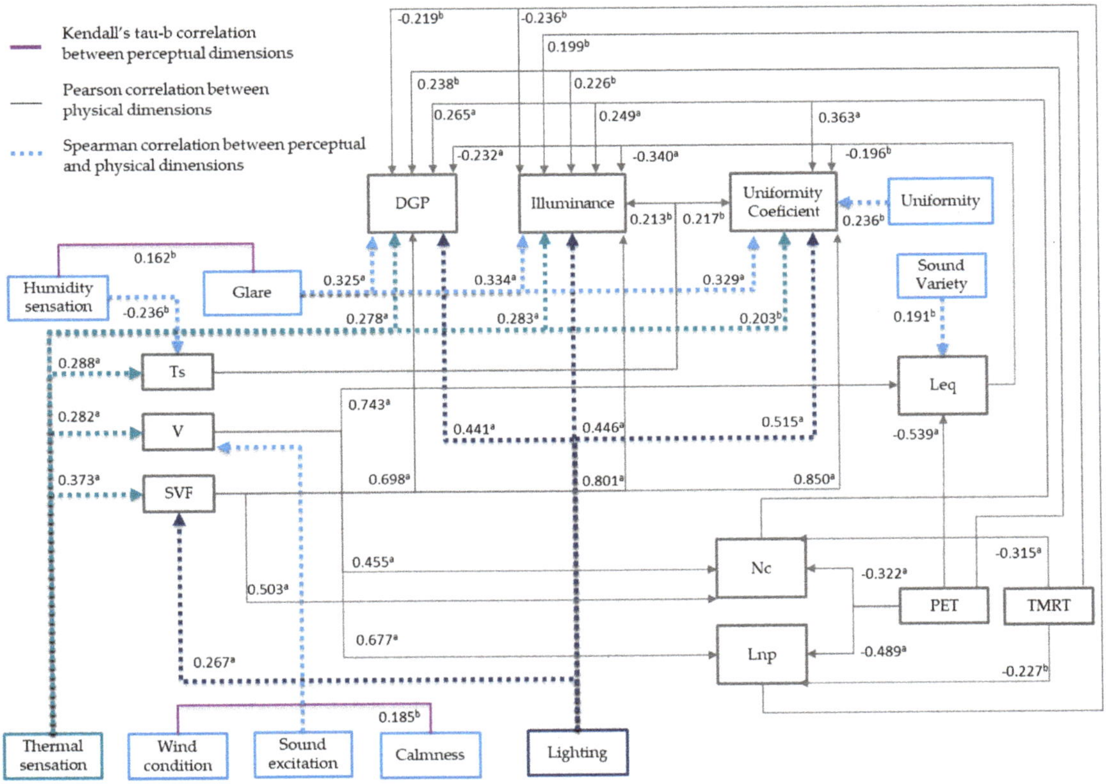

ᵃ Correlation is significant at the 0.01 level (two-tailed). ᵇ Correlation is significant at the 0.05 level (two-tailed)

Figure 7. Interrelationships network between the main influencer variables.

The results of Kendall's tau-b correlation analysis reveal statistically significant relationships between certain perceptual dimensions. A weak but significant correlation was found between humidity sensation and glare perception ($\tau = 0.162$, $p < 0.05$). Similarly, a weak yet statistically significant correlation was observed between calmness and wind conditions ($\tau = 0.185$, $p < 0.05$).

However, Pearson's correlation shows significant interactions between physical dimensions. Surface temperature (Ts) exhibited weak correlations with illuminance ($r = 0.213$, $p < 0.05$) and the uniformity coefficient (UC) ($r = 0.217$, $p < 0.05$), indicating limited influence on these luminous parameters. Similarly, wind velocity (V) strongly correlated with the equivalent sound pressure level (Leq) ($r = 0.743$, $p < 0.01$), noise climate (NC) ($r = 0.455$, $p < 0.01$), and noise pollution level (Lnp) ($r = 0.677$, $p < 0.01$), highlighting its role in shaping the acoustic environment. Illuminance (Illum) demonstrated a strong correlation with the sky view factor (SVF) (r = 0.801, $p < 0.01$) while also showing moderate associations with thermal variables such as the mean radiant temperature (tMRT) (r = 0.199, $p < 0.05$) and predicted mean vote (PET) (r = 0.226, $p < 0.05$). Interestingly, illuminance showed a negative correlation with Leq (r = −0.340, $p < 0.01$) and Lnp (r = −0.236, $p < 0.05$), indicating potential trade-offs between luminous and acoustic environments. Furthermore, SVF strongly correlated with the daylight glare probability (DGP) (r = 0.698, $p < 0.01$) and UC (r = 0.850, $p < 0.01$), suggesting its critical role in luminous uniformity and glare.

Thermal dimensions also displayed noteworthy relationships. The mean radiant temperature (tMRT) negatively correlated with NC (r = −0.315, $p < 0.01$) and Lnp (r = −0.227, $p < 0.05$), while PET exhibited strong negative correlations with Leq (r = −0.539, $p < 0.01$), NC (r = −0.322, $p < 0.01$), and Lnp (r = −0.489, $p < 0.01$). PET also had a weak positive correlation with DGP (r = 0.238, $p < 0.05$), linking thermal comfort to glare perception. Acoustic and luminous dimensions showed important interactions. Leq negatively correlated with DGP (r = −0.332, $p < 0.01$) and UC (r = −0.196, $p < 0.05$), indicating that high sound levels may reduce luminous comfort. NC was positively correlated with DGP (r = 0.265, $p < 0.01$) and UC (r = 0.363, $p < 0.01$), reinforcing its influence on visual parameters. Lastly, Lnp showed a weak negative correlation with DGP (r = −0.219, $p < 0.05$), further linking acoustic and luminous dimensions.

The results of Spearman's correlation analysis indicate several significant relationships between perceptual and physical dimensions. Thermal sensation was positively correlated with surface temperature (Ts) (r = 0.288, $p < 0.01$), wind velocity (r = 0.282, $p < 0.01$), illuminance (r = 0.283, $p < 0.01$), the sky view factor (SVF) (r = 0.373, $p < 0.01$), daylight glare probability (DGP) (r = 0.278, $p < 0.01$), and the uniformity coefficient (UC) (r = 0.203, $p < 0.05$). These correlations suggest that a higher surface temperature, wind velocity, and illuminance levels, as well as more open sky exposure and higher daylight glare probability, intensify thermal sensation. Uniformity also played a subtle role in influencing thermal sensation. Conversely, humidity perception had a negative correlation with surface temperature (r = −0.236, $p < 0.05$), indicating that higher surface temperatures may reduce the perception of humidity.

In the realm of sound perception, wind velocity exhibited a weak negative correlation with sound excitation (r = −0.237, $p < 0.05$), suggesting that higher wind velocity slightly dampens the perception of sound excitation. Additionally, the equivalent sound pressure level (Leq) showed a weak positive correlation with sound variety (r = 0.191, $p < 0.05$), indicating that higher sound levels marginally increase the perception of sound variety.

For glare perception, significant positive correlations were observed with illuminance (r = 0.334, $p < 0.01$), daylight glare probability (r = 0.325, $p < 0.01$), and uniformity coefficient (r = 0.329, $p < 0.01$). These findings indicate that higher illuminance, DGP, and uniformity contribute to increased glare perception. Meanwhile, lighting perception was strongly associated with illuminance (r = 0.446, $p < 0.01$), daylight glare probability (r = 0.441, $p < 0.01$), and the uniformity coefficient (r = 0.515, $p < 0.01$), with a moderate correlation with the sky view factor (r = 0.267, $p < 0.01$). These results suggest that uniformity, illuminance, and DGP are critical factors in shaping perceptions of lighting and glare.

4. Discussion

The physical dimensions of the environment are considered a significant criterion in assessing the quality of urban spaces [10,53]. The descriptive statistics of the physical dimensions suggest that the thermal perception based on PET balanced between comfortable without thermal stress and warm with moderate heat stress [45]. In addition, based on the measured illuminance values, the luminous environment in terms of daylight was classified as beneficial [54–56]. Concerning the audible environment, the soundscape was considered slightly noisy to very noisy due to the Leq values, which balanced between 63 dB and 90.8 dB [44]. The findings from the descriptive statistics of the participant responses mention that the three recreational spaces in the university were perceived differently in terms of general perception (suitability, comfort, and mood). Hence, the third area was perceived as more suitable, comfortable, and cheerful than the first and second spaces.

To provide a comprehensive clarification about the reasons for this difference, an ordinal regression was carried out to reveal the impact of the environmental components on the general perception. As a result, different factors contributed to the difference in the general perception of each area. Thermal sensation affected suitability and comfort in the first and second spaces, where the average was around very hot. In contrast to the third space, which was perceived as reasonably hot with an average near to a little hot, the thermal sensation effect was absent. A correlation between thermal sensation, the sky view factor, and surface temperature explains the variation in perception within the different spaces [23,57]. The audible perception significantly affected the general perception through the three areas by sound excitation and calmness, as well as the wind condition, which was perceived as an audible factor in the third space due to its relationship with calmness. The complex relationship among wind velocity, audible dimensions, calmness, sound excitation, and wind conditions can justify the effect of winds on audible perception [58,59]. Regarding the luminous perception, the light uniformity perception affected mood in the first area and suitability in the third area. Moreover, the correlation between the perceived and measured uniformity suggests a direct effect. Conforming to this, several studies have highlighted that light uniformity promotes relaxation and safety feeling [47,60].

The findings from the ordinal regression analysis also emphasize that the second area was affected by many factors, with numerous conditions causing its fragility. Spatial orientation may contribute to its vulnerability in terms of thermal and luminous environments owing to direct sun exposure [61]. The adjacency to the external wall of the faculty could also be a factor in this susceptibility [62]. Relating to the acoustic conditions, the proximity of the space to the university parking area and the direct and wide orientation to the high-traffic road W42 increase the possibility of its sensitivity [63]. This study revealed the significant impact of the audible environment in recreational space compared to the luminous and thermal environments. In support of this, previous studies in the literature have outlined similar results in assessments of indoor and outdoor environments [17,64].

Supporting the ordinal regression results, three types of correlations were conducted to explore the relationships between the various dimensions. For the physical dimensions, the sky view factor (SVF) was found to correlate with luminous dimensions (DGP, illuminance, and the uniformity coefficient), as well as with noise climate (NC). Similarly, surface temperature (Ts) exhibited a correlation with illuminance and the uniformity coefficient. The relationship between the SVF, luminous, and thermal dimensions has been demonstrated in previous studies [17,65]. Regarding wind velocity (v), its strong correlation with audible dimensions (NC, Lnp, and Leq) underscores its significant role in shaping the audible environment [58,59].

Regarding the perceptual dimensions, thermal sensation and humidity sensation had a correlation with thermal dimensions, the SVF, and luminous dimensions [66,67]. Additionally, luminous perception correlated with the measured luminous dimensions, particularly glare perception and uniformity perception, which were associated with daylight glare probability and the uniformity coefficient, respectively. In the audible environment, sound variety correlated with the equivalent sound pressure level (Leq), while sound excitation showed a correlation with sound variety and wind velocity. Calmness, on the other hand, was correlated with wind conditions. These findings suggest that wind plays a significant role in shaping the soundscape experience, as further supported by the strong correlation between wind velocity and all the aforementioned audible dimensions.

Due to the ambiguity and complexity of the multisensory interaction between the thermal, luminous, and audible environments, a multisensory fuzzy logic model was established to clarify the combined effect of these environments on the students' perception. Based on this study's results, this model predicted the users' perceived experience. It should be noted that the precision of the model's results relied on the number of rules.

The findings of this study offer several valuable insights for architects, urban planners, and designers. They provide a foundation for optimizing outdoor spaces on university campuses and beyond. By linking these results to the broader literature, we can highlight their practical significance and contribute to the evolution of user-centered design practices.

This study demonstrates the critical role of thermal perception in influencing user comfort and suitability, particularly in spaces with varying thermal sensations. Spaces with higher thermal stress (e.g., the first and second areas) were perceived as less comfortable, reinforcing the importance of mitigating thermal stress through design strategies. Practitioners can incorporate tree canopies, pergolas, or shade sails to reduce direct sun exposure, improving thermal comfort, as supported by Tablada et al. [68,69]. Orienting spaces to maximize natural ventilation could also be used to reduce heat buildup and enhance the thermal environment [50].

The findings also highlight the significant influence of the auditory environment on overall user perception, aligning with prior studies on soundscapes [14]. Noise from traffic and adjacent university facilities negatively impacted perceptions, while wind conditions in the third space were linked to improved calmness and auditory satisfaction. It is suggested that practitioners design buffers like vegetative screens or architectural barriers to mitigate noise pollution from high-traffic roads and parking areas. As Brown et al. noted, incorporating water features, wind chimes, or other calming auditory elements can enhance positive soundscapes [70]. Placing recreational areas farther from noise-generating facilities can improve auditory perception.

This study underscores the complexity of interactions between thermal, luminous, and auditory factors, with significant implications for the design of multisensory environments. Developing the fuzzy logic model provided an innovative tool for predicting user perceptions based on combined sensory inputs, addressing a critical gap in existing design practices.

A multisensory design is necessary to enhance the user experience in outdoor spaces, particularly within university campuses. This approach should integrate sensory evaluations, utilizing fuzzy logic models to predict and optimize user comfort by considering the complex interplay of thermal, luminous, and auditory stimuli. Furthermore, data-driven design should be employed. This involves collecting on-site data related to thermal comfort, lighting levels, and sound conditions. Complementing this quantitative approach, user engagement through participatory design processes is crucial.

While this research focused on university campuses, the findings have broader implications for the design of urban parks, plazas, and other recreational spaces. The identified principles advocate for incorporating urban greenery, such as trees and green infrastructure, to improve thermal and auditory conditions while simultaneously enhancing the aesthetic quality of the space and contributing to psychological well-being.

This research also stresses the importance of site-specific solutions. Recognizing the unique challenges of different spatial contexts, such as proximity to high-traffic roads, is crucial for developing tailored interventions. Finally, this study emphasizes the need for climate-sensitive design. Applying these findings across diverse climatic regions can enhance global urban design practices by providing strategies to address region-specific environmental challenges.

Limitations of the Study

This study significantly contributes to multisensory assessment in outdoor design, particularly in university recreational spaces. A notable strength lies in its mixed-methods approach, combining quantitative and qualitative methodologies to explore the influence of thermal, luminous, and auditory environments on students' perceptions. By integrating field surveys with on-site environmental measurements, this research provides a comprehensive understanding of how environmental factors shape user experiences.

Moreover, introducing a fuzzy logic model to predict user perceptions marks a novel contribution. While previous studies often focused on individual environmental factors, this research applied fuzzy logic to a multisensory context, setting a new benchmark for predictive modeling in outdoor design. Consistency with earlier studies in terms of the methodology and findings further validates the credibility of the results despite differences in context and analytical approaches.

While this study presents several strengths, it is crucial to acknowledge its limitations to provide a balanced perspective. This research was conducted within the recreational spaces at Bordj Bou Arreridj University. This localized context may limit the generalizability of the findings to other university campuses with differing spatial, cultural, or environmental characteristics. For instance, the design and use of recreational spaces in regions with contrasting climates or cultural practices may yield different results.

This research was conducted over a short period in late April 2023. This limited time frame captured only a snapshot of environmental conditions and user perceptions, potentially excluding seasonal variations that could influence thermal, luminous, and auditory environments. Extended studies across different times of the year could provide a more holistic understanding of these factors. Furthermore, this study focused exclusively on thermal, luminous, and auditory environments, neglecting other sensory factors impacting user experiences. For example, olfactory stimuli, such as vegetation or nearby odor sources, are critical in shaping comfort and mood in outdoor spaces. Previous research emphasizes the importance of smellscapes in environmental perception, suggesting that future studies should incorporate this dimension for a more comprehensive multisensory assessment [71].

The sample consisted of 106 participants distributed across specific areas of the campus. While sufficient for preliminary analysis, a more extensive and diverse sample could enhance the robustness of the findings, capturing a broader range of user perceptions. Although the fuzzy logic model is a novel contribution, it relies on the quality and range of input data. The exclusion of certain environmental variables, such as wind speed or tactile factors, may constrain the model's ability to fully predict user perceptions in complex outdoor settings.

5. Conclusions

This study aimed to provide a detailed understanding of students' multisensory perceptions in a university's recreational area, identifying which environments among the thermal, luminous, and audible environments have an interesting impact. The mixed method covers quantitative and qualitative approaches and was adopted to evaluate the relationship between physical dimensions and students' perceptions across three different spaces. A series of statistical tests were used to process the collected data, and the results reported that the three areas were experienced differently, in addition to the second space, which was considered vulnerable due to its direct exposure to sunlight and noise sources. This study's most influential environment was the audible environment, surpassing the thermal and luminous environments.

On the other hand, to explore the relationships between physical and perceptual dimensions, various types of correlation analyses were conducted, revealing a network of interrelationships between participants' sensations and physical dimensions. The findings highlighted key factors such as wind velocity, the sky view factor, and illuminance in shaping thermal, luminous, and acoustic perceptions. These results underscore the importance of integrating such analyses into outdoor spatial design to optimize user comfort and experience.

This paper also provides a multisensory fuzzy logic model to predict users' responses based on the results obtained from the data analysis. This model was developed using the MATLAB software, and the linguistic classifications of the membership functions were inspired by previous studies that utilized fuzzy logic. Finally, this research's findings provide meaningful information about the multisensory interaction between thermal, luminous, and audible environments and their impact on students' perceptions within the university's recreational area.

Author Contributions: Conceptualization, H.T., D.B. and T.A.K.B.; methodology, D.B., T.A.K.B. and H.T.; software, H.T.; validation, H.T.; formal analysis, H.T., D.B., T.A.K.B., S.K. and M.M.G.; investigation, H.T.; resources, H.T., D.B., T.A.K.B. and S.K.; data curation, H.T. and S.K.; writing—original draft preparation, H.T., D.B., T.A.K.B., S.K. and M.M.G.; writing—review and editing, H.T., D.B., T.A.K.B., S.K. and M.M.G.; visualization, H.T.; supervision, D.B., T.A.K.B. and S.K.; project administration, D.B., T.A.K.B., and S.K.; funding acquisition, H.T., D.B., T.A.K.B., S.K. and M.M.G. All authors have read and agreed to the published version of the manuscript.

Funding: We thank the Vice Presidency for Graduate Studies, Research & Business at Dar Al-Hekma University in Jeddah, Saudi Arabia for funding this research project and for offering their technical support.

Institutional Review Board Statement: Not applicable.

Informed Consent Statement: Not applicable.

Data Availability Statement: The data presented in this study are available on request from the corresponding author.

Conflicts of Interest: The authors declare no conflicts of interest.

References

1. Li, S.; Zhang, Y.; Zhang, Q.; Xue, P.; Wu, H.; Xu, W.; Ye, J.; Chen, L.; He, T.; Zheng, Y. Exploring the Multi-Sensory Coupling Relationship of Open Space on a Winter Campus. *Forests* **2024**, *15*, 876. [CrossRef]
2. Li, Y.; Wu, S.; Qiu, J.; Wei, T. Sustainable University Campuses: Temporal and Spatial Characteristics of Lightscapes in Outdoor Spaces. *Sustainability* **2024**, *16*, 7355. [CrossRef]
3. Ning, S.; Jing, W.; Ge, Z. Sunlight perception and outdoor thermal comfort in college campuses: A new perspective. *Sci. Rep.* **2023**, *13*, 16112. [CrossRef] [PubMed]
4. Ma, X.; Tian, Y.; Du, M.; Hong, B.; Lin, B. How to design comfortable open spaces for the elderly? Implications of their thermal perceptions in an urban park. *Sci. Total Environ.* **2021**, *768*, 144985. [CrossRef]
5. Spence, C. Senses of place: Architectural design for the multisensory mind. *Cogn. Res. Princ. Implic.* **2020**, *5*, 46. [CrossRef] [PubMed]
6. Vasilikou, C.; Nikolopoulou, M. Outdoor thermal comfort for pedestrians in movement: Thermal walks in complex urban morphology. *Int. J. Biometeorol.* **2020**, *64*, 277–291. [CrossRef]

7. Gomaa, M.M.; Othman, E.; Mohamed, A.F.; Ragab, A. Quantifying the Impacts of Courtyard Vegetation on Thermal and Energy Performance of University Buildings in Hot Arid Regions. *Urban Sci.* **2024**, *8*, 136. [CrossRef]
8. Li, G.; Liu, C.; He, Y. The effect of thermal discomfort on human well-being, psychological response and performance. *Sci. Technol. Built Environ.* **2021**, *27*, 960–970. [CrossRef]
9. Santos Nouri, A.; Costa, J.P.; Santamouris, M.; Matzarakis, A. Approaches to outdoor thermal comfort thresholds through public space design: A review. *Atmosphere* **2018**, *9*, 108. [CrossRef]
10. Xu, M.; Hong, B.; Jiang, R.; An, L.; Zhang, T. Outdoor thermal comfort of shaded spaces in an urban park in the cold region of China. *Build. Environ.* **2019**, *155*, 408–420. [CrossRef]
11. Shooshtarian, S.; Rajagopalan, P.; Sagoo, A. A comprehensive review of thermal adaptive strategies in outdoor spaces. *Sustain. Cities Soc.* **2018**, *41*, 647–665. [CrossRef]
12. Erfanian, M.; Mitchell, A.J.; Kang, J.; Aletta, F. The psychophysiological implications of soundscape: A systematic review of empirical literature and a research agenda. *Int. J. Environ. Res. Public Health* **2019**, *16*, 3533. [CrossRef] [PubMed]
13. Fiebig, A.; Jordan, P.; Moshona, C.C. Assessments of acoustic environments by emotions–the application of emotion theory in soundscape. *Front. Psychol.* **2020**, *11*, 573041. [CrossRef]
14. Kang, J.; Zhang, M. Semantic differential analysis of the soundscape in urban open public spaces. *Build. Environ.* **2010**, *45*, 150–157. [CrossRef]
15. Boiné, K.; Demers, C.M.; Potvin, A. Spatio-temporal promenades as representations of urban atmospheres. *Sustain. Cities Soc.* **2018**, *42*, 674–687. [CrossRef]
16. Berkouk, D.; Bouzir, T.A.K.; Boucherit, S.; Khelil, S.; Mahaya, C.; Matallah, M.E.; Mazouz, S. Exploring the Multisensory Interaction between Luminous, Thermal and Auditory Environments through the Spatial Promenade Experience: A Case Study of a University Campus in an Oasis Settlement. *Sustainability* **2022**, *14*, 4013. [CrossRef]
17. Nitidara, N.P.A.; Sarwono, J.; Suprijanto, S.; Soelami, F.N. The multisensory interaction between auditory, visual, and thermal to the overall comfort in public open space: A study in a tropical climate. *Sustain. Cities Soc.* **2022**, *78*, 103622. [CrossRef]
18. Sanchez, G.M.E.; Van Renterghem, T.; Sun, K.; De Coensel, B.; Botteldooren, D. Using Virtual Reality for assessing the role of noise in the audio-visual design of an urban public space. *Landsc. Urban Plan.* **2017**, *167*, 98–107. [CrossRef]
19. Nie, X.; Wang, Y.; Zhang, C.; Zhao, Y.; Kirkwood, N. The varied restorative values of campus landscapes to students' well-being: Evidence from a Chinese University. *BMC Public Health* **2024**, *24*, 487. [CrossRef]
20. Ribeiro, H.; Santana, K.V.d.S.; Oliver, S.L. Natural Environments in University Campuses and Students' Well-Being. *Int. J. Environ. Res. Public Health* **2024**, *21*, 413. [CrossRef] [PubMed]
21. Boucherit, S.; Maffei, L.; Masullo, M.; Berkouk, D.; Bouzir, T.A.K. Assessment of sighted and visually impaired users to the physical and perceptual dimensions of an oasis settlement urban park. *Sustainability* **2023**, *15*, 7014. [CrossRef]
22. Derdous, O.; Tachi, S.E.; Bouguerra, H. Spatial distribution and evaluation of aridity indices in Northern Algeria. *Arid Land Res. Manag.* **2021**, *35*, 1–14. [CrossRef]
23. Scarano, M.; Sobrino, J. On the relationship between the sky view factor and the land surface temperature derived by Landsat-8 images in Bari, Italy. *Int. J. Remote Sens.* **2015**, *36*, 4820–4835. [CrossRef]
24. Palinkas, L.A.; Aarons, G.A.; Horwitz, S.; Chamberlain, P.; Hurlburt, M.; Landsverk, J. Mixed method designs in implementation research. *Adm. Policy Ment. Health Ment. Health Serv. Res.* **2011**, *38*, 44–53. [CrossRef] [PubMed]
25. Sun, S.; Xu, X.; Lao, Z.; Liu, W.; Li, Z.; García, E.H.; He, L.; Zhu, J. Evaluating the impact of urban green space and landscape design parameters on thermal comfort in hot summer by numerical simulation. *Build. Environ.* **2017**, *123*, 277–288. [CrossRef]
26. Matallah, M.E.; Alkama, D.; Ahriz, A.; Attia, S. Assessment of the outdoor thermal comfort in oases settlements. *Atmosphere* **2020**, *11*, 185. [CrossRef]
27. Antoniadis, D.; Katsoulas, N.; Papanastasiou, D.K. Thermal environment of urban schoolyards: Current and future design with respect to children's thermal comfort. *Atmosphere* **2020**, *11*, 1144. [CrossRef]
28. Hu, S.; He, M.; Liu, G.; Lu, M.; Liang, P.; Liu, F. Correlation between the visual evoked potential and subjective perception at different illumination levels based on entropy analysis. *Build. Environ.* **2021**, *194*, 107715. [CrossRef]
29. Masullo, M.; Cioffi, F.; Li, J.; Maffei, L.; Ciampi, G.; Sibilio, S.; Scorpio, M. Urban Park Lighting Quality Perception: An Immersive Virtual Reality Experiment. *Sustainability* **2023**, *15*, 2069. [CrossRef]
30. Bellia, L.; Cesarano, A.; Iuliano, G.F.; Spada, G. Daylight glare: A review of discomfort indexes. *Vis. Qual. Energy Effic. Indoor Light. Today Tomorrow* **2008**, *31*, 1–10.
31. Yacine, S.M.; Noureddine, Z.; Piga, B.E.; Morello, E.; Safa, D. Towards a new model of light quality assessment based on occupant satisfaction and lighting glare indices. *Energy Procedia* **2017**, *122*, 805–810. [CrossRef]
32. Kalawapudi, K.; Singh, T.; Vijay, R.; Goyal, N.; Kumar, R. Effects of COVID-19 pandemic on festival celebrations and noise pollution levels. *Noise Mapp.* **2021**, *8*, 89–93. [CrossRef]

33. Hong, J.Y.; He, J.; Lam, B.; Gupta, R.; Gan, W.-S. Spatial audio for soundscape design: Recording and reproduction. *Appl. Sci.* **2017**, *7*, 627. [CrossRef]
34. Hong, J.Y.; Jeon, J.Y. Comparing associations among sound sources, human behaviors, and soundscapes on central business and commercial streets in Seoul, Korea. *Build. Environ.* **2020**, *186*, 107327. [CrossRef]
35. He, X.; An, L.; Hong, B.; Huang, B.; Cui, X. Cross-cultural differences in thermal comfort in campus open spaces: A longitudinal field survey in China's cold region. *Build. Environ.* **2020**, *172*, 106739. [CrossRef]
36. Li, B.; Liu, J.; Yao, R. Investigation and analysis on classroom thermal environment in winter in Chongqing. *J. Heat. Vent. Air Cond.* **2007**, *37*, 115–117.
37. Fang, Z.; Zheng, Z.; Feng, X.; Shi, D.; Lin, Z.; Gao, Y. Investigation of outdoor thermal comfort prediction models in South China: A case study in Guangzhou. *Build. Environ.* **2021**, *188*, 107424. [CrossRef]
38. Andrade, H.; Alcoforado, M.-J.; Oliveira, S. Perception of temperature and wind by users of public outdoor spaces: Relationships with weather parameters and personal characteristics. *Int. J. Biometeorol.* **2011**, *55*, 665–680. [CrossRef]
39. Castilla, N.; Llinares, C.; Bisegna, F.; Blanca-Giménez, V. Affective evaluation of the luminous environment in university classrooms. *J. Environ. Psychol.* **2018**, *58*, 52–62. [CrossRef]
40. Portnov, B.A.; Saad, R.; Trop, T.; Kliger, D.; Svechkina, A. Linking nighttime outdoor lighting attributes to pedestrians' feeling of safety: An interactive survey approach. *PLoS ONE* **2020**, *15*, e0242172. [CrossRef]
41. Axelsson, Ö.; Nilsson, M.E.; Berglund, B. A principal components model of soundscape perception. *J. Acoust. Soc. Am.* **2010**, *128*, 2836–2846. [CrossRef] [PubMed]
42. Hanoon, S.K.; Abdullah, A.F.; Shafri, H.Z.; Wayayok, A. Comprehensive vulnerability assessment of urban areas using an integration of fuzzy logic functions: Case study of Nasiriyah City in South Iraq. *Earth* **2022**, *3*, 699–732. [CrossRef]
43. Pedrycz, W.; Gomide, F. *Fuzzy Systems Engineering: Toward Human-Centric Computing*; John Wiley & Sons: Hoboken, NJ, USA, 2007.
44. Bressane, A.; Mochizuki, P.S.; Caram, R.M.; Roveda, J.A.F. A system for evaluating the impact of noise pollution on the population's health. *Cad. De Saude Publica* **2016**, *32*, e00021215.
45. Matzarakis, A.; Mayer, H.; Iziomon, M.G. Applications of a universal thermal index: Physiological equivalent temperature. *Int. J. Biometeorol.* **1999**, *43*, 76–84. [CrossRef] [PubMed]
46. Feriadi, H.; Hien, W.N. Modelling thermal comfort for tropics using fuzzy logic. In Proceedings of the Building Simulation, Eindhoven, The Netherlands, 11–14 August 2003; pp. 323–330.
47. Wänström Lindh, U.; Jägerbrand, A.K. Perceived lighting uniformity on pedestrian roads: From an architectural perspective. *Energies* **2021**, *14*, 3647. [CrossRef]
48. Al-Ghonamy, A. Analysis and evaluation of road traffic noise in Al-Dammam: A business city of the eastern province of KSA. *J. Environ. Sci. Technol.* **2010**, *3*, 47–55. [CrossRef]
49. Robinson, D.W. The concept of noise pollution level. *J. Occup. Environ. Med.* **1971**, *13*, 602. [CrossRef]
50. Nikolopoulou, M.; Steemers, K. Thermal comfort and psychological adaptation as a guide for designing urban spaces. *Energy Build.* **2003**, *35*, 95–101. [CrossRef]
51. Spagnolo, J.; De Dear, R. A field study of thermal comfort in outdoor and semi-outdoor environments in subtropical Sydney Australia. *Build. Environ.* **2003**, *38*, 721–738. [CrossRef]
52. Beranek, L.L.; Blackstock, D.T. *Acoustics, 1986 Edition by Leo L. Beranek*; Acoustical Society of America: Melville, NY, USA, 1988.
53. Yilmaz, S.; Külekçi, E.A.; Mutlu, B.E.; Sezen, I. Analysis of winter thermal comfort conditions: Street scenarios using ENVI-met model. *Environ. Sci. Pollut. Res.* **2021**, *28*, 63837–63859. [CrossRef] [PubMed]
54. Nabil, A.; Mardaljevic, J. Useful daylight illuminances: A replacement for daylight factors. *Energy Build.* **2006**, *38*, 905–913. [CrossRef]
55. Michael, A.; Heracleous, C. Assessment of natural lighting performance and visual comfort of educational architecture in Southern Europe: The case of typical educational school premises in Cyprus. *Energy Build.* **2017**, *140*, 443–457. [CrossRef]
56. Reinhart, C.F.; Mardaljevic, J.; Rogers, Z. Dynamic daylight performance metrics for sustainable building design. *Leukos* **2006**, *3*, 7–31. [CrossRef]
57. Liu, Y.; Yu, Z.; Song, Y.; Yu, X.; Zhang, J.; Song, D. Psychological influence of sky view factor and green view index on daytime thermal comfort of pedestrians in Shanghai. *Urban Clim.* **2024**, *56*, 102014. [CrossRef]
58. Fluitt, K.F.; Mermagen, T.J.; Letowski, S.; Letowski, T. *Auditory Perception in an Open Space: Detection and Recognition*; Army Research Laboratory (US): Adelphi, MD, USA, 2015.
59. Daugintis, R. Effect of Wind on Self-Audition of Human Voice. Master's Thesis, Aalto University, Espoo, Finland, 2021.
60. Stokkermans, M.; Vogels, I.; de Kort, Y.; Heynderickx, I. A Comparison of Methodologies to Investigate the Influence of Light on the Atmosphere of a Space. *Leukos* **2018**, *14*, 167–191. [CrossRef]

61. Hegazy, I.R.; Qurnfulah, E.M. Thermal comfort of urban spaces using simulation tools exploring street orientation influence of on the outdoor thermal comfort: A case study of Jeddah, Saudi Arabia. *Int. J. Low-Carbon Technol.* **2020**, *15*, 594–606. [CrossRef]
62. Yuan, J.; Masuko, S.; Shimazaki, Y.; Yamanaka, T.; Kobayashi, T. Evaluation of outdoor thermal comfort under different building external-wall-surface with different reflective directional properties using CFD analysis and model experiment. *Build. Environ.* **2022**, *207*, 108478. [CrossRef]
63. Paiva, K.M.; Cardoso, M.R.A.; Zannin, P.H.T. Exposure to road traffic noise: Annoyance, perception and associated factors among Brazil's adult population. *Sci. Total Environ.* **2019**, *650*, 978–986. [CrossRef]
64. Yang, W.; Moon, H.J. Combined effects of acoustic, thermal, and illumination conditions on the comfort of discrete senses and overall indoor environment. *Build. Environ.* **2019**, *148*, 623–633. [CrossRef]
65. Guedouh, M.S.; Zemmouri, N. Courtyard building's morphology impact on thermal and luminous environments in hot and arid region. *Energy Procedia* **2017**, *119*, 153–162. [CrossRef]
66. Chinazzo, G.; Pastore, L.; Wienold, J.; Andersen, M. A field study investigation on the influence of light level on subjective thermal perception in different seasons. In Proceedings of the tenth Windsor Conference, Windsor Great Park, UK, 12–15 April 2018; pp. 12–15.
67. Yang, W.; Moon, H.J. Cross-modal effects of illuminance and room temperature on indoor environmental perception. *Build. Environ.* **2018**, *146*, 280–288. [CrossRef]
68. Tablada, A.; Carmeliet, J.; Baelmans, M.; Saelens, D. Exterior louvers as a passive cooling strategy in a residential building. In Proceedings of the 26th Conference on Passive and Low Energy Architecture, Quebec City, QC, Canada, 22–24 June 2009.
69. Tablada, A.; De Troyer, F.; Blocken, B.; Carmeliet, J.; Verschure, H. On natural ventilation and thermal comfort in compact urban environments–the Old Havana case. *Build. Environ.* **2009**, *44*, 1943–1958. [CrossRef]
70. Brown, A.L. Advancing the concepts of soundscapes and soundscape planning. In Proceedings of the Conference of the Australian Acoustical Society (Acoustics 2011), Gold Coast, Australia, 2–4 November 2011.
71. Henshaw, V. *Urban Smellscapes: Understanding and Designing City Smell Environments*; Routledge: London, UK, 2013.

atmosphere

MDPI

Article

Humidity-Controlling Ceramic Bricks: Enhancing Evaporative Cooling Efficiency to Mitigate Urban Heat Island Effect

Xueli Jin [1], Junsong Wang [2], Kanghao Tan [2] and Zhenjie Zou [3,*]

1 College of Architectural Engineering, Guangzhou Panyu Polytechnic, Guangzhou 511400, China
2 State Key Laboratory of Subtropical Building and Urban Science, South China University of Technology, Guangzhou 510640, China
3 College of Civil Engineering and Architecture, Guangxi University, 100 University Road, Nanning 530004, China
* Correspondence: zou5di@126.com

Abstract: Passive evaporative cooling technology using the building envelope is a crucial measure to mitigate the urban heat island effect. This study aims to enhance the cooling efficiency of the surface of enclosure structures by utilizing volcanic ash, potassium–sodium stone powder, and silica-based mesoporous oxide (SMO) as primary materials. These components are incorporated into the ceramic brick production process to create innovative humidity-controlling ceramic bricks (HCCTs). This study extensively investigates the impact of SMO and the amount of applied glaze on the physical and mechanical characteristics of these HCCTs. Additionally, it examines the water absorption and evaporative cooling properties of the studied materials under optimal substitution conditions. Numerical calculations are used to determine the heat and moisture transfer properties of HCCTs. The results indicate that incorporating 2% SMO and applying 70 g/m^2 of glaze results in a moisture absorption capacity of 385 g/m^2 and a moisture discharge capacity of 370 g/m^2. These conditions also yield a notable flexural strength of 15.2 MPa. Importantly, the HCCTs exhibit significantly enhanced capillary water absorption and water retention capabilities. Increased water absorption reduces surface temperature by 2–3 °C, maintaining the evaporative cooling effect for 20 to 30 h. It is also found that the temperature of HCCTs decreases linearly with increasing water content and porosity, while the temperature difference gradually decreases with thickness. Water migration in HCCTs with greater thickness is notably influenced by gravity, with water moving from top to bottom. Therefore, it is recommended that brick thickness does not exceed 15 mm.

Keywords: urban heat island; humidity-controlling ceramic bricks; flexural strength; evaporative cooling; water retention capability

1. Introduction

Under the background of China's large-scale urbanization, the rapid expansion of urban infrastructure has led to an excessive amount of land being covered by artificial surfaces under cities, breaking the original balance of heat and moisture exchange between the natural soil interface and the atmosphere and resulting in many urban thermal environmental problems, such as the urban heat island (UHI) effect [1]. The Guangdong–Hong Kong–Macao Greater Bay Area is characterized by a typical humid and hot climate. One notable feature of this climate is the high solar radiation levels, particularly during the scorching summer season, wherein the cumulative solar radiation between June and August peaks at 545 MJ/m^2. This intense solar radiation serves as a primary heat source contributing to overheating in the built environment, making it an ideal candidate for UHI mitigation strategies such as evaporative cooling [2,3]. Furthermore, this region experiences strong monsoons, and the urban areas in the Greater Bay Area are well suited to harness wind energy resources. With over 6000–8000 h of wind speeds exceeding 3 m/s per year, both theoretical and experimental studies indicate that this abundant wind energy resource can enhance the evaporation of surface water. This, in turn, augments the effectiveness of evaporative cooling on building surfaces [4,5].

Notably, the evaporation ratio β (β = actual evaporation E/climatic evaporation EO) in this region is high. The value of β serves as a quantitative measure of the suitability for regional passive evaporative cooling technology. In the Guangdong–Hong Kong–Macao Greater Bay Area, the annual evaporation ratio stands at 0.8, a notable contrast to the lower values (ranging from 0.05 to 0.2) found in Northwest China. Additionally, this area experiences concentrated precipitation during the hot summer season (June, July, and August), further enhancing its potential for passive evaporative cooling. These climatic conditions create a robust foundation for passive evaporative cooling technology. The region can rely on the autonomous behavior of climate elements to ensure a guaranteed rate of over 80% during the cooling of the building environment when it is most needed. This rate can be further enhanced with the strategic addition of artificial water. Leveraging the unique attributes of the local natural environment, the seasonal climate energy advantage for passive evaporative cooling technology in this area is particularly promising [6].

Passive evaporative cooling, which is a cornerstone of the theory of natural harmony, plays a pivotal role in enhancing environmental quality and reducing artificial energy consumption in building design. This technology leverages a range of natural harmonious effects. Roofs and walls, serving as the primary interface between buildings and the natural environment, are key surfaces exposed to sunlight, heat, and precipitation [7]. In regions with hot summers and warm winters, these elements are vital in the design of building insulation. As construction technology advances rapidly, exterior walls are no longer confined to their traditional load-bearing role. There is a growing emphasis on comfort and ecological concerns, leading to heightened demands for additional functionalities in exterior-wall finishing materials. These demands have paved the way for the use of porous materials with water storage capabilities [4]. By applying such materials to the building envelope, evaporative cooling methods like surface water drenching can be employed to achieve effective water storage and cooling for building facades. The heat and moisture transfer attributes of porous materials delay water evaporation on the material surface, thereby enhancing the evaporative cooling effect [8,9].

A historical example of passive evaporative cooling through porous materials is the ancient oyster-shell wall from the Lingnan coastal area. This distinctive design utilizes the water-absorbing properties of oyster shells during rainy days. On sunny days, the stored water evaporates, contributing to natural cooling, resulting in a warm environment during winter and a cool one during summer. Despite its historical efficacy, the traditional oyster-shell wall faces limitations due to the unique characteristics of its materials and appearance, which have hindered its broader adoption in architectural practices [10]. Nevertheless, the concept of passive evaporative cooling has continued to inspire contemporary research and innovations in building design. For instance, Zhang et al. [6] utilized a thermal- and humid-climate wind-tunnel testing platform to investigate the thermodynamic properties of moistened aerated concrete blocks. This research also delved into the climatic suitability of porous bricks with varying thicknesses. Zhang proposed a suitable wall structure for exterior walls based on the passive evaporative cooling theory [11].

Expanding on this work, Pan et al. [12] conducted numerous experiments and established that porous-surface brick wall components with a high moisture content and significant solar radiation absorption coefficients exhibited superior moisture absorption and dissipation capabilities. Consequently, they demonstrated more pronounced passive evaporative cooling effects under identical conditions. Wang et al. [13] furthered this research by experimenting with the drenching and cooling of porous brick walls under natural conditions. Their research revealed that the effectiveness of evaporative cooling in decorative brick walls was closely tied to the water absorption rate, monitored at two-hour intervals. Their results illustrated that higher water absorption rates led to more effective evaporative cooling. Additionally, differences in the evaporative cooling performance due to various drenching methods were identified. These findings provide essential theoretical insights for the practical application of porous tiles in engineering projects. Furthermore, Zhang et al. [11] explored the dynamic surface heat transfer coefficient of porous building materials during the evaporation process using a wind tunnel in a hot and humid climate. Their work included proposing an experimental calculation method for determining the external surface heat transfer coefficient of porous building materials during water storage and evaporation.

Currently, porous tiles are considered to be an eco-friendly material and are crafted from primary resources like clay, fly ash, and silt subjected to high-temperature roasting. However, the effectiveness of sintered porous bricks in managing wetness is hampered by the propensity of the mesopores to become obstructed by the glass phase formed during high-temperature firing. This limitation reduces their humidity-controlling performance. Furthermore, the selection of sintered porous bricks is somewhat restricted when it comes to their color, texture, and overall appearance, making them less appealing to designers. In contrast, ceramic tiles/bricks represent a widely used material for exterior wall decoration, boasting attributes such as moisture resistance, resistance to pollutants, and ease of cleaning. Ceramic tiles, with their extensive variety of colors, textures, and finishes, can cater to diverse architectural styles [14]. However, the evaporative cooling potential of ceramic tiles has often been overlooked. Previous studies have demonstrated that diatom ceramic bricks with a compressive strength of 4.6 MPa, a density of 1.35 g/cm^3, a hygrometric capacity of 32 mg/g, and a discharge capacity of 30 mg/g can be successfully prepared at a calcination temperature of 900 °C [15]. In a novel approach, Vu et al. [16] utilized volcanic ash and waste glass slag subjected to calcination at 800 °C to create humidity-controlling ceramic bricks (HCCTs). These bricks exhibited a moisture absorption capacity of 275 g/m^2 and a moisture discharge capacity of 235 g/m^2. However, their average flexural strength measured at only 5.6 MPa, falling considerably short of the national standards' requirements for the flexural strength of ceramic bricks. Consequently, the development of high-strength, humidity-controlling ceramic tiles with evaporative cooling capabilities holds significant importance.

To address a critical research gap, this study leverages volcanic ash's abundance and cost-effectiveness as a primary raw material. In conjunction, silicon-based mesoporous oxide (SMO) serves as an additive to enhance performance. SMO synthesis is achieved through an acid–base etching process. Subsequently, these materials are seamlessly integrated into the ceramic matrix, resulting in the successful fabrication of HCCTs with remarkable humidification capabilities. This research scrutinizes the impact of SMO and the quantity of glaze-spraying on these HCCTs' inherent properties. The ensuing analysis includes an assessment of the innovative HCCTs' hygrothermal properties. In terms of simulation, based on COMSOL Multiphysics software, the heat and moisture transfer characteristics of the HCCTs are numerically calculated, and the influence of optimizing the parameters of the HCCTs on their heat and moisture performance is further discussed.

2. Materials and Methods

2.1. Raw Materials

In this study, we utilized four key raw materials: silica-based mesoporous oxide (SMO), volcanic ash, potassium–sodium stone powder, and a low-temperature melting block. The chemical composition of each primary material is delineated in Table 1. The SMO was synthesized from volcanic ash through a template-free method, meticulously engineered to enhance the strength and moisture control characteristics of the ceramic tiles. For a detailed description of the production process, please consult our previously published research [17].

2.2. Preparation of HCCTs

In this study, each HCCT consisted of a green body and a glaze layer. The green body was composed of raw materials in the following weight percentages: 80% volcanic ash with a nano-mesoporous structure, 15% borax, and 3.5% silica-based mesoporous oxide.

Fumed silica with a specific surface area of 200 m^2/g was used as an additive. Slurries of volcanic powder with 0–4% fumed silica were prepared with distilled water and 0.5% sodium polyacrylate, resulting in a mixture mass ratio of 1:1. The slurries were planetary-milled for 30 min. The resultant powders with 4.2–15.1% water content were prepared by drying, granulating, and sieving through a 30-mesh sieve. Green bodies (dimensions: 8 × 45 × 95 mm) with different SMO contents were formed by dry-pressing the resultant powders at 9 MPa. The green bodies were dried at 120 °C in a drying oven and fired in an electrically heated muffle furnace (Nabertherm P330) with an air temperature heating rate of 5 °C/min and a holding time of 30 min. The glaze layer included the following weight percentages: 35% SiO_2, 20% Al_2O_3, 18% B_2O_3, 12% Li_2O, 13% Na_2O, 5% K_2O, and 3% $CaCO_3$. Our investigation aimed to explore the influence of varying proportions of SMO and glaze coating quantities on the final product. Subsequently, two different glaze colors were prepared and each one was applied to an HCCT to control the reflectivity of its specific surface: one HCCT specimen was coated in red, named HCCT-R, and another was coated in white, named HCCT-W. A comprehensive depiction of the HCCT preparation process is illustrated in Figure 1.

Table 1. Chemical compositions of the primary materials (wt%).

Materials	SiO_2	Al_2O_3	Fe_2O_3	TiO_2	CaO	MgO	K_2O	Na_2O
Volcanic powder	67.92	12.44	1.12	0.15	3.14	4.12	0.43	1.15
Potassium–sodium powder	72.43	13.38	0.06	0.08	0.43	0.01	8.06	5.29
Low-temperature melt	27.32	1.05	0.13	3.12	3.45	1.12	2.89	11.63
SMO	91.11	1.92	0.08	0.14	0.5	0.01	0.28	0.22

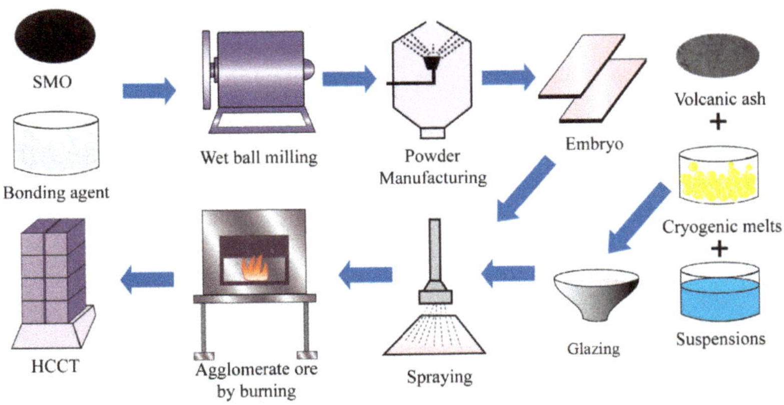

Figure 1. Schematic representation of the preparation process for an innovative HCCT.

In this paper, two types of wet tiles are designed. The reflectivity of their specific surfaces is mainly controlled by the color of the glaze layer on the surface of the tile.

2.3. Experimental Methods

2.3.1. Flexural Strength and Water Absorption Tests

According to ASTM C78/C78M-2015a [18], the determination of flexural strength was conducted on specimens sized 8 × 45 × 95 mm through a three-point bending test employing an INSTRON-5567 apparatus from Instron Engineering Corporation, Norwood, MA, USA. The test configuration featured a lower span of 60 mm, and the test was executed under a crosshead speed of 0.5 mm/min for accurate and controlled measurement. The water absorption of the ceramic samples was tested according to ASTM C1585 [19].

2.3.2. Pore Structure Test

For pore size distribution analysis, the HCCT samples underwent drying in an oven at 200 °C for a duration of 4 h. The nitrogen adsorption and desorption properties of the inorganic silica-based mesoporous material were assessed using a specific surface area and porosity analyzer (TriStar II 3flex, Micromeritics, Norcross, GA, USA) at 77 K, and N_2 adsorption–desorption isotherms were generated for comprehensive analysis.

2.3.3. Microstructural Analysis

Preceding testing, the HCCT samples underwent a series of preparatory steps, including drying, grinding, and sieving through a 200-mesh sieve. Subsequently, a Rigaku SmartLab SE X-ray diffractometer equipped with a Cu Kα radiation source (λ = 1.5418 Å) and a D/te X Ultra detector facilitated the analysis of their crystal structure and phase composition. The scanning parameters were set at a speed of 2°/min and an angle range of 5° to 80°. Further elucidation of the energy spectrum was conducted using X-ray photoelectron spectroscopy (K-Alpha*, Thermo Fisher Scientific, Waltham, MA, USA) with a monochromatized AlKa source (energy of 1486.6 eV). The full-spectrum scan featured a pass energy of 100 eV and a step size of 1 eV, while the narrow-spectrum scan employed a pass energy of 30 eV and a step size of 0.1 eV. Microstructural observations at both the micrometer and nanometer scales were carried out utilizing a field emission scanning electron microscope (TESCAN MIRA LMS, Brno, Czech Republic) and an ultra-high-resolution cold field scanning electron microscope (SU8220, Hitachi, Tokyo, Japan), respectively.

2.3.4. Evaporation Performance Test

In pursuit of a comprehensive analysis of the hygrothermal characteristics inherent to HCCTs, the primary investigation focused on the initial test section within a wind tunnel expressly designed for hot and humid climates. This innovative wind tunnel bifurcates into two distinct sections: the former is dedicated to scrutinizing the thermal and humid milieu of the surrounding materials and the latter is dedicated to analyzing the atmospheric boundary layer, as delineated in Figure 2. The inaugural test section adeptly emulates conditions of heightened thermal and humid extremity, thereby facilitating a nuanced evaluation of building materials, components, and modules across diverse environmental scenarios. In contrast, the latter test section emulates both neutral and non-neutral atmospheric boundary layers, rendering it particularly apt for inquiries into phenomena such as urban heat islands and regional planning. The dimensional specifications of the primary test section are calibrated at 3.0 × 2.5 × 6.0 m and it is equipped with a suite of instruments, including a temperature stratification device, a turbulence generator, a hot-and-cold radiant plate, and a model turntable with a 2.4 m diameter aligned along the airflow direction. This sophisticated platform excels in the meticulous replication of complex meteorological parameter environments. For further elucidation on the intricate design parameters of the wind tunnel, we direct readers to our antecedent work [13].

Meteorological parameters were calibrated to mimic the prevalent weather conditions in Guangzhou, as delineated in Figure 3, in strict adherence to the design standards articulated in JGJ 286-2013 [20]. In order to monitor the evaporation during the evaporation process, a BW32KH-series electronic scale (measuring range: 0–32 kg, accuracy: 0.01 g) was used to continuously and automatically record the quality changes in the samples (Figure 4). The sample to be tested was placed in a square box made of acrylic, and its four sides and bottom were wrapped with polyethylene foam plates to cut off the influence of local heat sources around the evaporation of the specimen, so as to ensure heat transfer from the surface of the specimen to the interior. The sample surface was adjusted to be consistent with the illumination of the light source by adjusting the height supported by the type II stainless-steel bracket under the acrylic box. The electronic balance, intricately linked to a computer via a professional transmission line, systematically recorded alterations in specimen mass on a minute-by-minute basis. Temperature alterations in the specimens were monitored with a high-precision T-type thermocouple line affixed to the specimen's surface, securely bonded with heat-conducting silica gel. The opposing terminus of the thermocouple line was interfaced with a data acquisition instrument. The entire duration of the evaporation experiment spanned 48 h, during which all samples were preconditioned by saturation through immersion in distilled water.

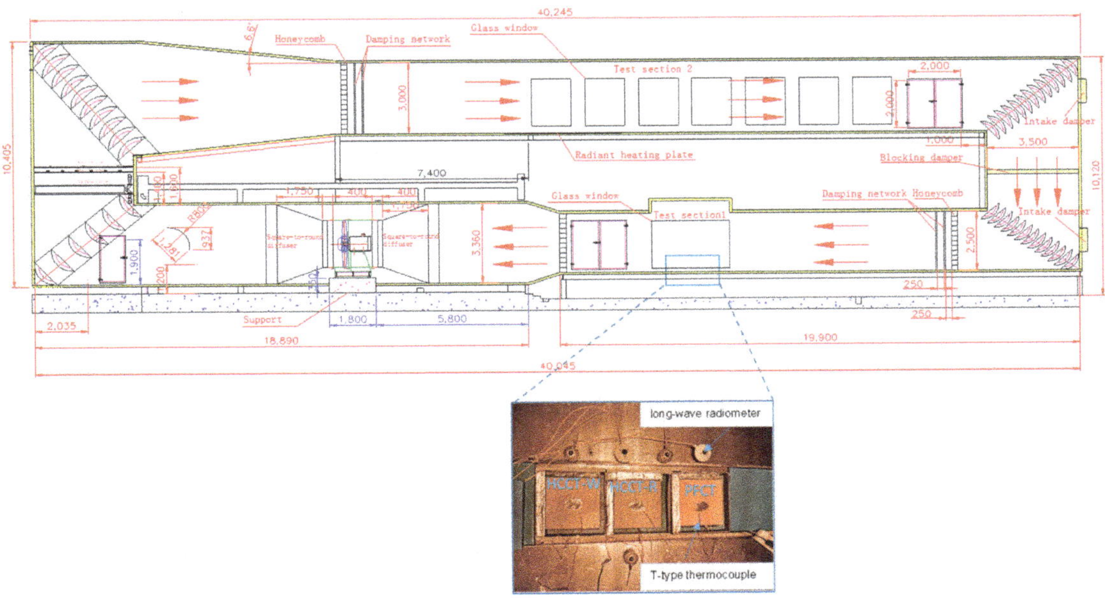

Figure 2. Test specimens were subjected to an evaporation cooling experiment in a wind-tunnel device.

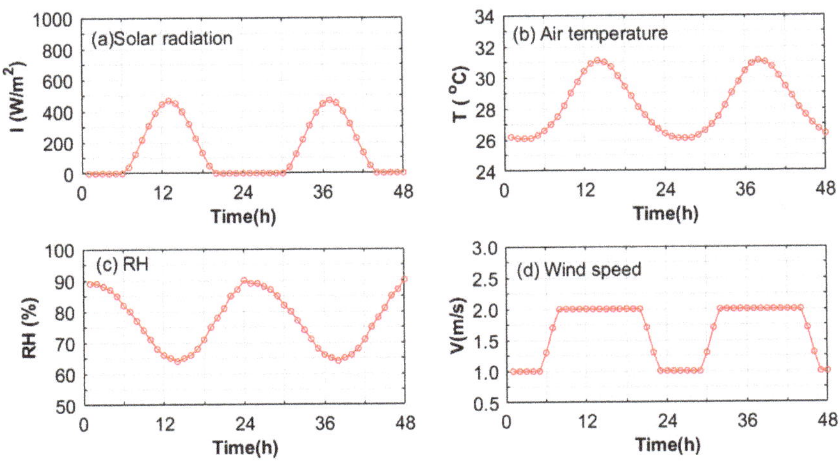

Figure 3. The wind tunnel reproduced typical meteorological daily parameters of Guangzhou.

2.4. Numerical Simulation

When studying the evaporative cooling characteristics of an HCCT, it is only feasible to select one or more measuring points on its surface for measurement. The temperature and humidity distributions along its internal one-dimensional line cannot be measured. Under certain specific meteorological conditions, the heat and moisture transfer process of an HCCT in a complex environment cannot be predicted. In this study, we employed a two-dimensional (2D) numerical model to simulate the influence of humidity and porosity variations on HCCTs. The internal pore structure of the actual materials is highly complex and heterogeneous. Accurately capturing this complexity in a 3D model would require extremely detailed input data and significant computational resources. Given the current scope and objectives of our study, the use of a 2D model provided a feasible and practical approach for investigating the factors influencing the heat and moisture transfer of HCCTs.

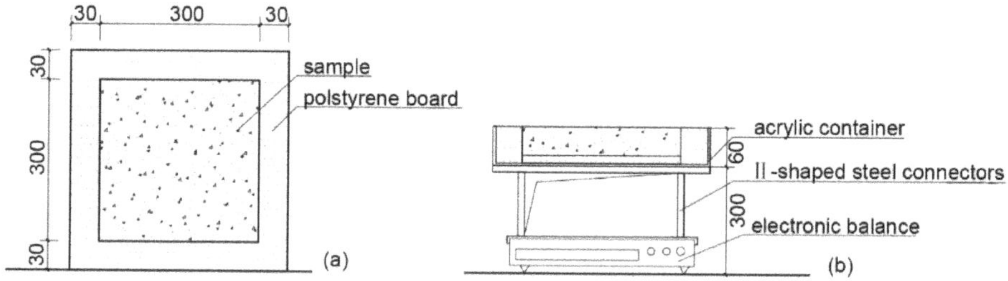

Figure 4. Sample evaporation measurement unit: (**a**) top view; (**b**) section view.

2.4.1. Model Foundation

In this section of our study, COMSOL Multiphysics 6.0 was used to simulate the heat and moisture transfer characteristics of the HCCTs, and then the wind-tunnel test results were used to verify the reliability of the simulation results, and then the structural parameters of the HCCT samples were optimized using the simulation software. A 2D model of the HCCT structure was developed, as shown in Figure 5. The typical parameters of an HCCT is shown in Table 2.

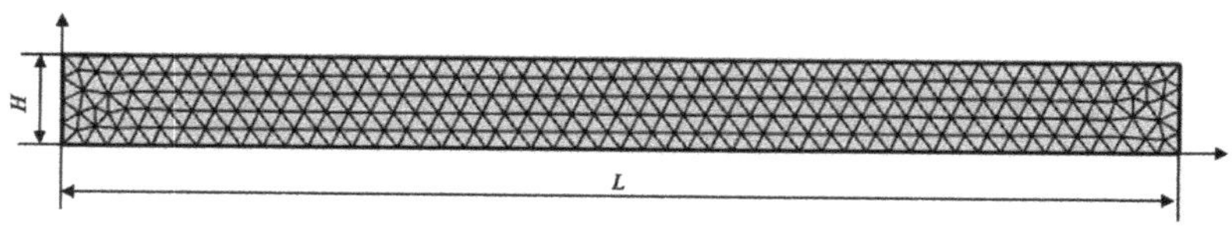

Figure 5. Geometric physical model of an HCCT.

Table 2. Thickness and physical parameters of an HCCT.

Material	Thickness (mm)	Density (kg/m^3)	Heat Capacity (J/(kg·K))	Thermal Conductivity (W(m·°C))	Porosity (%)
HCCT	12	1700	900	1.5	20

2.4.2. Governing Equations

As a typical porous medium, an HCCT's structure is usually composed of three phases: solid, liquid, and gas. In unsaturated porous media, wet fractions not only exist in the porous media's skeleton pores in the form of liquid and vapor, but also consider the movement of non-condensate gasses (such as air) inside the pores. In order to reveal the transmission mechanism of moisture in the HCCT structure, we mainly focused on the water vapor diffusion movement in the process of evaporation and drying. Under the node "Water Transport in Building Materials" in COMSOL Multiphysics, the moisture control equation includes both water transfer and gas diffusion. For the solid skeleton part of an HCCT, the main established moisture control equation is as follows:

$$\omega\frac{\partial\phi}{\partial t} - \nabla\cdot(\omega D_w\nabla\phi + \delta_p\nabla(\phi P_{sat})) = G \quad (1)$$

where ω represents water storage, kg/m^3; D_ω represents water diffusivity, m^2/s; G is the water vapor source, kg/(m^3·s); and ϕ is porosity.

The heat transfer process of an HCCT includes mutual contact between the solid skeleton (particles) and the heat conduction process of fluids in the void space, convective heat transfer of fluids in the void space (forced convection or natural convection), phase-change heat transfer such as evaporation, and radiation heat transfer between solid particles (the skeleton) or gasses. To facilitate accurate modeling, the following assumptions were made:

(i) The expansion and contraction deformation of the HCCT skeleton were not considered, and the surface material was assumed to be homogeneous and isotropic;

(ii) Heat radiation and air convection heat transfer in the void were not considered;

(iii) Viscosity dissipation and pressure variation were not considered;

(iv) The satisfaction of local thermal equilibrium was assumed, that is, $T_s = T_f = T$, where T_s and T_f denote the temperature of the solid phase and that of the fluid phase, respectively. According to the local thermal equilibrium assumption, the thermal governing equation for an HCCT is as follows:

$$(\rho c)_m \frac{\partial T}{\partial t} + (\rho c_p)_f V \cdot \nabla T = \nabla \cdot (\lambda_m \nabla T) + \phi q''_m \tag{2}$$

$$\lambda_m = (1-\phi)\lambda_s + \phi\lambda_f \tag{3}$$

$$q''_m = (1-\phi)q''_s + \phi q''_f \tag{4}$$

$$(\rho c)_m = (1-\phi)(\rho c)_s + \phi(\rho c_p)_f \tag{5}$$

where the subscripts *s* and *f* denote solid phase and fluid phase, respectively, and q''_m, λ_m, and $(\rho c)_m$ denote the apparent heat generation rate, apparent thermal conductivity, and apparent heat capacity of the internal heat source, respectively. c_p is the specific heat of the fluid at constant pressure (J/(kg·K)); c is the specific heat of the solid (J/(kg·K)); λ is the thermal conductivity (W(m·°C)); ρ is the density of the solid or fluid phase (kg/m^3); and q'' is the heat per unit volume generated by the internal heat source (W/m^3).

2.4.3. Boundary Conditions

The perimeter boundary and bottom surface of the HCCT were set to thermal insulation, and the initial temperature was set to 20 °C. The ambient temperature, solar radiation, humidity, and wind speed were consistent with the wind-tunnel meteorological data, and the convective heat flux on the surface of the HCCT was calculated as follows:

$$q_0 = h_c(T_s - T_a) \tag{6}$$

where h_c is the heat transfer coefficient, W/(m^2·K); T_s and T_a are, respectively, the initial specimen temperature and air temperature, K.

The convective water vapor flux of HCCT was calculated as follows:

$$G_0 = \beta(\theta_{ext} p_{sat}(T_{ext}) - \theta p_{sat}(T)) \tag{7}$$

where β is the water transport coefficient, 2×10^{-8} s/m; θ and θ_{ext} are the initial and external relative humidity, respectively.

3. Results and Discussion

3.1. Flexural Strength and Water Absorption

Figure 6a illustrates the impact of the SMO content on the flexural strength of HCCTs. The results reveal a discernible trend wherein the flexural strength experienced an incremental rise with the increasing SMO content until reaching an optimal threshold, beyond which a decline can be observed. Notably, at a 2% SMO content, the flexural strength of the HCCTs exhibits a substantial improvement from 9.5 MPa to 11.3 MPa, constituting a noteworthy 19% augmentation compared to SMO-free HCCTs. This enhancement is ascribed to the robust interfacial bonding between SMO and the ceramic matrix, fostering enhanced cohesion and crack resistance at the optimum SMO content. However, an excessive presence of SMO is observed to have induced heightened porosity, thinning of the pore walls, and an unstable pore structure within the HCCTs, detrimentally impacting their mechanical properties.

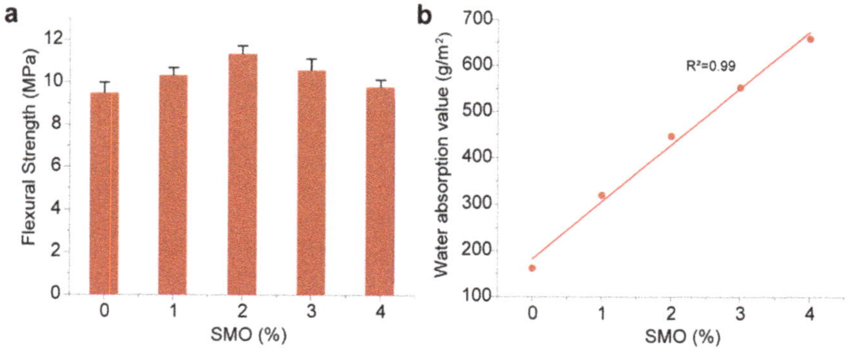

Figure 6. Effect of different SMO contents on the properties of HCCTs: (**a**) flexural strength; (**b**) water absorption.

Figure 6b portrays the water absorption behavior of the HCCTs under high-humidity conditions (humidity: 75–90%). A discernible positive correlation emerges between the SMO content and the water absorption of the HCCTs. This underscores the pivotal role played by SMO, rich in mesopores, in significantly enhancing the moisture-absorbing capabilities of these ceramic tiles. The impact of SMO on water absorption is more pronounced than that on flexural strength primarily due to the extensive mesoporous structure introduced by SMO, which significantly increases the overall porosity and water retention capacity of HCCTs.

To explore the mechanism by which SMO improves HCCT strength, we measured the XRD spectra, presented in Figure 7, of samples prepared without and with the 2% addition of SMO. This figure indicates that the diffraction pattern of the HCCTs prepared with SMO has characteristic diffraction peaks corresponding to hydrated calcium silicate, confirming the formation of calcium silicate hydrate (C-S-H). Further insight is provided by Figure 8, presenting the full-spectrum X-ray photoelectron spectroscopy (XPS) scans of the samples. Seven prominent photoelectron lines of O 1s, Si 2p, Mg 2p, Al 2p, C 1s, Ca 2p, and Na 2p indicate the presence of O, Si, Mg, Al, C, Ca, and Na elements in the samples. Peak-fitting of the Si 2p photoelectron line revealed binding energies of 103.3 eV, 102.6 eV, and 101.7 eV, aligning with the phases of silica, silicate, and tobermorite, respectively, as established by Black et al. [21]. The existence of tobermorite, a calcium silicate hydrate mineral with a Ca/Si ratio of 5:6, further confirms the formation of C-S-H phases in the samples with the addition of 2% SMO.

Figure 9 presents the SEM micrograph results for HCCTs that were fabricated with either 0% or a 2% addition of SMO. It can be clearly seen that HCCTs with the 2% addition of SMO exhibit loose lamellar aggregates compared to those without this addition. These lamellae structures, aligned with the pore size distribution curve in Figure 9a,b, accommodate numerous mesopores, facilitating the condensation and evaporation of water vapor. In the absence of SMO, the pore sizes span from 4.0 to 100.0 nm. Contrastingly, a 2% addition expands this range to 4.0 to 1000.0 nm, with a predominant presence of mesopores in the 5.0–10.0 nm range. This indicates that the introduction of inorganic silica-based mesoporous materials augments the mesopore count and overall tile porosity. Additionally, Figure 9c reveals that the adsorption–desorption isotherms of these tiles exhibit a Type V hysteresis loop, indicative of a combination of Type D and Type E hysteresis loops. This suggests that the primary pore types in these ceramic tiles are slit-like and ink-bottle-shaped. This structural arrangement empowers the ceramic tiles to effectively regulate humidity in response to environmental changes.

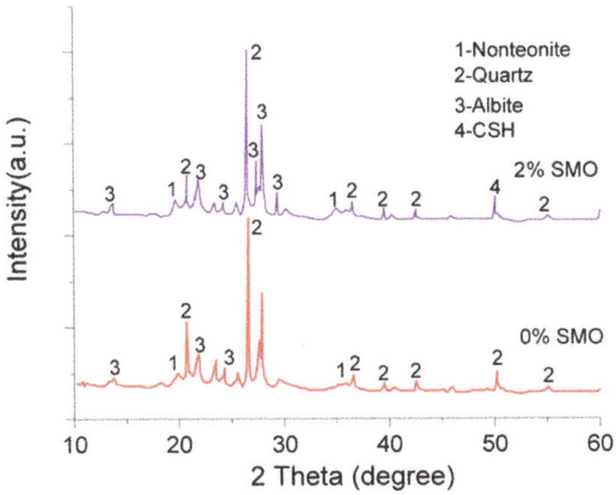

Figure 7. Effect of different SMO contents on the XRD patterns of HCCTs.

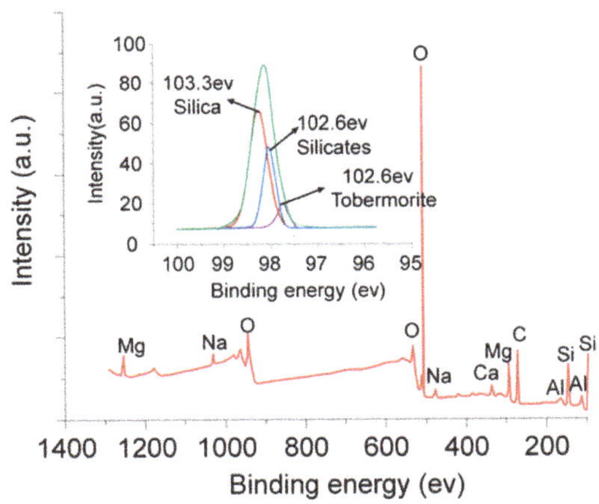

Figure 8. XPS spectra of HCCTs prepared with 2% SMO.

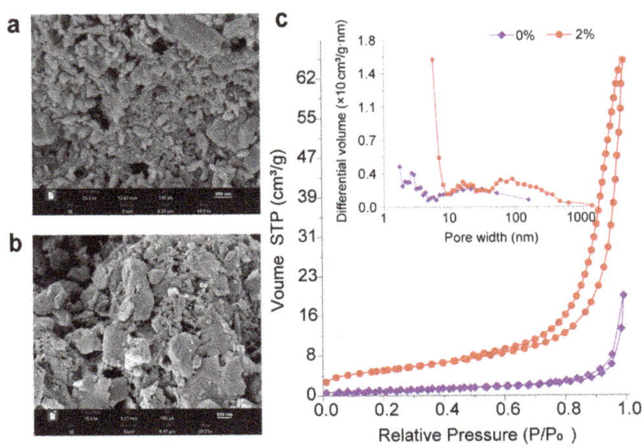

Figure 9. Microstructural analysis of HCCTs: (a) SEM images of HCCTs without SMO addition; (b) SEM images of HCCTs with 2% SMO addition; (c) N_2 adsorption–desorption curves and pore size distribution for HCCTs with 2% SMO and without SMO.

3.2. Impact of Glaze Quantity

Under standard conditions, applying glaze to the surface of an HCCT is crucial. This dual-purpose method enhances the decorative elements of the moistened ceramic brick while simultaneously serving functions such as anti-fouling, concealing the color of the unfired body, and facilitating the adhesion of decorative ceramic ink. To assess the impact of different amounts of sprayed surface glaze on the characteristics of HCCTs, a glaze-spraying machine was employed to apply low-temperature glaze of different qualities to the surface of the dry HCCTs. Following this, the HCCTs underwent calcination at 700 °C for 30 min, resulting in the production of glazed HCCTs. The hygrometric properties of these glazed HCCTs were then subjected to testing. Figure 10 illustrates the hygroscopic capacity of the glazed HCCTs prepared using different quantities of sprayed glaze.

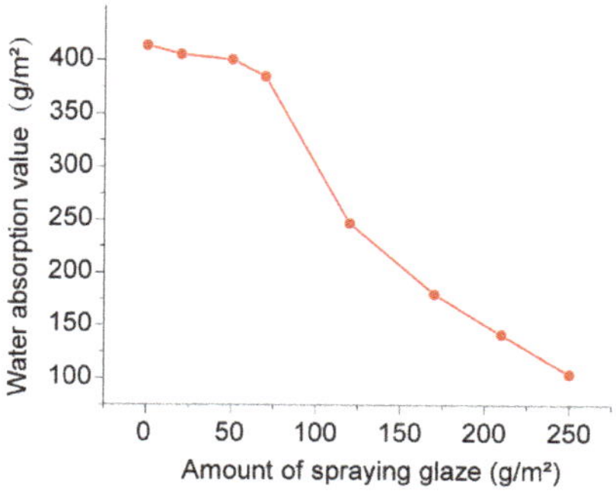

Figure 10. Adsorbed moisture content curve for glazed HCCTs prepared with varied amounts of sprayed glaze.

Observations of this figure reveal that, as the amount of glaze sprayed on the ceramic brick body's surface increases, the water absorption of the glazed HCCTs within the high humidity range (70–95%) gradually diminishes. When the spraying amount is below 70 g/m^2, the reduction in moisture absorption is relatively gradual. However, exceeding the threshold of 70 g/m^2 leads to a rapid decline in moisture absorption. This highlights a direct correlation between the quantity of sprayed glaze and the hygroscopic properties of the ceramic tiles. Excessive glaze on the surface obstructs nano-pores, diminishing the hygroscopic properties of the ceramic tiles through the melting and blocking of the glaze.

3.3. Hygrothermal Properties

The analyses conducted above reveal that the optimal conditions for the fabrication of glazed HCCTs involve a 2% addition of SMO and a glaze-spraying amount of 70 g/m^2. Subsequently, two HCCTs were re-prepared with different glaze colors: one in red, named HCCT-R, and the other in white, named HCCT-W. These innovative HCCTs exhibited an average flexural strength of 15.2 MPa, surpassing the stipulated requirement of the Chinese National Standard [22] (GB/T4100-2015) for ceramic tiles. Furthermore, the water absorption of these tiles was measured at 303 g/m^2, significantly exceeding the standard set by JC/T 2082-2011 [23].

The primary objective of this study was to evaluate the viability of these innovative HCCTs as functional decorative materials for external building walls. A comprehensive assessment of the hygrothermal properties of these tiles was conducted, and a comparative analysis was made with the properties of a porous fired ceramic tile (PFCT) available in the market.

3.3.1. Water Absorption Profile

The capillary water absorption coefficient serves as a crucial parameter for characterizing the water absorption capacity of porous materials. According to ASTM standard

C1585 [19], the capillary water absorption process of porous materials can be divided into two distinct stages. The first stage covers the initial 6 h of water absorption, while the second stage encompasses the subsequent 6 to 192 h. To calculate the capillary water absorption coefficient, the slope of the curve relating the cumulative water absorption per unit area to the square of time is fitted through regression analysis.

Figure 11 illustrates the water absorption profiles of the three porous tile specimens. These profiles demonstrate a common trait, featuring a rapid water absorption rate during the initial stages, followed by a relatively slower rate in the later stages. The outcomes of the single-side immersion test indicate that the two types of humidity-controlling ceramic tiles exhibit similar water absorption process. Table 3 further compares the water absorption coefficient of the three porous tile specimens. The results from the single-side immersion test revealed that the first-stage water absorption coefficients (S_1) of HCCT-W and HCCT-R were 3.69×10^{-2} and 3.65×10^{-2} mm/s$^{0.5}$, respectively. These values were approximately 30% lower than that of the PFCT. This disparity is attributed to the capillary water absorption's connection to the pore-related structural characteristics of porous tile, including factors like porosity, pore size, and pore shape.

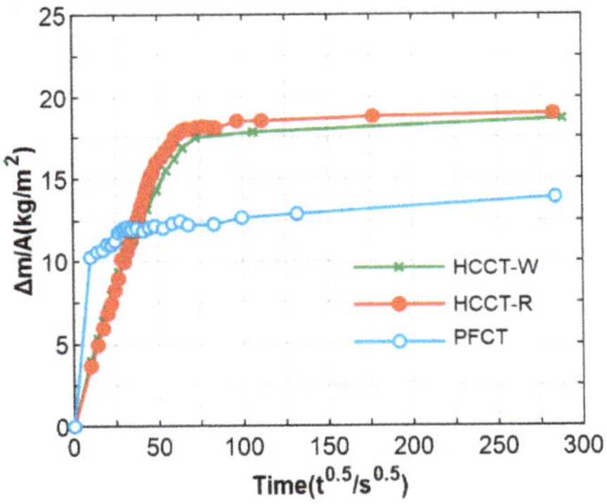

Figure 11. Cumulative water absorption of three different porous tile specimens.

Table 3. Water absorption properties of the three porous tile specimens.

Material	S_1 ($\times 10^{-2}$ mm/s$^{0.5}$)		S_2 ($\times 10^{-2}$ mm/s$^{0.5}$)		W_{vap} (kg/m^3)		Open Porosity (%)	
	Mean	**SD***	**Mean**	**SD***	**Mean**	**SD***	**Mean**	**SD***
PFCT	4.92	0.251	0.04	0.034	130.85	4.205	35.05	0.124
HCCT-W	3.69	0.312	0.07	0.052	173.39	5.031	36.98	0.261
HCCT-R	3.65	0.123	0.07	0.047	177.41	4.936	36.24	0.173

S_1—initial rate of sorptivity; S_2—secondary rate of sorptivity; W_{vap}—total water absorption capacity; SD*—standard deviation.

The results of the vacuum saturation test reveal that the three specimen types display comparable levels of open porosity. Consequently, the distinctions in water absorption are primarily attributed to variations in pore distribution. In contrast to the PFCT, the HCCTs exhibit loosely structured lamellar aggregates, with a notable abundance of mesopores in the range of 10–30 nm, as previously found in the porosity analysis. These mesoporous structures contribute significantly to enhancing the capillary water absorption capacity and volumetric water content of the ceramic bricks. The impact of pore structure on the second-stage capillary water absorption coefficients (S_2) of the three specimen types is relatively marginal, as the water absorption capacity of porous materials tends to become saturated during the later stage. In summary, the final total water absorption capacities (W_{cap}) of the three materials can be ranked as follows: PFCT < HCCT-W < HCCT-R.

3.3.2. Evaporation Profile

The evaporation process of porous materials after their capacity for water absorption reaches saturation can be categorized into two stages. In the first stage, evaporation is primarily driven by the vapor pressure differences in the surrounding environment, and water is transported from the saturated zone to the evaporation surface through capillary action. The second stage of evaporation predominantly occurs within the material medium. As free water on the surface evaporates, the free water within the pores also initiates an evaporation process. Internal water vapor diffuses to the surface through molecular diffusion and subsequently evaporates into the air.

Figure 12 compares the hourly evaporation rates and cumulative evaporation over time for the aforementioned three saturated porous tile samples. It is evident that the hourly evaporation rates for the porous tiles closely follow the fluctuations in solar radiation. The specimens exhibited substantial evaporation during the daytime but minimal evaporation at night, underscoring the predominant influence of solar radiation intensity on water evaporation. At noon on the first day after water spraying, both the HCCT-W and HCCT-R specimens reached similar peak evaporation rates, at approximately 0.63 kg/(m^2·h) and 0.61 kg/(m^2·h), respectively, while the evaporation rate of the PFCT specimen was only 0.42 kg/(m^2·h). However, these rates decreased to nearly zero during the night. This is attributed to the porous ceramics absorbing moisture from the near-surface air at night. On the second day, the evaporation rates of all specimens declined to varying degrees. The HCCT-R specimen peaked at an evaporation rate of about 0.61 kg/(m^2·h), while the HCCT-W specimen was only approximately 0.52 kg/(m^2·h), with the difference between the two gradually widening. This is attributable to the fact that when their initial water retention volumes are equal, the HCCT-R specimen has a lower surface solar radiation reflectance; thus, road surfaces built with this material would be more prone to water evaporation due to its increased capacity to absorb solar radiation. Regarding 48 h cumulative evaporation, the HCCT-R specimen sustained evaporative cooling for a longer duration than the HCCT-W specimen (Figure 12b).

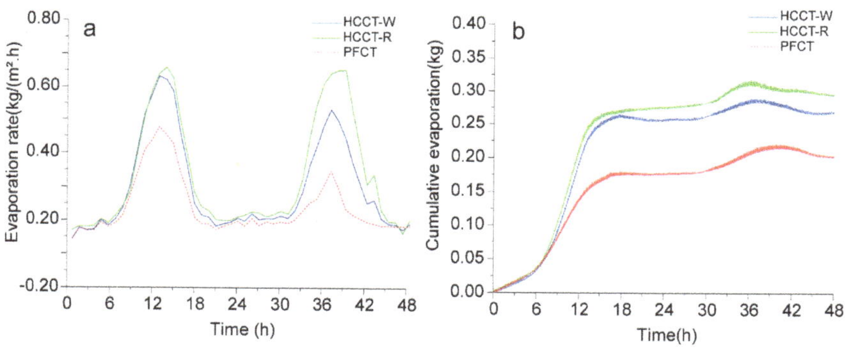

Figure 12. Evaporation profiles of three different porous tile specimens: (**a**) hourly evaporation rate; (**b**) cumulative evaporation.

3.3.3. Temperature Profile

Figure 13 shows that the temperature at various points within the three porous tile specimens exhibits periodic fluctuations due to the influence of outdoor weather conditions. The surface temperature exhibits the following characteristics:

In the dry state (Figure 13a,b), the PFCT specimen had the highest temperature (52.0 °C), followed by the HCCT-R specimen (51.8 °C), and then the HCCT-W specimen (48.9 °C). This variation may be related to surface reflectivity, as surfaces with higher reflectivity tend to absorb less solar radiation. At night, the surface temperatures of the HCCT specimens were slightly lower than that of the PFCT. This is due to the high emissivity of porous ceramic, allowing for the dissipation of surface heat through thermal radiation heat transfer at night, resulting in a decrease in the nighttime surface temperature.

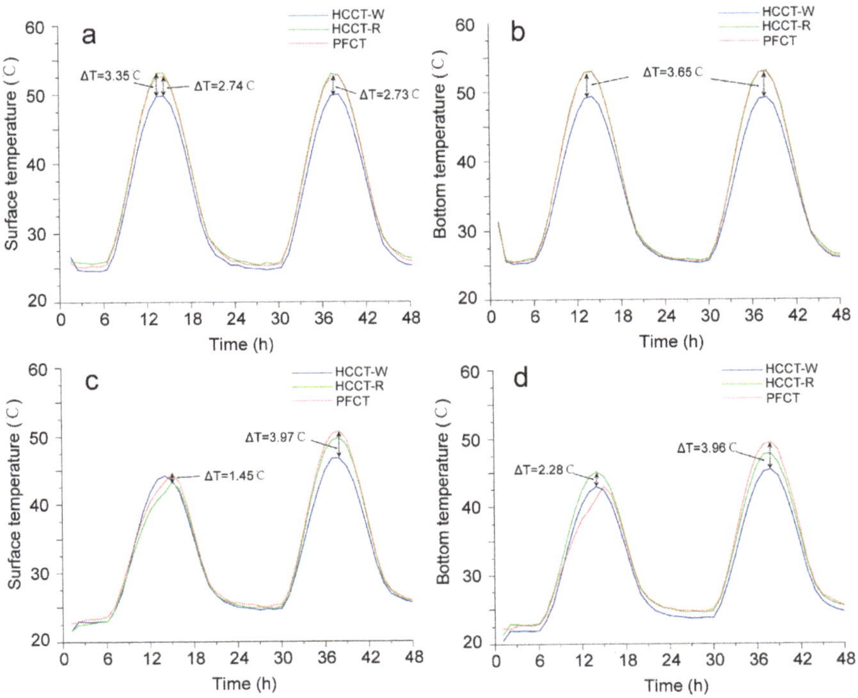

Figure 13. Surface temperature of three different porous tile specimens: (**a**,**b**) under dry conditions; (**c**,**d**) under wet conditions.

In the wet state (Figure 13c,d), the temperatures of the three porous tile specimens decreased by 5–8 °C compared to the dry state, and this cooling effect lasted for 20–30 h. This reduction is the outcome of forced heat exchange due to the latent heat of water evaporation and the temperature difference between the water and the porous tiles. This suggests that water spraying can significantly reduce the surface temperature of ceramic-tile walls during the day, mitigating the overall building's energy consumption. Notably, in the high-speed evaporation stage, water is continuously transported from the saturated layer to the material surface, and the evaporative cooling characteristics are mainly affected by meteorological conditions. Therefore, the temperature curves of the three porous ceramics show almost the same trend. This result is consistent with the findings reported by Wang et al. [13]. They suggested that in the whole evaporation process, about 60–70% of heat is lost in the first stage of evaporation so that the material can maintain a low temperature for a certain amount of time. At this time, the amount of evaporation that occurs is controlled by the heat distribution. With the depletion of water inside the porous material, the heat flow distribution during evaporation in the second stage gradually changes to be controlled by reflectivity. On the second day of our experiment, the maximum temperatures of HCCT-W and HCCT-W rose slower than that of the PFCT under the same heat source, with maximum temperature drops of 3.9 °C and 3.2 °C, respectively. A similar trend could be observed in the minimum temperature data, which are presented in Figure 13d. Hence, the temperature results confirm that our innovative HCCTs can increase the length of the first stage of evaporation, thereby increasing the evaporation rate and prolonging the evaporative cooling effect.

3.4. Numerical Calculation of Heat and Moisture Transfer Process in HCCTs

3.4.1. Validation Model

As shown in Figure 14, the surface and volume moisture content changes of the HCCTs were simulated, wherein the ambient temperature of the HCCTs in the wind tunnel was 33 °C, the ambient wind speed was 1.5 m/s, the ambient relative humidity was 60%, and the net radiant heat flow was 500 W/m^2. It can be seen that the two curves of temperature change have a high degree of fitting: R^2 = 0.98; Std = 1.034 °C. In addition, there is a certain deviation between the simulated calculations of the volumetric moisture content of the HCCTs and the measured values, which may be due to fluctuations in the meteorological parameters in the wind-tunnel environment resulting in the measured values being slightly higher than the simulated values.

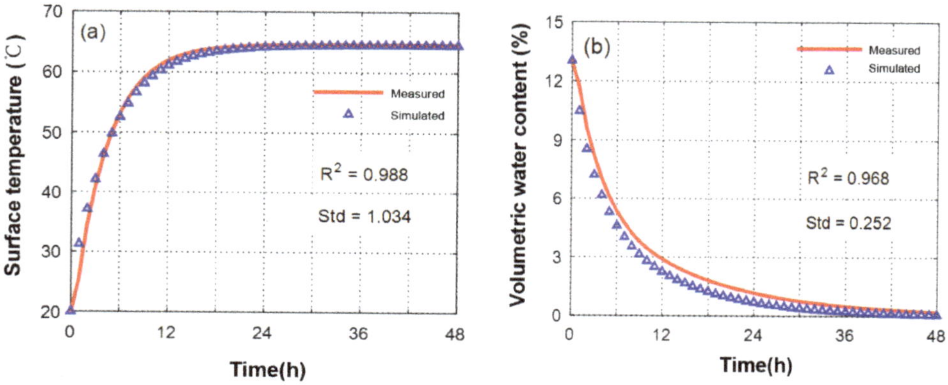

Figure 14. Simulated and measured values in the evaporation process of HCCTs: (**a**) surface temperature; (**b**) volumetric water content.

3.4.2. Factors Influencing Heat and Moisture Transfer of HCCTs

(1) Volumetric water content

Figure 15a illustrates the results pertaining to the influence of the volumetric water content on the internal temperature field of the HCCTs. It can be seen from this figure that the temperature variations reveal a gradual decrease in the temperature gradient in accordance with the thickness of the material. HCCTs with a higher water content maintain a larger temperature gradient, contributing to their enhanced heat transfer capacity. For example, the temperatures of the HCCTs with a 16% volumetric water content at thicknesses of 0 mm, 6 mm, and 12 mm were 39.1 °C, 36.8 °C, and 36.0 °C, respectively. In comparison, at a 28% volumetric water content, the corresponding temperatures were 41.3 °C, 39.1 °C, and 38.2 °C, representing increases of 2.25 °C, 2.19 °C, and 2.15 °C, respectively. Furthermore, in HCCTs with a higher water content, the capillary force of water, unable to overcome gravity, initiates a water redistribution process, leading to increased liquid-phase movement speed and higher convection intensity. Simultaneously, strong phase transitions at the liquid-gas interface enhance heat transfer speed in HCCTs, resulting in a slower temperature rise. Figure 15b further details the temperature distribution among HCCTs of different thicknesses after 3 h of evaporation. The observations presented in this figure reveal a gradual decrease in the temperature gradient as the thickness increases. Notably, HCCTs with a higher water content demonstrate an ability to maintain a larger temperature gradient, contributing to their substantial heat transfer capacity. This phenomenon is a key factor influencing the enhanced heat transfer capacity of HCCTs with a high initial moisture content.

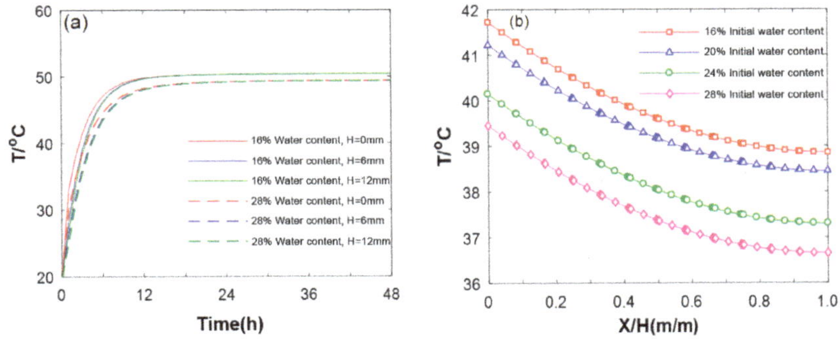

Figure 15. Temperature field of HCCTs: (**a**) temperature variation at different thicknesses; (**b**) temperature distribution across different initial water contents.

(2) Porosity

Figure 16a shows the results pertaining to the influence of different porosity levels on the internal temperature field of HCCTs in the early stage of evaporation (3 h). The temperature at each point within the brick exhibits a linear decrease with increasing porosity. However, a nuanced observation reveals a gradual temperature decrease as the thickness increases, resulting in an increasing gap between the three points. For instance, in permeable bricks with 25% porosity, the maximum temperatures at 0 cm, 3 cm, and 6 cm are 41.1 °C, 39.2 °C, and 38.5 °C, respectively. In comparison, HCCTs with 10% porosity have maximum temperatures of 40.5 °C, 37.9 °C, and 36.7 °C at the corresponding points. The former values are lower by 0.6 °C, 1.3 °C, and 1.8 °C, respectively. This temperature difference arises due to the higher thermal diffusivity of air compared to that of HCCTs. Consequently, a larger air content results in a faster temperature change in the HCCTs. Additionally, the heat capacity of air is smaller than that of permeable bricks, and as their porosity increases, the heat capacity of HCCTs decreases, leading to larger temperature fluctuations. Figure 16b depicts changes in the volumetric water content of the HCCTs alongside their varying porosity. Remarkably, porosity has minimal influence on the changes in bulk moisture content within the brick, exhibiting a consistent trend. The overall moisture change results effectively reflect the pattern of water migration. However, the distribution of volumetric moisture content in the HCCTs along the depth-wise direction demonstrates notable non-monotonic changes; further investigations are required to elucidate the underlying factors causing these changes.

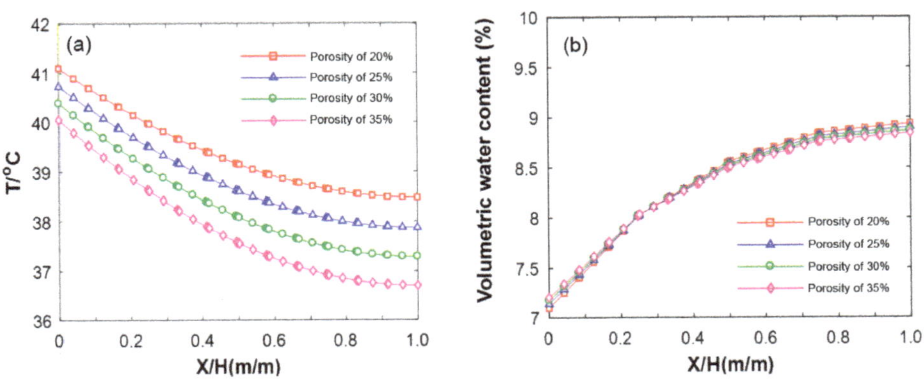

Figure 16. Influence of porosity on temperature–moisture field in HCCTs: (**a**) temperature; (**b**) moisture.

(3) Thickness

Figure 17a presents the results illustrating the impact of different thicknesses on the internal temperature field of HCCTs. Notably, the surface temperature of the brick decreases with an increase in thickness, and this temperature difference intensifies with greater depth. For instance, during the early stages of evaporation (3 h), the surface layer temperatures of bricks with thicknesses of 6 mm, 9 mm, 12 mm, and 15 mm are 41.0 °C, 37.8 °C, 36.4 °C, and 35.8 °C, respectively. In comparison, the corresponding temperatures of the bottom layer are 38.2 °C, 32.2 °C, 28.1 °C, and 25.2 °C, respectively. The former are lower than the latter by 2.7 °C, 5.6 °C, 8.3 °C, and 10.6 °C, respectively. This observation indicates that water migrates from the lower layer to the surface layer, emphasizing the significant role of gravity in influencing water migration. When the brick thickness reaches 15 mm, the moisture in the HCCT can no longer overcome the internal resistance to reach the surface for evaporation. This implies that further increasing the thickness of the brick contributes negligibly to the cooling effect. A similar phenomenon is also evident in its impact on the volumetric water content of the HCCTs (Figure 17b). Considering both thermal and moisture-related performance, the HCCT thickness correlates with the evaporative cooling effect of the road surface. Consequently, it is recommended that the thickness of each brick does not exceed 15 mm.

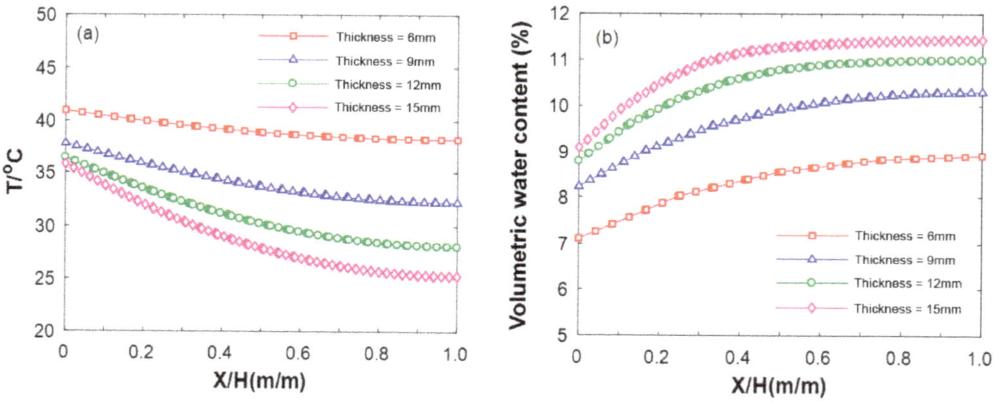

Figure 17. Influence of thickness on temperature–moisture field in HCCTs: (**a**) temperature; (**b**) moisture.

4. Conclusions

This study investigates the preparation and performance of HCCTs using volcanic ash with a nanoporous structure and potassium–sodium powder with a high alkali content as the primary raw materials, as well as SMO as an additive. This research explores the impact of SMO addition and the amount of glaze coating on the humidity-controlling capabilities of HCCTs, resulting in the following key findings:

(1) The inclusion of SMO substantially improves the flexural strength and water absorption of HCCTs. Specifically, with a 2% SMO content, our HCCTs achieved a flexural strength of 11.6 MPa and a water absorption rate of 480 g/m^2. Microstructural analysis revealed the formation of CSH and an increase in mesopores within the HCCTs, contributing significantly to the enhancement of both their mechanical properties and their humidity-controlling properties.

(2) With a 2% SMO content and a glaze coating of 70 g/m^2, HCCTs sintered at 700 °C exhibit an average flexural strength of 15.2 MPa and a water absorption rate of 385 g/m^2 under high-humidity conditions. These results align with the standards outlined in GB/T 4100-2015 [23].

(3) Our innovative HCCTs have a rich porous-structure network that can improve their water storage capacity, and their cumulative water absorption capacity is up to 40% higher than that of a PFCT. Under dry conditions, the surface temperature of the porous tiles during the day is controlled by their reflectance; in contrast, humidified tiles have better heat dissipation at night. In the wet condition, the temperature of HCCT-W and HCCT-R pavement is 5–8 °C lower than that of the PFCT material, and HCCT-W pavement has a longer cooling time.

(4) Under the action of temperature gradient, the temperature of a high-moisture-content HCCT is less than that of a low-moisture-content one. The temperature of an HCCT decreases linearly with the increase in its porosity, but this temperature difference decreases gradually with increases in the HCCT's thickness. Water migration of HCCTs with higher thicknesses is obviously affected by gravity, and water migrates from top to bottom under the action of gravity. It is recommended that the thickness of an HCCT should not be more than 15 mm.

Future work should focus on the long-term performance and durability of HCCTs under various environmental conditions. Optimizing the SMO content and exploring additional additives can enhance both mechanical and humidity-controlling properties. Developing advanced 3D numerical models will provide deeper insights into HCCTs' heat and moisture transfer mechanisms. A comprehensive environmental impact assessment, including life-cycle analyses, is crucial for evaluating the sustainability-related benefits of HCCTs. Scaling-up HCCT production and applying it in real-world scenarios will be vital for encouraging their broader adoption. Exploring different glaze formulations to enhance reflectivity, thermal properties, and aesthetic appeal can increase HCCTs' versatility. Despite promising results, this study could be improved through efforts to more directly correlate HCCT performance with urban heat island mitigation, quantifying potential temperature reductions and their impacts on urban microclimates. Addressing these areas will further improve the effectiveness of HCCTs in mitigating urban heat islands.

Author Contributions: Investigation, K.T.; Writing—original draft, X.J.; Writing—review & editing, Z.Z.; Supervision, J.W. All authors have read and agreed to the published version of the manuscript.

Funding: The author would like to thank the National Natural Science Foundation of China (No. 51878287), Guangdong Provincial Department of Education Innovation Team Project (No. 2023KCXTD071), Guangzhou Yangcheng Scholar Project (No. 202235323) and the China Postdoctoral Science Foundation (No. 2024M750943).

Data Availability Statement: The original contributions presented in the study are included in the article, further inquiries can be directed to the corresponding author.

Conflicts of Interest: The authors declare no conflict of interest.

References

1. Maka, A.O.; Alabid, J.M. Solar energy technology and its roles in sustainable development. *Clean Energy* **2022**, *6*, 476–483. [CrossRef]
2. Zhang, L.; Zhang, R.; Zhang, Y.; Hong, T.; Meng, Q.; Feng, Y. The impact of evaporation from porous tile on roof thermal performance: A case study of Guangzhou's climatic conditions. *Energy Build.* **2017**, *136*, 161–172. [CrossRef]
3. Wang, J.; Meng, Q.; Yang, C.; Ren, P.; Santamouris, M. Spray optimization to enhance the cooling performance of transparent roofs in hot-humid areas. *Energy Build.* **2023**, *286*, 112929. [CrossRef]
4. Zhang, Y.; Zhang, L.; Pan, Z.; Meng, Q.; Feng, Y.; Chen, Y. Hydrological properties and solar evaporative cooling performance of porous clay tiles. *Constr. Build. Mater.* **2017**, *151*, 9–17. [CrossRef]
5. Li, L.; Xu, F.; Tang, H.; Wang, J.; Meng, Q. Experimental and theoretical study on cooling performance of membrane roof with circulated water film. *Constr. Build. Mater.* **2023**, *403*, 133115. [CrossRef]
6. Zhang, L.; Zhang, R.; Hong, T.; Zhang, Y.; Meng, Q. Impact of post-rainfall evaporation from porous roof tiles on building cooling load in subtropical China. *Appl. Therm. Eng.* **2018**, *142*, 391–400. [CrossRef]
7. Mao, H.; Meng, Q.; Liu, Z.; Wang, S.; Wang, J. A comprehensive evaluation of water spray system on transparent roof towards optimal operation schedule. *Build. Environ.* **2023**, *242*, 110590. [CrossRef]
8. He, J.; Hoyano, A. Experimental study of practical applications of a passive evaporative cooling wall with high water soaking-up ability. *Build. Environ.* **2011**, *46*, 98–108. [CrossRef]

9. He, J.; Hoyano, A. Experimental study of cooling effects of a passive evaporative cooling wall constructed of porous ceramics with high water soaking-up ability. *Build. Environ.* **2010**, *45*, 461–472. [CrossRef]
10. Hall, M.R. Assessing the environmental performance of stabilised rammed earth walls using a climatic simulation chamber. *Build. Environ.* **2007**, *42*, 139–145. [CrossRef]
11. Zhang, Y.; Zhang, L.; Meng, Q.; Feng, Y.; Chen, Y. A dynamic experimental study on the evaporative cooling performance of porous building materials. *Heat Mass Transf.* **2017**, *53*, 2651–2662. [CrossRef]
12. Pan, Z.; Meng, Q.; Li, Q.; Xie, J.; Liu, J. Evaporative cooling of porous tiles with seawater in a tropical climate with salty humid air. *Constr. Build. Mater.* **2019**, *204*, 727–739. [CrossRef]
13. Wang, J.; Santamouris, M.; Meng, Q.; He, B.-J.; Zhang, L.; Zhang, Y. Predicting the solar evaporative cooling performance of pervious materials based on hygrothermal properties. *Sol. Energy* **2019**, *191*, 311–322. [CrossRef]
14. Tian, W.; Shui, A.; Ke, S.; Huang, L.; Xi, X.; He, C.; Chen, W.; Du, B. Low-temperature preparation of humidity self-regulating porous ceramics with high strength performance. *Mater. Lett.* **2019**, *243*, 128–131. [CrossRef]
15. Dal, S.; Sutcu, M.; Gok, M.S.; Gencel, O. Characteristics of lightweight diatomite-based insulating firebricks. *J. Korean Ceram. Soc.* **2020**, *57*, 184–191. [CrossRef]
16. Vu, D.-H.; Wang, K.-S.; Bac, B.H. Humidity control porous ceramics prepared from waste and porous materials. *Mater. Lett.* **2011**, *65*, 940–943. [CrossRef]
17. Tian, W.; Du, B.; Shui, A.; Ke, S.; Shan, Q.; He, C.; Ma, J. Template-free preparation of humidity self-regulating silica-based mesoporous oxide from volcanic ash. *J. Sol-Gel Sci. Technol.* **2020**, *94*, 416–424. [CrossRef]
18. *ASTM C78/C78M-2015a*; Standard Test Method for Flexural Strength of Concrete (Using Simple Beam with Third-Point Loading). ASTM: West Conshohocken, PA, USA, 2015.
19. *ASTM C1585e13*; Standard Test Method for Measurement of Rate of Absorption of Water by Hydraulic-Cement Concretes. ASTM: West Conshohocken, PA, USA, 2013.
20. *JGJ 286-2013*; Design Standard for Thermal Environment of Urban Residential Areas. Ministry of Housing and Urban-Rural Development: Beijing, China, 2013.
21. Black, L.; Stumm, A.; Garbev, K.; Stemmermann, P.; Hallam, K.R.; Allen, G.C. X-ray photoelectron spectroscopy of aluminium-substituted tobermorite. *Cem. Concr. Res.* **2005**, *35*, 51–55. [CrossRef]
22. *GB/T 4100-2015;* Ceramic Tiles. China National Standardization Administration: Beijing, China, 2015.
23. *JC/T 2082-2011;* Humidification Function Interior Building Decoration Materials. Ministry of Science and Technology of the People's Republic of China: Beijing, China, 2011.

atmosphere

MDPI

Systematic Review

Climate Change and Occupational Risks in Outdoor Workers: A Systematic Review of the Health Effects of Extreme Temperatures

Maria Francesca Rossi *, Raimondo Leone and Umberto Moscato

Department of Health Surveillance and Bioethics, Università Cattolica del Sacro Cuore, 00168 Rome, Italy
* Correspondence: mariafrancesca.rossi@unicatt.it

Abstract: Climate change is one of the most important current threats to global health. Outdoor workers are among the most vulnerable people to its effects. The aim of this systematic review is to assess the occupational risks related to climate change, investigating health outcomes in outdoor workers and estimating its impact in the occupational context. The review was performed following PRISMA guidelines, screening three databases (PubMed, Web of Science, and Scopus). Studies written in English or Italian languages, performed on outdoor workers, assessing occupational risks linked to climate change, and reporting on health outcomes were included. A quality assessment was performed using the Newcastle–Ottawa Scale. Thirteen studies were included in the review, performed mostly on construction (seven studies, 53.8%) and agricultural (five studies, 38.5%) workers. Twelve of the included studies (92.3%) reported on occupational risks related to heat stress, one on the effects of cold weather. Four studies (30.8%) reported a high prevalence of heat-related symptoms, ranging from 64.0% to 90.3% of workers. This systematic review highlights heat-related stress in outdoor workers as an important occupational risk, but it also underlines an important gap in scientific knowledge regarding other occupational risks relating to climate change.

Keywords: climate change; global health; heat stress; occupational health; outdoor workers

1. Introduction

Climate change is one of the most significant current threats to global health due to its impact on all aspects of human life, including the social, economic, psychological, and health-related dimensions [1]. Furthermore, climate change may have an impact on health, work and employment, housing, safety, and the ability to grow food [2]. According to World Health Organization (WHO) estimates, climate change could cause approximately 250,000 additional deaths per year by 2030–2050 due to malnutrition, malaria, diarrhoea, and heat stress alone; the direct health-related damage costs are estimated to range between two and four billion dollars per year by 2030 [3].

The economic impact of climate change has been reported in the scientific literature [4,5]. In particular, it has been highlighted that higher temperatures substantially reduce economic growth (a temperature increase of 1 °C leads to a reduction of 1% in economic growth in the long term) [6], with an impact on growth rates as well, and an alarming impact on agricultural output, industrial output, and political stability [5]. Moreover, it must be considered that the economic impact of climate change has been reported to be worse for poorer countries, further emphasizing the social inequities underlining this growing issue [4,5,7].

From an occupational health perspective, outdoor workers are among the most vulnerable populations to the effects of climate change [8,9]. Considering the profound changes that the occupational world has undergone in the COVID-19 and post-COVID-19 era [10–13], identifying and assessing emerging occupational risks is of paramount importance to implement prevention strategies and ensure the health and safety of workers [14]. In this context, preparedness strategies based on a One Health approach are essential to identify emerging environmental changes and counteract these challenges through a preventive approach [15,16].

Climate change's effects on outdoor temperatures have been thoroughly documented [17]. It has been showcased that individuals working under heat stress, especially if performing physically strenuous tasks, are more likely to experience heat-related strain and illnesses, which can lead to heat stroke or even death, even in young and healthy workers [18]. Furthermore, a 2018 meta-analysis by Flouris et al., which included studies performed on a global scale, reported that individuals working a single shift under heat stress conditions were four times more likely to experience occupational heat strain [19].

However, heat risk is not the only occupational hazard affected by climate change. Climate change has an effect on the ozone layer, along with other contributing factors; a recent report from the European Commission has reported that during heatwaves the ozone levels are consistently above European Union limits (registered levels > 120 $\mu g/m^3$ compared to the cut-off of 100 $\mu g/m^3$) [20]. Moreover, the thinning of the ozone layer contributes to the increase in intensity of ultraviolet radiation (UV), with an increase in type A and B ultraviolet radiation of up to 35% during heatwaves [21]. Outdoor workers are, therefore, more exposed to ultraviolet radiation, leading to an increased risk of adverse eye effects and skin cancer [22]. A clear estimation of the occupational risk present in outdoor work is very difficult to generate, especially considering that skin cancer caused by occupational exposure seems to be underreported [23]; however, current scientific evidence indicates an increased risk of skin cancer and other dermatological issues for outdoor workers [24–26].

Furthermore, climate change is contributing to the increase in air pollution levels; according to the World Meteorological Organization, air quality and pollution worsen during heatwaves, with registered ozone levels above the cut-off (100 $\mu g/m^3$) and an increase in particulate matter ≤ 2.5 μm ($PM_{2.5}$) [27]. This could exacerbate the health issues that have been associated with air pollution in outdoor workers [28], and as their occupational exposure levels increase these problems could become more severe, with higher prevalence of occupational allergies and respiratory diseases [29–31].

In the context of global health and climate change, the aim of this systematic review is to assess the occupational hazards related to climate change, investigating health outcomes in outdoor workers in order to give an estimation of the impact of climate change in the occupational context.

2. Materials and Methods

The review was performed following the Preferred Reporting Items for Systematic Reviews and Meta-Analyses (PRISMA) Statement [32], screening three databases (PubMed, ISI Web of Knowledge, and Scopus). The P(E)CO model [33] was used to formulate the query: the *Population* identified was outdoor workers, the *Exposure* was climate change, and the *Outcome* was any health outcome correlated with occupational risks identified in the study. Due to the aim of the systematic review, no *Control* was chosen for the query.

Therefore, to make the research as comprehensive as possible, the query was structured to include synonyms and Medical Subject Headings (MeSH) terms for the following keywords: "outdoor workers", "climate change", and "occupational risk".

The records found through this search were screened for duplicates through the online site Rayyan [34]. The screening by titles and abstracts was performed using triple-blind methodology through the same website in order to reduce selection bias [34]. The records unanimously included or excluded were not discussed further, while the articles for which the website showed conflict were openly discussed between the three researchers, and a unanimous decision about inclusion or exclusion was eventually reached. A full-text screening was then performed for all included records.

The study was carried out in accordance with the Declaration of Helsinki. The study protocol was not submitted to an Institutional Review Board, as this is a systematic review and does not involve human subjects.

2.1. Inclusion and Exclusion Criteria

Studies published up until the initial search was performed were included in the systematic review if they were performed on outdoor workers, assessed occupational risks related to climate change, and reported on health outcomes associated with these risks. Only manuscripts in English or Italian languages were included (languages spoken by the researchers), while manuscripts written in other languages were excluded from the review. Studies that did not assess climate change-related occupational risks or did not consider health outcomes were also excluded from the review. Study designs that did not report on primary data (editorials, literature reviews, commentaries, etc.) were excluded.

2.2. Data Extraction

Data was extracted from the included studies and reported in a Microsoft Excel sheet. The following data was gathered from each article: the first author, publication year, country, study timeframe, study design, work tasks of the workers included, sample size, gender ratio, mean age of the workers, type of occupational risk assessed in the study, health outcomes considered, and main results.

A quality assessment was performed on all included studies using the Newcastle–Ottawa Scale [35]. This quality assessment scale is divided into three sub-scales, evaluating the selection, comparability, and outcome criteria used in the study. The selection sub-scale comprises four questions assessing the representativeness of the sample (maximum 1 point), sample size (maximum 1 point), non-respondent reports (maximum 1 point), and methods used to assess the exposure or risk factor (maximum 2 points) for a maximum of 5 points. The comparability sub-scale comprises one question with a maximum of 2 points, assessing whether confounding factors were considered appropriately in the study. Finally, the outcome sub-scale comprises two items concerning outcome assessment (maximum 1 or 2 points) and the statistical tests used (maximum 1 point) for a maximum of 3 points. Therefore, each study can score from 0 to 10 points on this scale [35].

3. Results

The systematic search resulted in 1306 relevant articles from the three databases (PubMed, Web of Science, and Scopus). After removing duplicates, the initial search resulted in 965 eligible articles. Three researchers screened the articles by title and abstract using a blinded methodology, including 113 full-text records. The final full-text screening resulted in the 13 included studies [36–48] (Figure 1).

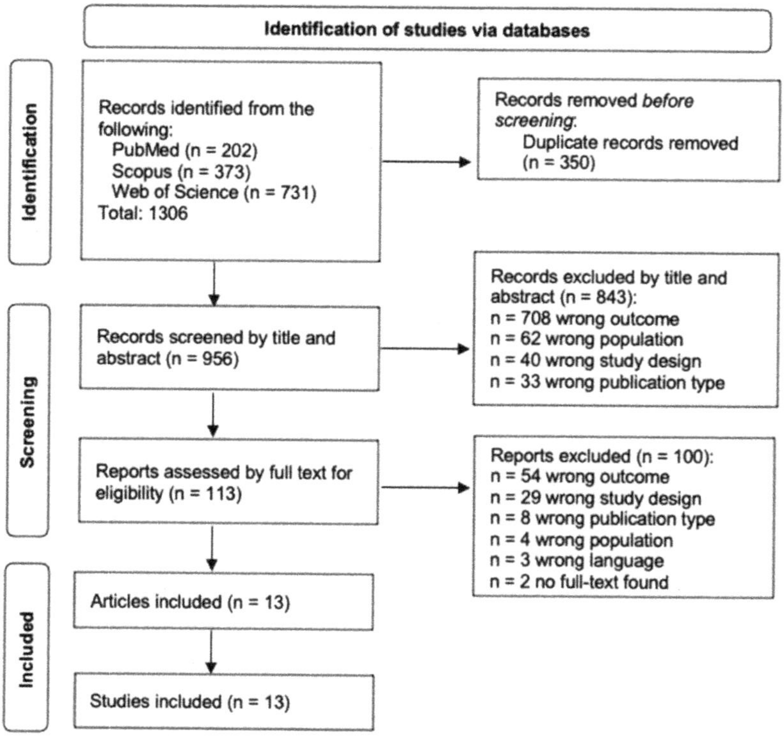

Figure 1. A PRISMA flowchart of the inclusion and exclusion processes.

The included studies were performed in various countries and included various types of workers, mainly construction workers (seven studies, 53.8%) [36–39,42,44,46] and agricultural workers (five studies, 38.5%) [39,41,42,46,47]. Concerning study design, nine studies were cross-sectional studies (69.2%) [36,37,39,40,42,44–47], one was a case–control [48], one was a case series [43], one was a cohort study [38], and one was a trial [41] (each representing 7.7%). The largest study included over a thousand workers [39], while most studies included between a hundred and a thousand workers (nine studies, 69.2%) [36–38,40,41,45–48]; two studies (15.4%) [43,44] included fewer than one hundred workers [43,44], and one study did not report the number of participants [42].

Ten studies (76.9%) had a majority of male workers [36–39,41,43,45–48], while two (15.4%) had a majority of female workers [40,44], and one study did not report the gender ratio [42]. Concerning age, the studies included all working ages, meaning a differentiation based on age range or mean age was not possible.

Finally, all included studies except for one (12 studies, 92.3%) reported on the occupational risks related to heat stress [37–45,47,48]. One study investigated the effects of cold weather caused by climate change on the occupational hazards of outdoor workers [36].

The main results for each study included in this systematic review are reported in Table 1. A quality assessment using the Newcastle–Ottawa Scale was performed on all included studies. All studies showcased at least fair quality (six to seven points total) (Table 2).

Table 1. The data extracted from the included studies, ordered by year of publication and first author's name (NR = not reported; SD = standard deviation; OR = odds ratio; CBT = core body temperature).

Citation	Country	Study Design	Study Timeframe	Work Tasks	Sample Size	Gender Ratio (*n*,%)	Age (Mean ± SD or Range)	Occupational Risks	Outcomes Considered	Results	*p* Value
Karthick (2022) [36]	USA	Cross-sectional	NR	Construction	100	M = 87 (87.0%)	Range from 21 to ≥60	Cold weather	Respiratory issues	OR 1.141	0.041
									Freezing of exposed parts	OR 1.217	0.032
Han (2021) [37]	China	Cross-sectional	July–September 2020	Construction	318	M = 239 (75.2%)	Range from ≤24 to ≥55	Heat stress	Heat-related illness	37 (11.6%)	NR
									Heat-related injury	35 (11.0%)	NR
Phanprasit (2021) [38]	Thailand	Cohort	March–June 2016	Construction	90	M = 70 (77.8%)	Range 20–60	Heat stress	Heat-related symptoms	20 (22.0%)	0.028
				Foundry	78	M = 75 (96.2%)	Range 20–46		Heat-related symptoms	30 (39.0%)	0.028
Venugopal (2021) [39]	India	Cross-sectional	NR	Agriculture, construction, brick making, salt pans	1053	M = 532 (50.5%)	NR	Heat stress	Self-reported heat-related illness	951 (90.3%)	<0.001
									Rise in CBT > 1 °C	189 (17.9%)	<0.001
									Sweat rate > 1/h	282 (26.8%)	<0.001
									Urine specific gravity	300 (28.5%)	0.020
									Reduced kidney function	268 (48.2%)	0.007
Mansor (2019) [40]	Malaysia	Cross-sectional	March–April 2016	Cleaning, gardening	320	M = 148 (46.2%)	43 ± 9.49	Heat stress	Heat-related illness	152 (47.5%)	<0.001
Suter (2019) [41]	Indonesia	Trial	October–November 2017	Agriculture	331	M = 331 (100%)	42.3 ± 11.0	Heat stress	Skin rush	3 (1.7%)	NR
									Muscle cramps	19 (11.0%)	
									Dizziness	14 (8.1%)	
									Heavy sweating	146 (84.9%)	
									Extreme weakness	3 (1.7%)	
									Confusion	6 (3.5%)	
Riley (2018) [42]	USA	Cross-sectional	2005–2010	Construction	NR	NR	NR	Heat stress	Emergency department visits	OR 1.023	NR
									Hospitalizations	OR 1.035	NR
				Agriculture, forestry, fishing, hunting, mining				Heat stress	Emergency department visits	OR 1.080	NR
									Hospitalizations	OR 0.907	NR

Table 1. *Cont.*

Citation	Country	Study Design	Study Timeframe	Work Tasks	Sample Size	Gender Ratio (n,%)	Age (Mean ± SD or Range)	Occupational Risks	Outcomes Considered	Results	p Value
Meade (2017) [43]	USA	Case series	Two consecutive days	Work involving electrical utilities	4	M = 4 (100%)	38 ± 12	Heat stress	Body core temperature, dehydration, urine specific gravity	NR	NR
Rahman (2016) [44]	Bangladesh	Cross-sectional	May 2014	Construction	8	F = 8 (100%)	24.67 ± 8.0	Heat stress	Tympanic Temperature	37.2 ± 0.4	NR
									Core temperature	41.2	NR
									Metabolic rate (W/m^2)	280	NR
Xiang (2016) [45]	Australia	Cross-sectional	August–November 2012	Miscellaneous (completely outdoor work)	82	M = 96.0%	Range < 24 to >55	Heat stress	Concern for extreme heat	OR 71.6	NR
				Miscellaneous (mostly outdoor work)	300					OR 53.2	NR
Venugopal (2015) [46]	India	Cross-sectional	2012–2013	Metal work, security, agriculture, construction	442	M = 314 (71%)	35.8 ± 12.7	Heat stress	Heat-related symptoms	421 (96%)	<0.001
Bethel (2014) [47]	USA	Cross-sectional	July–August 2013	Agriculture	100	M = 60 (60.0%)	31.8 ± 10.1	Heat stress	Heat-related illness	27 (27.3%)	NR
									Heat-related symptoms (past week)	64 (64.0%)	NR
Luo (2013) [48]	China	Case–control	2003–2010	Spray painting	37	M = 788 (83%)	Range 20–59	Heat stress	Urolithiasis	OR 4.4	NR
				Smelter work	48					OR 4.0	NR
				Welding	178					OR 3.7	NR
				Production security and quality inspection	159					OR 2.7	NR
				Planing machine operation	12					OR 4.0	NR
				Gas cutting work	16					OR 2.6	NR
				Assembly	203					OR 2.2	NR

Table 2. A quality assessment of the included studies, performed using the Newcastle–Ottawa Scale.

First Author (Year)	Selection	Comparability	Outcome	Total
Karthick (2022) [36]	2	2	2	6
Han (2021) [37]	3	2	1	6
Phanprasit (2021) [38]	3	2	2	7
Venugopal (2021) [39]	4	2	3	9
Mansor (2019) [40]	4	2	2	8
Suter (2019) [41]	4	2	3	9
Riley (2018) [42]	3	2	2	7
Meade (2017) [43]	2	2	3	7
Rahman (2016) [44]	2	2	3	7
Xiang (2016) [45]	4	2	3	9
Venugopal (2015) [46]	3	2	2	7
Bethel (2014) [47]	3	2	2	7
Luo (2013) [48]	4	2	3	9

3.1. *Heat-Related Symptoms*

Eight of the included studies (61.5%) investigated the presence of subjective symptoms related to heat stress.

Overall, the presence of heat-related symptoms in outdoor workers was reported as very high in four of the included studies: Venugopal et al. [46] reported symptoms in 96% of their included workers, and self-reported heat-related illnesses occurred in 90.3% of the workers [39] included in another study by the same authors; Bethel et al. [47] reported symptoms in 64.0% of workers in the week preceding their study; and Suter et al. [41] reported heavy sweating (the most common symptom) in 84.9% of their included workers. One study by Mansor et al. [40] reported heat-related illness in 47.5% of workers. The study by Phanprasit et al. highlighted that 39.0% of foundry workers and 22.0% of construction workers reported heat-related symptoms. Han et al. [37] reported heat-related illnesses in only 11.6% of workers.

Three of these studies investigated heat-related symptoms [38,41,46]. Phanprasit et al. [38] investigated working conditions related to heat in a population of construction and foundry workers in Thailand; the authors investigated body temperature and heart rate, relating these parameters to the workload. The study was limited by the sample (mostly males) and the short study period. The study by Venugopal et al. [46] analysed the health and productivity impacts of occupational heat stress in 442 Indian outdoor workers. The authors highlighted the fact that employees with higher workloads reported lower productivity and higher rates of heat-related issues. The outcomes considered, however, were not evaluated with objective parameters, but were based on self-reported symptoms. Furthermore, the samples for each sector were small and comprised mostly male workers (71%). Finally, the study by Suter et al. [41] examined heat-related health outcomes in 331 Indonesian workers. This study, however, had an exclusively male sample and the data gathering period was very short.

One of the studies measured heat-related illnesses. Mansor et al. [40] conducted a study to evaluate the role of hydration practices in determining the severity of heat-related illnesses in outdoor workers (gardeners and cleaners). The authors concluded that hydration practices were related to the severity of heat-related illness, highlighting the importance of drinking water throughout the day rather than only during breaks or when feeling thirsty. The study included workers who had reported heat-related symptoms; therefore, all the participants had reported symptoms in the previous week, so the study lacked a proper control group concerning hydration practices; in Table 1 we report heat-related illnesses before the study's timeframe.

The study by Bethel et al. [47] measured both heat-related illnesses and symptoms. This study considered a population of migrant farmworkers, investigating whether they had experienced heat-related illnesses or symptoms. Interestingly, the authors also reported on the availability of water sources nearby (reported by 76.0% of workers), as well as the possibility of moving to cooler spots or shadowed areas when the heat was perceived as too high (60.0% reported that shaded areas were available). Furthermore, the authors reported that 52 participants had received heat-related hazard training. However, the sample size was relatively small (100 participants) and the survey was only performed on migrant workers, lacking a control sample.

The study by Han et al. [37] evaluated heat-related illnesses and injuries in construction workers. In this study, the authors considered possible prevention strategies enacted by the workers, who adjusted based on heat exposure, as well as the possible usage of personal protective equipment by the workers. The study, due to its cross-sectional nature, did not allow for the experimental measurement of injuries or illnesses caused by heat stress, but it did describe the reported heat-related injuries or illnesses that had occurred to the workers in the past; furthermore, the study had a mostly male sample (75.2%).

The study by Venugopal et al. [39] assessed multiple health-related outcomes, both self-reported and clinical. The study had a large sample and both genders were well represented. Furthermore, the assessment of body temperature rise, sweat rate, urine specific gravity, and reduced kidney function was performed through objective methods rather than self-reported assessments. Participation in the study was on a voluntary basis, which may have produced a selection bias; furthermore, heat-related symptoms were self-reported and were not correlated with other factors that may have influenced these symptoms (diet, genetic factors, and other non-occupational exposures).

Xiang et al. [45] conducted a study to investigate heat risk perception in 749 outdoor workers, evaluating four different aspects (i.e., concerns, policy, guidelines, preventive measures): 51.2% of the respondents reported concerns about workplace heat stress and 68.8% reported they would like to change their work habits to reduce the risk of negative health effects. However, most of the participants were males and were recruited on a voluntary basis, which may have introduced potential selection bias.

3.2. Heat-Related Clinical Outcomes

Six studies included in this review (46.1%) considered objective clinical outcomes [36,39,42–44,48]. One of these studies, by Venugopal et al. [39], was discussed in the previous section of this manuscript.

The study by Karthick et al. [36] was the only one that investigated health outcomes related to cold weather. The authors highlighted that outdoor workers had a 14.1% higher chance (OR 1.14, p value = 0.041) of developing respiratory issues due to cold weather, and a 21.7% (OR 1.217, p value = 0.032) higher chance of experiencing freezing of exposed parts compared to indoor workers. Furthermore, the authors highlighted that male workers had a 69.7% higher risk of respiratory issues and a 43.9% higher risk of freezing of exposed parts than female workers. However, the sample size was relatively small (100 participants) and the response rate was less than 50%. Furthermore, the cross-sectional design did not allow for tracking issues developed after the survey's administration. Moreover, while the authors mentioned investigating some preventive habits in the population considered (i.e., wearing personal protective equipment against cold weather), they did not perform a correlation analysis concerning these responses.

Riley et al. [42] investigated two types of workers strongly affected by work-related heat stress (construction and agriculture) and evaluated two important outcomes, emergency department visits and hospitalizations. The study also evaluated data collected over five years, which is a valid period for an optimal evaluation of the considered outcomes. However, the large sample did not allow for consideration of additional factors, which can often be co-causes of heat-related illness. Furthermore, the authors did not report a precise sample size or a gender ratio, limiting the generalization of their results, since gender can greatly impact occupational risks [49].

Meade et al. [43], in their case report, analysed heat stress parameters (body core temperature, urine specific gravity) in four workers in the field of electrical utilities on two consecutive days in hot outdoor conditions, and highlighted a thermal strain increase in hot conditions even with decreased work activity. The study had a small, exclusively male sample and the period was limited to two consecutive days, so it did not consider the long-term effects of chronic exposure.

The study by Rahman et al. [44] investigated the tympanic temperature increase in eight pregnant outdoor workers before, during, and after heat exposure, to estimate the risk of miscarriages or foetal anomalies. The study was the only one that analysed this category of women. The authors highlighted how female outdoor workers had higher body temperatures than indoor ones. However, the sample size was very small and the body temperature measuring method may not have been very accurate due to factors (i.e., ear conformation, thermometer position, environmental factors) that can alter the measurement.

In their case–control study, Luo et al. [48] investigated the association between heat exposure and outdoor workers' urolithiasis in a shipbuilding company in a subtropical city of China, analysing health data and adjusting important confounding variables (sex, nutrition, pre-existing heat-related health conditions). The results highlighted a strong connection between heat exposure and the considered outcome, with a higher incidence in outdoor workers and a dose–response relationship. This study did not collect information on family history, obesity, weight, BMI, social class, metabolic diseases, or smoking and drinking behaviours. Furthermore, the study had a mostly male sample (83%).

4. Discussion

The results of this systematic review highlight the fact that heat-related stress is well documented as an occupational risk associated with climate change. Twelve of the included studies investigated the effects of heat-related stress on outdoor workers, while one investigated the effects of cold weather. Four of the included studies reported a very high prevalence of heat-related symptoms, ranging from 64.0% to 90.3% of workers [39,41,46,47]. These results highlight occupational heat stress as a global public health concern that affects the well-being of different types of outdoor workers worldwide.

Higher temperatures have been highlighted as affecting the well-being of certain occupational groups of outdoor workers more than others: construction and agriculture workers have reportedly had a higher incidence of heat-related illness and death [50–52]. This is consistent with the direct exposure to sunlight and strenuous physical activities performed by these categories of workers, even on days with extreme temperatures. Furthermore, recent projections estimate that the number of days with extreme heat stress may double by 2050 in Southern Europe [53], a trend that would further increase the vulnerability of these groups of outdoor workers in the near future.

The studies included in this review highlight the fact that heat can also increase the risk of injuries. This may be due to factors such as fogging of protective glasses, sweating hands, dizziness, and reduced concentration, leading to additional hazards [22,23]. These results are consistent with the currently available literature and are reflected in the international guidelines available [50–52]. Furthermore, it has been demonstrated that heat-related illnesses are directly proportional to personal risk factors such as heat acclimation, dehydration, an age over 60 years, previous heat-related illness, specific medications, and a high body mass index [54].

The results of this systematic review are largely consistent with previous literature reviews that examined the effects of occupational heat exposure on outdoor workers [19,55]. Previous literature reviews have reported that environmental heat exposure is associated with an increased risk of heat-related illnesses, reduced work capacity, and elevated risk of occupational injuries in outdoor workers, particularly in the agriculture, construction, and mining sectors. Similarly, this systematic review found a high prevalence of heat-related symptoms, confirming the significant health burden associated with occupational heat stress.

However, our review adds to the existing evidence by including more recent studies and analyzing the role of climate change-related variables. In addition, some of the studies included in our review focused on biometeorological aspects, which are less frequently addressed in earlier reviews. Moreover, while most previous reviews have focused primarily on heat, our review also included one study addressing the potential occupational risks associated with cold exposure, highlighting the broader range of climate-related hazards affecting outdoor workers.

One of the studies included in this systematic review analyzed the impact of extreme cold temperatures on occupational health issues. While the majority of studies focused on heat-related occupational risks, it is important to acknowledge that climate change may also contribute to the occurrence of extreme cold events in certain regions [56]. Cold exposure can pose significant health and safety risks for outdoor workers, including hypothermia, frostbite, decreased manual dexterity, impaired cognitive function, and an increased risk of injuries [57–59]. Given this very limited assessment of the effects of extreme cold temperatures on climate change-related occupational risks, future studies should explore how climate variability may alter the frequency, duration, and severity of cold exposure, and should aim to identify vulnerable worker populations and implement adaptation strategies for occupational health protection in colder environments.

It has to be acknowledged that the use of the keyword "climate change", which was used to unequivocally identify studies correlating occupational risks to this issue, may have contributed to the selection of studies reporting on the effects of climate change (i.e., extreme temperatures) that were perceived as having a direct impact on occupational risks (i.e., heat stress), and may have contributed to the exclusion of the indirect effects on occupational health that were highlighted in the Introduction section (i.e., occupational allergies).

As previously mentioned, no studies included in the review investigated occupational allergies, dermatologic illnesses, or ocular disorders, although an association with climate change has been highlighted in previous narrative studies [23–26,29]. This review highlights the need to perform studies that gather primary data on occupational risks related to climate change beyond temperature-related risks, showcasing a possible gap in the knowledge of scientific studies regarding pulmonary, dermatologic, and ocular disorders in outdoor workers in relation to the growing issue of climate change.

Although the Introduction section of this review highlighted other climate-related occupational hazards, such as ultraviolet radiation and air pollution, no eligible studies that specifically investigated these exposures met the inclusion criteria. This research gap may reflect several factors. First, while the health risks related to these factors are well established in the general population, their investigation in occupational settings and outdoor occupations, within the context of climate change, remains limited. This may be due to the complex interplay between environmental exposures, variability across geographic regions, and difficulties in accurately assessing individual occupational exposure levels over time (since these exposures are related to long-term health issues rather than short-term ones, contrary to heat stress). This gap in the current knowledge underlines the need to perform long-term evaluations of the effects of outdoor occupational exposures related to climate change in order to assess the possible health issues related to these factors.

This review should be considered in light of some limitations and strengths. The use of the keyword "climate change" and its synonyms may have led to the exclusion of studies that did not use this keyword but considered effects (dermatological, respiratory, etc.) that may be indirectly correlated with climate change. Furthermore, most of the included studies were cross-sectional and, therefore, could not assess the long-term effects of climate change-related occupational risks, nor was a causal inference possible. Future research should include cohort and intervention-based studies in order to enrich the current scientific knowledge available in this field. Despite these limitations, the systematic approach ensured the methodological soundness of this study. The review included studies performed in many different countries, giving a global overview of the issue. The growing interest in climate change underlines the need to provide a comprehensive overview of the topic, and this review provides insight into the direct effects of climate change on the health of outdoor workers worldwide.

5. Conclusions

This systematic review highlights heat-related stress in outdoor workers as an important occupational risk, but it also highlights an important gap in scientific knowledge regarding certain age or gender categories, with the majority of studies being performed on male workers and not stratified for age. Future studies should investigate the impact of climate change on the most vulnerable occupational categories while including at-risk populations (i.e., older workers, female workers, migrant workers, etc.). Furthermore, future scientific research should evaluate the knowledge and attitudes of these vulnerable workers regarding climate change, heat stress, and occupational exposure.

Mitigation and prevention strategies should be developed and implemented at the company, national, and supra-national levels, to ensure the safety and well-being of outdoor workers worldwide. These interventions should focus on educating workers about the occupational risks caused by heat stress, as well as climate change and heatwaves in general; furthermore, training programs on the preventive measures already available to mitigate these risks (sunscreen, water intake, appropriate work wear, etc.) should be implemented.

Author Contributions: Conceptualization: M.F.R. and U.M.; data curation: M.F.R.; investigation: M.F.R., R.L. and U.M.; methodology: M.F.R.; supervision: U.M.; writing—original draft: M.F.R.; writing—review and editing: U.M. All authors have read and agreed to the published version of the manuscript.

Funding: This research received no external funding.

Data Availability Statement: No new data were created or analyzed in this study.

Conflicts of Interest: The authors declare no conflicts of interest.

References

1. Crowley, R.A. Climate Change and Health: A Position Paper of the American College of Physicians. *Ann. Intern. Med.* **2016**, *164*, 608–610. [CrossRef]
2. What Is Climate Change? | United Nations. Available online: https://www.un.org/en/climatechange/what-is-climate-change (accessed on 17 July 2023).
3. Climate Change. Available online: https://www.who.int/health-topics/climate-change (accessed on 19 July 2023).
4. Nordhaus, W.D. Geography and Macroeconomics: New Data and New Findings. *Proc. Natl. Acad. Sci. USA* **2006**, *103*, 3510–3517. [CrossRef]
5. Dell, M.; Jones, B.F.; Olken, B.A. Temperature Shocks and Economic Growth: Evidence from the Last Half Century. *Am. Econ. J. Macroecon.* **2012**, *4*, 66–95. [CrossRef]
6. Gospodinov, N.; Gaffney, I.L.; Ng, S. The Economic Impact of Low- and High-Frequency Temperature Changes. *arXiv* **2025**, arXiv:2505.08950. [CrossRef]
7. Dong, J.; Tol, R.S.J.; Wang, J. The Effects of Climate and Weather on Economic Output: Evidence from Global Subnational Data. *arXiv* **2025**, arXiv:2505.17946. [CrossRef]
8. Gibb, K.; Beckman, S.; Vergara, X.P.; Heinzerling, A.; Harrison, R. Extreme Heat and Occupational Health Risks. *Annu. Rev. Public Health* **2024**, *45*, 315–335. [CrossRef]
9. Hunt, A.P.; Brearley, M.; Hall, A.; Pope, R. Climate Change Effects on the Predicted Heat Strain and Labour Capacity of Outdoor Workers in Australia. *Int. J. Environ. Res. Public Health* **2023**, *20*, 5675. [CrossRef]
10. Formica, S.; Sfodera, F. The Great Resignation and Quiet Quitting Paradigm Shifts: An Overview of Current Situation and Future Research Directions. *J. Hosp. Mark. Manag.* **2022**, *31*, 899–907. [CrossRef]
11. Borrelli, I.; Santoro, P.E.; Gualano, M.R.; Moscato, U.; Rossi, M.F. Assessing the Great Resignation Phenomenon: Voluntary Resignation of Young Italian Workers during the COVID-19 Pandemic. *Ann. Ig. Med. Prev. E Comunita.* **2024**, *36*, 88–98. [CrossRef]
12. Xu, M.; Dust, S.B.; Liu, S. COVID-19 and the Great Resignation: The Role of Death Anxiety, Need for Meaningful Work, and Task Significance. *J. Appl. Psychol.* **2023**, *108*, 1790–1811. [CrossRef]
13. Rossi, M.F.; Beccia, F.; Gualano, M.R.; Moscato, U. Quiet Quitting: The Need to Reframe a Growing Occupational Health Issue. *Soc. Work* **2024**, *69*, 313–315. [CrossRef] [PubMed]
14. Chirico, F.; Taino, G. Climate Change and Occupational Health of Outdoor Workers: An Urgent Call to Action for European Policymakers. *Environ. Dis.* **2018**, *3*, 77–79. [CrossRef]
15. Ferrari, G.N.; Leal, G.C.L.; Thom de Souza, R.C.; Galdamez, E.V.C. Impact of Climate Change on Occupational Health and Safety: A Review of Methodological Approaches. *Work Read. Mass.* **2023**, *74*, 485–499. [CrossRef]
16. Aristei, L.; D'Ambrosio, F.; Villani, L.; Rossi, M.F.; Daniele, A.; Amantea, C.; Damiani, G.; Laurenti, P.; Ricciardi, W.; Gualano, M.R.; et al. Public Health Regulations and Policies Dealing with Preparedness and Emergency Management: The Experience of the COVID-19 Pandemic in Italy. *Int. J. Environ. Res. Public Health* **2022**, *19*, 1091. [CrossRef]
17. Carlon, O. Cambiamenti Climatici: Una Minaccia al Benessere Delle Persone e Alla Salute del Pianeta. Agire ora Può Mettere al Sicuro il Nostro Futuro. IPCC—Focal Point Italy. 2022. Available online: https://ipccitalia.cmcc.it/cambiamenti-climatici-una-minaccia-al-benessere-delle-persone-e-alla-salute-del-pianeta-agire-ora-puo-mettere-al-sicuro-il-nostro-futuro/ (accessed on 20 June 2025).
18. Gubernot, D.M.; Anderson, G.B.; Hunting, K.L. Characterizing Occupational Heat-Related Mortality in the United States, 2000–2010: An Analysis Using the Census of Fatal Occupational Injuries Database. *Am. J. Ind. Med.* **2015**, *58*, 203–211. [CrossRef]
19. Flouris, A.D.; Dinas, P.C.; Ioannou, L.G.; Nybo, L.; Havenith, G.; Kenny, G.P.; Kjellstrom, T. Workers' Health and Productivity under Occupational Heat Strain: A Systematic Review and Meta-Analysis. *Lancet Planet. Health* **2018**, *2*, e521–e531. [CrossRef] [PubMed]
20. Copernicus: High Temperatures Trigger Seasonal Ozone Pollution Across Europe. Available online: https://atmosphere.copernicus.eu/copernicus-high-temperatures-trigger-seasonal-ozone-pollution-across-europe (accessed on 13 June 2025).
21. Kobza, J.; Dul, L.; Geremek, M. The Influence of NOx, Temperature, Wind and Total Radiation on the Level of Ozone Concentration in the Upper Silesian Agglomeration. *Front. Public Health* **2024**, *12*, 1485333. [CrossRef]
22. Schulte, P.A.; Chun, H. Climate Change and Occupational Safety and Health: Establishing a Preliminary Framework. *J. Occup. Environ. Hyg.* **2009**, *6*, 542–554. [CrossRef]
23. Modonese, A.; Gobba, F. Occupational risk related to natural optical radiation exposure and skin cancers. *G. Ital. Med. Lav. Ergon.* **2020**, *42*, 329–331.
24. de Troya Martín, M.; Aguilar, S.; Aguilera-Arjona, J.; Rivas-Ruiz, F.; Rodríguez-Martínez, A.; de Castro-Maqueda, G.; Cambil-Martín, J.; de Gálvez-Aranda, V.; Blázquez-Sánchez, N. Risk Assessment of Occupational Skin Cancer among Outdoor Workers in Southern Spain: Local Pilot Study. *Occup. Environ. Med.* **2023**, *80*, 14–20. [CrossRef]

25. Ruppert, L.; Ofenloch, R.; Surber, C.; Diepgen, T. Occupational Risk Factors for Skin Cancer and the Availability of Sun Protection Measures at German Outdoor Workplaces. *Int. Arch. Occup. Environ. Health* **2016**, *89*, 1009–1015. [CrossRef] [PubMed]
26. Parks, C.G.; Cooper, G.S. Occupational Exposures and Risk of Systemic Lupus Erythematosus. *Autoimmunity* **2005**, *38*, 497–506. [CrossRef] [PubMed]
27. WMO Bulletin: Heatwaves Worsen Air Quality and Pollution. Available online: https://public.wmo.int/news/media-centre/wmo-bulletin-heatwaves-worsen-air-quality-and-pollution (accessed on 13 June 2025).
28. Schulte, P.A.; Jacklitsch, B.L.; Bhattacharya, A.; Chun, H.; Edwards, N.; Elliott, K.C.; Flynn, M.A.; Guerin, R.; Hodson, L.; Lincoln, J.M.; et al. Updated Assessment of Occupational Safety and Health Hazards of Climate Change. *J. Occup. Environ. Hyg.* **2023**, *20*, 183–206. [CrossRef] [PubMed]
29. D'Ovidio, M.C.; Annesi-Maesano, I.; D'Amato, G.; Cecchi, L. Climate Change and Occupational Allergies: An Overview on Biological Pollution, Exposure and Prevention. *Ann. Ist. Super. Sanita.* **2016**, *52*, 406–414. [CrossRef]
30. Berhane, K.; Kumie, A.; Samet, J. Health Effects of Environmental Exposures, Occupational Hazards and Climate Change in Ethiopia: Synthesis of Situational Analysis, Needs Assessment and the Way Forward. *Ethiop. J. Health Dev. Yaltyopya Tena Lemat Mashet.* **2016**, *30*, 50–56.
31. Raulf, M.; Annesi-Maesano, I. Occupational Allergy and Climate Change. *Curr. Opin. Allergy Clin. Immunol.* **2025**, *25*, 83–87. [CrossRef]
32. Page, M.J.; McKenzie, J.E.; Bossuyt, P.M.; Boutron, I.; Hoffmann, T.C.; Mulrow, C.D.; Shamseer, L.; Tetzlaff, J.M.; Akl, E.A.; Brennan, S.E.; et al. The PRISMA 2020 Statement: An Updated Guideline for Reporting Systematic Reviews. *BMJ* **2021**, *372*, n71. [CrossRef]
33. Eriksen, M.B.; Frandsen, T.F. The Impact of Patient, Intervention, Comparison, Outcome (PICO) as a Search Strategy Tool on Literature Search Quality: A Systematic Review. *J. Med. Libr. Assoc. JMLA* **2018**, *106*, 420–431. [CrossRef]
34. Ouzzani, M.; Hammady, H.; Fedorowicz, Z.; Elmagarmid, A. Rayyan—A Web and Mobile App for Systematic Reviews. *Syst. Rev.* **2016**, *5*, 210. [CrossRef]
35. Wells, G.; Shea, B.; O'Connell, D.; Peterson, J.; Welch, V.; Losos, M.; Tugwell, P. *The Newcastle-Ottawa Scale (NOS) for Assessing the Quality of Nonrandomised Studies in Meta-Analyses*; Ottawa Hospital Research Institute: Ottawa, ON, Canada, 2014; Available online: https://www.ohri.ca/programs/clinical_epidemiology/oxford.asp (accessed on 20 June 2025).
36. Karthick, S.; Kermanshachi, S.; Loganathan, K. Effect of Cold Temperatures on Health and Safety of Construction Workers. In *Tran-SET 2022*; ASCE: Reston, VA, USA, 2022; pp. 237–247. [CrossRef]
37. Han, S.-R.; Wei, M.; Wu, Z.; Duan, S.; Chen, X.; Yang, J.; Borg, M.A.; Lin, J.; Wu, C.; Xiang, J. Perceptions of Workplace Heat Exposure and Adaption Behaviors among Chinese Construction Workers in the Context of Climate Change. *BMC Public Health* **2021**, *21*, 2160. [CrossRef]
38. Phanprasit, W.; Rittaprom, K.; Dokkem, S.; Meeyai, A.C.; Boonyayothin, V.; Jaakkola, J.J.K.; Näyhä, S. Climate Warming and Occupational Heat and Hot Environment Standards in Thailand. *Saf. Health Work* **2021**, *12*, 119–126. [CrossRef] [PubMed]
39. Venugopal, V.; Shanmugam, R.; Kamalakkannan, L.P. Heat-Health Vulnerabilities in the Climate Change Context—Comparing Risk Profiles between Indoor and Outdoor Workers in Developing Country Settings. *Environ. Res. Lett.* **2021**, *16*, 085008. [CrossRef]
40. Mansor, Z.; Rosnah, I.; Ismail, N.H.; Hashim, J.H. Effects of Hydration Practices on the Severity of Heat-Related Illness among Municipal Workers during a Heat Wave Phenomenon. *Med. J. Malays.* **2019**, *74*, 275–280.
41. Suter, M.K.; Miller, K.A.; Anggraeni, I.; Ebi, K.L.; Game, E.T.; Krenz, J.; Masuda, Y.J.; Sheppard, L.; Wolff, N.H.; Spector, J.T. Association between Work in Deforested, Compared to Forested, Areas and Human Heat Strain: An Experimental Study in a Rural Tropical Environment. *Environ. Res. Lett.* **2019**, *14*, 084012. [CrossRef]
42. Riley, K.; Wilhalme, H.; Delp, L.; Eisenman, D.P. Mortality and Morbidity during Extreme Heat Events and Prevalence of Outdoor Work: An Analysis of Community-Level Data from Los Angeles County, California. *Int. J. Environ. Res. Public Health* **2018**, *15*, 580. [CrossRef]
43. Meade, R.D.; D'Souza, A.W.; Krishen, L.; Kenny, G.P. The Physiological Strain Incurred during Electrical Utilities Work over Consecutive Work Shifts in Hot Environments: A Case Report. *J. Occup. Environ. Hyg.* **2017**, *14*, 986–994. [CrossRef]
44. Rahman, J.; Fakhruddin, S.H.M.; Rahman, A.K.M.F.; Halim, M.A. Environmental Heat Stress Among Young Working Women: A Pilot Study. *Ann. Glob. Health* **2016**, *82*, 760–767. [CrossRef]
45. Xiang, J.; Hansen, A.; Pisaniello, D.; Bi, P. Workers' Perceptions of Climate Change Related Extreme Heat Exposure in South Australia: A Cross-Sectional Survey. *BMC Public Health* **2016**, *16*, 549. [CrossRef]
46. Venugopal, V.; Chinnadurai, J.S.; Lucas, R.A.I.; Kjellstrom, T. Occupational Heat Stress Profiles in Selected Workplaces in India. *Int. J. Environ. Res. Public Health* **2015**, *13*, 89. [CrossRef]

47. Bethel, J.W.; Harger, R. Heat-Related Illness among Oregon Farmworkers. *Int. J. Environ. Res. Public Health* **2014**, *11*, 9273–9285. [CrossRef]
48. Luo, H.; Turner, L.R.; Hurst, C.; Mai, H.; Zhang, Y.; Tong, S. Exposure to Ambient Heat and Urolithiasis among Outdoor Workers in Guangzhou, China. *Sci. Total Environ.* **2014**, *472*, 1130–1136. [CrossRef]
49. Santoro, P.E.; Borrelli, I.; Gualano, M.R.; Amantea, C.; Tumminello, A.; Daniele, A.; Rossi, M.F.; Moscato, U. Occupational hazards and gender differences: A narrative review. *J. Sex-Gend.-Specif. Med.* **2022**, *8*, 154–162.
50. Dinuoscio, C. Heat Stress and Strain. ACGIH. 2022. Available online: https://www.acgih.org/heat-stress-and-strain-2/ (accessed on 20 June 2025).
51. Jacklitsch, B.; Jon Williams, W.; Musolin, K.; Coca, A.; Kim, J.H.; Turner, N. *Criteria for a Recommended Standard: Occupational Exposure to Heat and Hot Environments—Revised Criteria 2016*; NIOSH: Atlanta, GA, USA, 2023. [CrossRef]
52. *ISO 7243:2017*; Ergonomics of the Thermal Environment—Assessment of Heat Stress Using the WBGT (Wet Bulb Globe Temperature) Index. ISO: Geneva, Switzerland, 2017. Available online: https://www.iso.org/standard/67188.html (accessed on 19 January 2024).
53. Europe Faces the Growing Threat of Heat Stress. Euractiv. 2023. Available online: https://www.euractiv.com/section/eet/opinion/europe-faces-the-growing-threat-of-heat-stress/ (accessed on 20 June 2025).
54. Park, J.; Kim, Y.; Oh, I. Factors Affecting Heat-Related Diseases in Outdoor Workers Exposed to Extreme Heat. *Ann. Occup. Environ. Med.* **2017**, *29*, 30. [CrossRef]
55. Levi, M.; Kjellstrom, T.; Baldasseroni, A. Impact of Climate Change on Occupational Health and Productivity: A Systematic Literature Review Focusing on Workplace Heat. *Med. Lav.* **2018**, *109*, 163–179. [CrossRef]
56. Cohen, J.; Agel, L.; Barlow, M.; Garfinkel, C.I.; White, I. Linking Arctic Variability and Change with Extreme Winter Weather in the United States. *Science* **2021**, *373*, 1116–1121. [CrossRef]
57. Karthick, S.; Kermanshachi, S.; Pamidimukkala, A.; Namian, M. A Review of Construction Workforce Health Challenges and Strategies in Extreme Weather Conditions. *Int. J. Occup. Saf. Ergon.* **2023**, *29*, 773–784. [CrossRef]
58. Ray, M.; King, M.; Carnahan, H. A Review of Cold Exposure and Manual Performance: Implications for Safety, Training and Performance. *Saf. Sci.* **2019**, *115*, 1–11. [CrossRef]
59. Farbu, E.H.; Höper, A.C.; Brenn, T.; Skandfer, M. Is Working in a Cold Environment Associated with Musculoskeletal Complaints 7–8 Years Later? A Longitudinal Analysis from the Tromsø Study. *Int. Arch. Occup. Environ. Health* **2021**, *94*, 611–619. [CrossRef]

MDPI AG
Grosspeteranlage 5
4052 Basel
Switzerland
Tel.: +41 61 683 77 34

Atmosphere Editorial Office
E-mail: atmosphere@mdpi.com
www.mdpi.com/journal/atmosphere

www.ingramcontent.com/pod-product-compliance
Lightning Source LLC
Chambersburg PA
CBHW040835250926
51850CB00007BA/1320
9783725877195